BROADBAND TELECOMMUNICATIONS TECHNOLOGIES AND MANAGEMENT

BROADBAND TELECOMMUNICATIONS TECHNOLOGIES AND MANAGEMENT

Riaz Esmailzadeh
Carnegie Mellon University, Adelaide, Australia

To Izumi, Amin Asad, Kian and Nima

Contents

Preface

The primary reason for writing this book was my need for a textbook for a telecommunications technology management course that I have been teaching at Carnegie Mellon University. As with many technology management topics, analysis and management of a telecommunications company necessitates a knowledge of the utilised technology, an understanding of the competitive landscape and business issues, and policy issues pertaining to the roll out and operation of the business. While there exist a number of books to address these aspects, each book generally focuses on any one of the technology, business or policy issues. To my knowledge, there existed no book that addresses the broadband telecommunications topic holistically from the technology, business and policy points of view. This book does take a holistic view and presents models by which such companies and organisations may be analysed or managed considering all three viewpoints.

This book is valuable for two groups of readers. The first group is post-graduate students on an inter-disciplinary technology management course, with a career outlook to become technology analysts or to manage technology companies as chief technology officers (CTOs) or chief information officers (CIOs). Those who choose a career in information communication technology (ICT) with a telecommunications focus will find this book of particular value as it provides an overall broadband telecommunication technology, business and policy analysis coverage. Those whose career choice is in other areas of ICT will find the book useful as it describes the technologies, business and policy issues which underpin the telecommunications industry and ICT services. The book also trains students to analyse technology products using a technology, business and policy framework.

The second group is business analysts, or technology managers focusing on telecommunications companies. The book will help these professionals to understand the basics of new broadband telecommunications technologies, and how these technologies impact their industry. The book further benefits this group by covering the business and policy factors. These factors contribute to, and in many instances determine, the success of a technology product/company in the market place. Again, the technology–business–policy framework discussion is educational and of benefit in a diverse range of industries.

While a telecommunications technology background is beneficial, it is not a prerequisite for readers of this book. Technology is discussed from a sufficiently basic level to allow the reader to understand and follow the topics. Furthermore, while the focus of the book is broadband telecommunications, much of the discussion is applicable to information technology students. This is in line with the objective of overall technology management courses of which broadband telecommunications is a part.

Book Format and Structure

The book is primarily written for an educational setting, and therefore each chapter starts with a set of preview questions. These questions are designed to help the reader understand the chapter's main topics and their main import. For example, the following preview questions can be stated for the book as a whole:

1. How may a broadband content and service provision business be defined in terms of its eco-system components?
2. What are the technological differences between 3G and 4G mobile communications systems?
3. What are the technologies that enable Fibre-to-the-Node and Fibre-to-the-Premise systems, and what are the economic considerations in choosing one over another?
4. How does the roll-out of broadband services impact on economic development?
5. Should national governments fund the roll-out of fixed broadband infrastructure?

These questions are followed by a list of learning objectives. These further help the reader with an overview for the topics discussed in the chapter and what may be learned. Some of the learning objectives for this book as a whole are:

1. What are the latest technological developments in telecommunications?
2. Which technology is more likely to survive the competition?
3. Who will (or should) fund the building of national broadband telecommunications infrastructure?

To illustrate the concepts discussed in each chapter, a number of case studies and historical notes are included. These provide a context where technical, managerial and policy issues may be discussed, both in a classroom setting and also for individual study and reflection. Further learning tools include on-line simulation programs, and recordings of the author's lectures for the Broadband Telecommunications Technologies and Management course at Carnegie Mellon University. A dedicated website includes a large number of case studies on various telecommunications businesses. These links are available from Wiley's website and general search engines. Chapters end with a set of review questions and references. Teaching notes and presentation materials, as well as videos of our class presentations, are available for course instructors.

Although our scope covers technology, business and policy aspects of telecommunications, a number of chapters have a particular focus on telecommunications technology. In these and other chapters, technology–business–policy synthesis is achieved through examples, case studies and review questions, where the reader can apply the analysis framework.

As discussed above, telecommunications can be defined in terms of how information is handled. Chapter 1 is an introduction and discusses the telecommunications evolution and adoption. Chapter 2 introduces the technology–business–policy analysis model and framework and the broadband telecommunications infrastructure. Chapter 3 introduces Voice Communications and is followed by Chapter 4 on Information Theory, a fundamental background to the digital broadband telecommunications system in use today. Chapters 5–8 are about how information is processed and communicated digitally. Three principal broadband communications systems: fixed, wireless and satellite are discussed in Chapters 9–11, followed by a discussion on complementary personal telecommunications systems in Chapter 12. Chapter 13 discusses a terrestrial wireless network infrastructure, and is followed by a discussion on how content business provides the traffic that flows over these networks in Chapter 14.

Case studies and historical notes are used throughout the book to illustrate the technology, business and policy issues, and synthesise their relative interdependence and importance. These are further intended to enhance discussion in a classroom setting, as well as assist individual learning.

1

Introduction

Preview Questions

- How do technology, business and policy considerations matter to the telecommunications industry?
- How does the telecommunications industry feature in the global economy?
- In recent years, which industry segment has filed the largest number of patents?
- What are the reasons for the phenomenal growth in the number of broadband subscribers in recent years?
- How may a broadband content/service provision functions and structure be characterised?

Learning Objectives

- An historical overview of the telecommunications industry
- An overview of the current state of the telecommunications industry
- An introduction to the technology–business–policy framework
- An introduction to the content/service, retail, infrastructure model for analysing broadband telecommunications businesses
- An introduction to telecommunications management and information and information technology management fields

Broadband Telecommunications Technologies and Management, First Edition. Riaz Esmailzadeh.
© 2016 Riaz Esmailzadeh. Published 2016 by John Wiley & Sons, Ltd.
Companion Website: www.wiley.com/go/BTTM

Historical Note

An early example of a *telecommunications system* is the Royal Mail of the Persian Empire circa 500 BC. The Empire ruled over the region between the Indus River (present day India/ Pakistan) and Thrace in the present day European Turkey. It also extended as far as Libya in North Africa. The Empire was divided into different 'Satraps' or provinces, which were ruled by governors appointed by the central government. To rule effectively, fast and reliable messaging between Susa the capital and provincial centres was necessary. However, the distances were prohibitively long, and the means of travel were slow. For example, the distance between Susa and Sardis, a major Empire centre in Lydia, was some 2700 km. Travelling this distance on foot would take 90 days. A faster way of travelling was by horse but the animal and the rider needed food, water and rest. A messenger travelling by horse could expect to cover perhaps 100 or so kilometres per day, which would have been a great improvement but this still required 27 days of travelling. The Empire needed to respond to emergencies, whether local unrest or natural disasters, and clearly such long delays were not acceptable as they would greatly reduce the chances of a successful response.

The Persians had constructed a network of roads and relay stations to improve travelling time. At these relay stations fresh horses and couriers were ready to receive a message and take it to the next station, thereby removing the need to rest riders and horses. The ancient Greek historian Herodotus (484–425 BC) wrote of these Persian mounted postal carriers: 'Neither snow nor rain nor heat nor gloom of night stays these couriers from the swift completion of their appointed rounds'. Interestingly, this is the unofficial motto of the US Postal Service [1].

The main Royal Highway that connected Susa and Lydia comprised of 111 'relay' stations, as recounted below by Herodotus describing the road, the courier resting places and the horse exchange system as a marvel of its time. This relay system enabled the Royal Mail mounted couriers to travel the distance of 2700 km in 7 days. This was the fastest method of 'telecommunications' for its time and was an important tool in the governance and security of the Empire [2] (Figure 1.1).

Now the true account of the road in question is the following: Royal stations exist along its whole length, and excellent caravanserais; and throughout, it traverses an inhabited tract, and is free from danger. In Lydia and Phrygia there are twenty stations within a distance of 94½ parasangs.[1] On leaving Phrygia the Halys has to be crossed; and here are gates through which you must needs pass ere you can traverse the stream. A strong force guards this post. When you have made the passage, and are come into Cappadocia, 28 stations and 104 parasangs bring you to the borders of Cilicia, where the road passes through two sets of gates, at each of which there is a guard posted. Leaving these behind, you go on through Cilicia, where you find three stations in a distance of 15½ parasangs. The boundary between Cilicia and Armenia is the river Euphrates, which it is necessary to cross in boats. In Armenia the resting-places are 15 in number, and the distance is 56½ parasangs. There is one place where a guard is posted. Four large streams intersect this district, all of which have to be crossed by means of boats. The first of these is the Tigris; the second and the third have both of them the same name, though they are not only different rivers, but do not even run from the same place. For the one which I have called the first of the two has its source in Armenia, while the other flows afterwards out of the country of the Matienians. The fourth of

---

[1] A parasang is a measure of distance and equals approximately 6 km.

Figure 1.1 Persian Empire circa 500 BC, and the Royal Mail route. Reproduced with permission of the University of Texas [3]

the streams is called the Gyndes, and this is the river which Cyrus dispersed by digging for it three hundred and sixty channels. Leaving Armenia and entering the Matienian country, you have four stations; these passed you find yourself in Cissia, where eleven stations and 42½ parasangs bring you to another navigable stream, the Choaspes, on the banks of which the city of Susa is built. Thus the entire number of the stations is raised to one hundred and eleven; and so many are in fact the resting-places that one finds between Sardis and Susa [4].

Development and operation of the Royal Highway and Royal Mail depended on a number of technological advances. These included better tools for building roads; better methods for horse breeding; training programmes for horse riding couriers and so on. Technological development for other applications contributed as well. For example, the development of papyrus as a medium for writing contributed to the operational simplicity as paper was lighter to carry than baked clay and less susceptible to breakage. The location of relay stations was likely determined based partly on a number of these technological parameters. For instance, the inter-station distance would have been a factor of the terrain, as well as the fastest speed a horse could typically travel.

Of similar importance were business and operational issues including the costs of operation and maintenance of stations, the number of couriers and their rotation, the security of the stations and so on. These issues and associated costs would have been weighed against the value of telecommunications speed: how important was a 7-day end-to-end travel compared with an 8-day or 6-day travel? In modern days a cost–benefit analysis (CBA) or a cost–effectiveness analysis (CEA) is usually undertaken to arrive at an optimal solution. While it is not known whether the Persians ever undertook such a formal analysis, experience from other courier systems should have given them a guideline for 'good' – if not the 'best' – practice. As the cost of operation was borne by the Empire, it is difficult to determine how much each individual message transfer cost. Private usage of the courier system, perhaps for trade purposes, is not recorded. The cost of sending such a private message could have given insight into the overall cost of the operation.

The Royal Highway and the Royal Mail clearly had a great strategic and security importance for the Persian Empire. Business considerations would have included the high cost of constructing, securing and operating the stations, and of maintaining the roads but this must have been justified by the strategic benefits that the system provided. The Empire's policy in funding, building and maintaining the road was also important in the continued operation of the system.

Communication has always been a basic human need. We primarily used it for security: to protect ourselves and our kin. As we evolved, we have used it to learn about others and convey information about ourselves such as what we do and will, how we wish to do business, what our interests are, and so on. Development of speech (our primary communication method), and as we evolved, writing at the dawn of civilization, were major advances which have laid the foundation of an ever advancing civilization in which we live today. Such information may be communicated in person, or through means and devices that carry information to remote locations. We define this transfer of information across time and place as telecommunications, and consider that it has been with us from the very beginning of civilisation, as illustrated in the Royal Mail example.

'Telecommunications' is a recent term (first used in a book in 1904 [5]) and generally means the science and technology of communicating over a distance. An alternative definition may focus on the *utility* of telecommunications, which is communication of *information* over a distance. The terminology may only be a century old, however the need for communicating and conveying information to remote locations (both physically and temporally) predates civilization and has existed since the dawn of humankind. An in-depth discussion and analysis of telecommunication methods of early societies is outside the scope of this text. The brief history below is intended to establish a context within which we may analyse present day broadband telecommunication systems.

A major reason for the development of telecommunications in early human history was security. Smoke signalling or sound signalling would have alerted a friend to the approach of a foe. The means by which information was conveyed, the 'technology' in the above example, were fire and associated smoke or sound. These techniques were still in use by many indigenous societies when they came into contact with explorers from the 'old' world. These techniques are obviously limited by the extent of human hearing and sight, but relay techniques could be used to extend the telecommunications range. The extent of these beacons and their reliability was a function of terrain and weather, and therefore their utility was limited.

As human societies evolved there was another important application for telecommunications. As humans transitioned from a hunter–gatherer existence into complex farming communities, early forms of trade appeared. There was a need to convey messages between trading partners residing in remote places. Initially the messages were carried orally. Soon however, letters were being exchanged: a letter written on clay dates back to 7000 years ago [6]. As societies became more complex, the sending of messages between centres of governance and trade became more frequent and the speed of transmission more important. As a result, conveying of messages in written form using couriers became common. The Royal Mail of the Persian Empire was an advanced form of such a telecommunications system.

Security and trade have continued to remain a major reason for the development and maintenance of telecommunications infrastructure. Increasingly trade, for example across the Silk

Road, required a reliable messaging system. In contrast with the internal security-focused telecommunications systems, such as that of the Royal Mail, many of these trade-focused communications were across 'national' boundaries. It should be noted that the benefits of this telecommunications system were experienced by all and not just the senders and receivers. The society at large benefits from improved security, enhanced efficiency of trade and so on.

Very few telecommunications technology advances are recorded until the late 18th century and the development of Visual Telegraphy by Claude Chappe in 1794 (see Historical Note in Chapter 2). This development owed its existence to a number of advances including architecture and building construction technologies and the invention of optical telescopes. It also depended on social/policy developments such as re-emergence of strong central governance or 'empire' which could build and benefit from a telecommunications infrastructure. The discoveries of electricity and electromagnetic field in the 19th century were further theoretical foundations which led to the invention of electrical telegraphy and telephony and the modern systems we use today.

Some 2500 years after the Royal Mail was constructed and operated by the Persian Empire, many governments around the world are making decisions on how to build, and to what extent fund, their national telecommunications infrastructure. Again, construction and operating costs of a telecommunications system are weighed against business and strategic benefits of a national robust infrastructure which benefits most if not all of society. From a government policy point of view, telecommunications is a public need as well as a tool for state security. Broadband telecommunications has been called the infrastructure of the 21st century, and is considered as important as electricity, water, roads and other national infrastructure. Many of our needs are delivered over this infrastructure such as education, health, information, entertainment, security, social connectivity and so on. Efficient development, management and operation of broadband telecommunications are high on the agenda of governments of developed and developing countries.

Most national governments generally have aimed to ensure access to high quality telecommunications services are provided to the society on an equitable basis. An infrastructure to 'connect people' and facilitate fast and reliable messaging continues to be a multifaceted strategic asset.

This book focuses on broadband telecommunications as an ecosystem designed not only to connect people to each other, but also as a means to connect people to content and services such as information, entertainment, work, trade, goods, and so on. It is more than just an infrastructure: it also contains means of customer service, data collection and content delivery. An end-user connects to a broadband telecommunications infrastructure managed by a business entity that ensures high quality information transfer. The subscriber connectivity needs are managed by a service provision retailer who not only connects the users to others, but also delivers content from countless sources worldwide. This industry ecosystem can be drawn as a three-layer diagram as shown in Figure 1.2. We will demonstrate the complexity of the underlying technologies, business structure, and policy requirement of constructing and operating this ecosystem, and describe models which can be used for its analysis.

For this ecosystem to operate efficiently several factors need to be considered. Each layer uses a range of different technologies, each of which contributes to system efficiency. Moreover, diverse business models and industry alliances may exist at and across layers. From a public policy point of view, broadband telecommunications is vital to a national

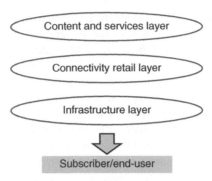

Figure 1.2   Broadband telecommunications industry ecosystem

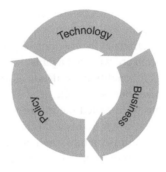

Figure 1.3   Technology–business–policy framework

economy. Furthermore, there are monopoly considerations, as well as privacy and confiden-tiality issues which require government attention and often regulation. Analysing a broadband telecommunications business is complex; it requires examination of technology, business and policy issues.

The Royal Mail *telecommunications system* is a good example of the analysis model we use extensively in this book. In this model three important elements contribute to the success of a product: technology, business and policy [7]. As illustrated in Figure 1.3, the three elements influence each other: the technology value proposition provides business advantage, which exists within a policy milieu, which may be subject to adjustment and government intervention. Policy direction may in turn influence the technology selection process, thereby closing the loop. We demonstrate that an overall evaluation of all three elements is necessary in analysing a telecommunications technology product. Clearly different products depend to a different degree on each aspect, but nevertheless all three will impact on a product's success.

The technology–business–policy framework is valuable to telecom analysts as it enables them to first examine whether a telecommunications technology is reliable and efficient. Next it requires them to consider costs of development of the product, provision of service, maintenance and so on, and determines whether a product can be sold to a customer for a price to produce a profit. It then needs to review the national and global policy issues that govern product development and service delivery, and determines product compliance; or, if necessary, what policy modification is needed.

Why Broadband Telecommunications?

The second half of the 20th century witnessed the evolution of the information communication technology (ICT) industry. The industry grew from serving niche applications to becoming a vital tool in helping to improve and even facilitate government and business operations and processes.

Early ICT systems were standalone computers which were mainly used in process automation and number-crunching applications. ICT systems have since evolved to facilitate not only automation, but also facilitate connectivity between suppliers and customers in business value chains. Information gathering from customers and suppliers is possible on an unprecedented scale. Furthermore, information gathering from ubiquitous sensor networks is forecast to become a major source of data collection in the future. Information so gathered is being used to obtain business intelligence to further enhance operational efficiency and to control business functions. Because of these developments, the ICT industry is now one of the most important sectors of the world economy, and employs an increasingly large number of technology and managerial ICT professionals. Figure 1.4 shows the employment trends in the ICT industry for 1995 and 2010 and how the share of ICT jobs in the economies of Canada, USA, Australia and the EU has grown. Note that this is a narrow definition which only considers ICT specialists as those, 'who have the capabilities to develop, operate and maintain ICT systems. ICT

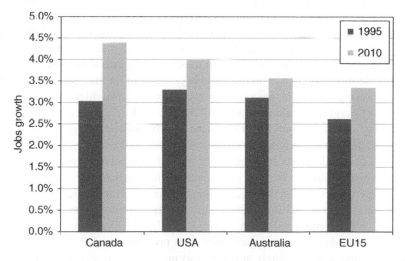

Figure 1.4 Sample job number growth within the ICT industry. Reproduced with permission of Organisation for Economic Co-operation and Development (OCED)

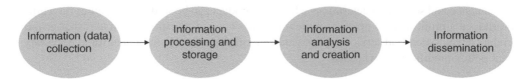

Figure 1.5   Information flow model

constitutes the main part of their job'. It does not include workers who use ICT products in their everyday jobs [8].

'Information' is at the centre of ICT industry activities, whose processes deal with how this 'information' is gathered, stored, analysed and disseminated. This information flow, as illustrated in Figure 1.5, is one way to analyse how effectively an organisation manages and uses the information it has. Among these processes, telecommunications can be defined as the technology and business processes associated with gathering and disseminating information.

'Modern' telecommunications, defined by its use of mechanical and electrical equipment and provision of a much faster transmission of messages, is a relatively recent development and its history can be traced to the late 18th century. Developments of visual and electrical telegraphy and telephony have significantly transformed how we *tele*-communicate. Modern telecommunications has been instrumental in the coming of a new age of global cooperation (and at times conflict) across national boundaries, leading to our present-day world order. In particular, the digital communications revolution of the last two decades of the 20th century has been the platform on which the ICT industry has been built.

Telecommunications has been the subject of research in multinational companies as well as major universities and government-funded institutions for more than two centuries. According to the World Intellectual Property Organisation (WIPO) 6.8% of all patents filed under the Patent Cooperation Treaty (PCT) in 2013 were in the telecommunications field. Indeed electrical machinery, apparatus, energy (14 897 filings), computer technology (14 684 filings), and digital communication (14 059 filings) were the top three areas of patent filings [9]. Moreover, the top four and all of the top 15 patent applicants were either purely telecommunications companies or had telecommunications units (Figure 1.6).

Telecommunications is also a major industry segment in its own right with many companies including operators, manufacturers and service providers among the world's most valuable. In addition, telecommunications has traditionally been one of the most regulated industries both nationally and internationally: indeed, the incumbent monopoly telephone operator was government owned until recent years. Although the industry has been deregulated, still the telecommunications portfolio is usually represented by a senior government minister, and one or more departments who regulate and monitor telecommunications service provision. At the international level, telecommunications technologies were one of the earliest to be standardised across national boundaries as interoperability was necessary for international trade. Nowadays, the development and ratification of international standards is generally accepted as a prerequisite for the success of a new telecommunications system. Such standardisation is done under the auspices of the International Telecommunications Union (ITU).

Modern telecommunications, if marked by the inventions of telegraphy and telephony, date back to the mid-to-late 19th century. Despite this, usage of telecommunication services was

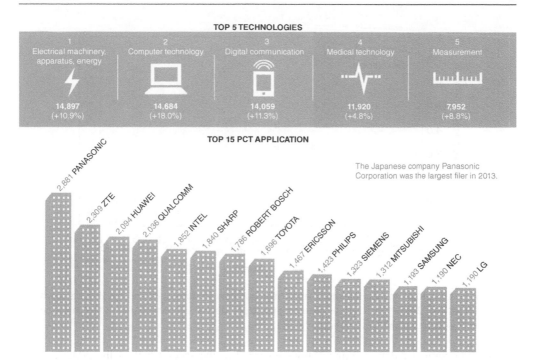

Figure 1.6 Patent application filed under the PCT in 2013. Reproduced with permission of World Intellectual Property Organization [9]

quite limited until very recently. Access to these services became widespread in developed countries by the second half of the 20th century. However, the vast majority of the population in developing countries had little or no access to telephony services, nor much use for telegraphy. Universal access to voice and message transmission only became possible with the introduction of inexpensive mobile telephony and internet systems and services. Figure 1.7 shows the growth in the number of mobile phone subscribers in developed and developing countries. The number of subscribers grew by a factor of 7.5 over a period of 12 years, reaching a global 100% penetration rate in 2013, a compound annual growth rate (CAGR) of 18.3%. The growth rate and rapid adoption in developing countries for these services is indicative of the basic need for telecommunications services. If prices are affordable, everyone will want to have access. This rapid adoption also demonstrates the effectiveness of national and international policies in standardisations, allocation of resources such as frequency spectrum, and deregulation of the telecommunications industry [10].

The technological and business transformations noted above have led to the rise of broadband telecommunications. Broadband telecommunications may be defined as information transmission at 'very high' rates. This 'very high' has been changing over the years: while a 256 kilobits per second (kbps) link was considered broadband in 2000; broadband today (2014) refers to several tens of megabits per second (Mbps). Some analysts even define broadband telecommunications systems as those capable of supporting transmission rates of 100 Mbps to 1 gigabit per second (Gbps) and more.

In contrast to telephony, the history of data communication and subsequently broadband telecommunications is very recent and began only in the late 1980s. Technologies to enable

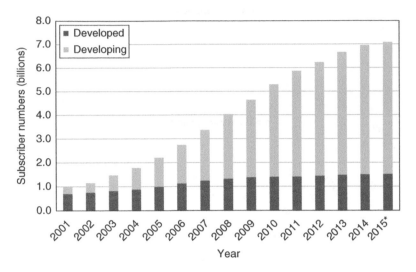

Figure 1.7 Growth of mobile telephone subscriber numbers in developed and developing countries during 2001–2015. 2013–*2015 figures are estimates. Reproduced with permission of Organisation for Economic Co-operation and Development (OCED)

computer-to-computer data communications emerged in the second half of the 20th century. Initially such services were provided to large companies, governments and universities using dedicated links such as coaxial and optical fibre cables. Data communications from subscriber premises also became possible at very low transmission rates using telephone lines and dial-up modems.

Large scale broadband telecommunications service delivery to subscriber premises only gained popularity with services such as email and the World Wide Web. Higher transmission rates became possible as technologies such as digital subscriber line (DSL), and optical fibre based systems such as Fibre-to-the-Home (FttH) and hybrid fibre-cable (HFC) were developed in the late 1990s. The introduction of wireless broadband technologies within the 3G standards in the early 2000s made it possible for nearly a third of the world's population to access the data communications network from individual subscriber premises. In most OECD countries the total fixed and wireless broadband subscription numbers are at near saturation levels (Figure 1.8). The main reason for such growth is the economic benefits of broadband connectivity. Efficient provision of broadband telecommunications services is an important business and policy issue in virtually every country around the world.

The growth of broadband telecommunications services has profoundly impacted many facets of our lives. It has significantly changed the way we study, work, play, entertain, take care of our health, share knowledge, socialise, and even wage war. The way we do business has changed to an extent that a whole new field of business – electronic commerce – enabled by the advancements in the telecommunications technology has emerged, and given rise to many successful new companies over the past two decades. The impact of the telecommunications industry may be measured by the fact that while the costs of making long distance phone calls has fallen to nearly zero, investment in infrastructure equipment as a share of global Gross Domestic Product (GDP) has risen significantly. The OECD reports a CAGR of 5.7% over the 2000–2009 decade in the ICT industry. As a share of GDP, the ICT industry has been growing as shown in Figure 1.9. It is expected to grow further over the coming decade as our life and work changes more as a result of information and communications technology development [11].

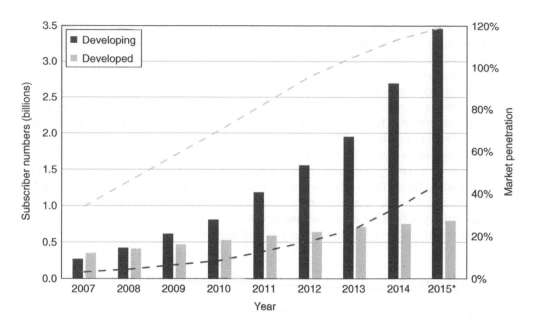

Figure 1.8 Fixed and wireless broadband subscriber numbers (columns, left-hand vertical axis) and the total penetration ratio (dotted line, right-hand vertical axis) in developed and developing countries. ITU data [10] and author analysis. 2013–*2015 figures are estimates. Source: Informa Telecoms and Media, July 2011. Reproduced with permission of Organisation for Economic Co-operation and Development (OCED)

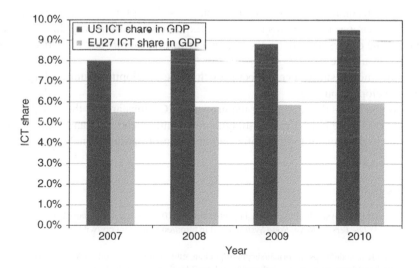

Figure 1.9 ICT share of GDP in the US and EU. Source: http://www.oecd.org/sti/broadband/oecdkeyictindicators.htm, from 'Share of ICT-related occupations in the total economy in selected countries, narrow definition', accessed 4 February 2014. Reproduced with permission of Organisation for Economic Co-operation and Development (OECD)

Further demonstrating the importance of telecommunications is the significant share of ICT expenditure and revenue in national GDPs. In the OECD alone the total telecom sector revenue stood at some $1.36 trillion. It has been growing steadily at a CAGR of nearly 7% over the past three decades [12, 13].

A large number of services depend on broadband telecommunications. These include tele-medicine, remote manufacturing, tele-education, and telepresence conferencing. Broadband telecommunications has facilitated a global infrastructure that supports hitherto unforeseen services and business activities. The coming decades are expected to witness the roll-out of a network of sensors which collect information on our environment and surroundings, our health, our work and many other aspects of our lives and share it across a network referred to as the Internet of Things (IoT). These systems are expected to connect over a ubiquitous tele-communications network, development and roll-out of which is forecast to be a major economic activity in both the developed and developing countries of the world in the near to medium future. Broadband telecommunications will remain a significant part of our lives.

Review Questions

1. Why did the Persians build the Royal Highway?
2. The cost of building the Royal Highway must have been quite high. Why do you see the trade-off considerations for the Persian Empire vis-à-vis this cost at the design stage?
3. What technological, business and policy issues had to be addressed in building the Royal Mail system?
4. How can the technology, business and policy framework be useful in analysing telecom-munications systems?
5. What are the three layers in a broadband content/service provision system? How do these relate to each other?
6. Although telecommunications comprises only 3–4% of global GDP, the share of global patent applications in 2013 was nearly 7%. Explain two reasons why this is the case.
7. What are the drivers behind the phenomenal growth of mobile and broadband subscriber numbers around the world?
8. Why is the mobile subscriber number growth in the developing countries larger than that of the developed countries?
9. What do you see as the reasons for the growth of ICT jobs around the world?
10. Why do you see the importance of studying telecommunications management? How does this relate to the broader topic of information and information technology management?

References

[1] http://www.postalmuseum.si.edu/museum/1e_faqs.html#history10, accessed 4 February 2014.
[2] Graf, D.F. (1994) *The Persian Royal Road System*, cited in http://en.wikipedia.org/wiki/Royal_Road, accessed 20 August 2015.
[3] http://www.lib.utexas.edu/maps/historical/shepherd/persian_empire.jpg, accessed 20 August 2015.
[4] http://www.thelatinlibrary.com/historians/herod/herodotus7.html, accessed 20 August 2015.
[5] Estuanie, E. (1904) *Traite Pratique de Telecommunications Electrique*, cited in Huurdeman, A.A. (2003) The Worldwide History of Telecommunications, Wiley Interscience.

[6]   Anonymous (1928) World's Oldest Letter, The Canberra Times (31 May), p. 4.

[7]   Krishnan, R. (2007) Lecture slides on E-Commerce course, Heinz College, Carnegie Mellon University.

[8]   http://www.oecd.org/sti/broadband/oecdkeyictindicators.htm, accessed 20 August 2015.

[9]   http://www.wipo.int/export/sites/www/ipstats/en/docs/infographics_patents_2013.pdf, accessed 20 August 2015.

[10] http://www.itu.int/en/ITU-D/Statistics/Pages/stat/default.aspx, accessed 20 August 2015.

[11] ICT R&D, Innovation and Growth. http://ec.europa.eu/digital-agenda/sites/digital-agenda/files/KKAH12001ENN-chap4-PDFWEB-4_0.pdf, accessed 20 August 2015.

[12] www.oecd.org/sti/ieconomy/46471775.xls, accessed 20 August 2015.

[13] OECD. OECD Communications Outlook 2013. http://www.oecd.org/sti/broadband/oecd-communications-outlook-19991460.htm

2

Technology, Business and Policy

Preview Questions

- Why do technology, business and policy parameters matter to the telecommunications industry?
- How can broadband telecommunications systems be analysed and compared using a technology–business–policy framework?
- How do infrastructure, retail and service businesses interact in the context of broadband telecommunications?
- What are multi-sided networks?
- What is value and how can it be analysed?

Learning Objectives

- Analysis of broadband telecommunications businesses using the technology–business–policy model
- Analysis of broadband telecommunications businesses' ecosystem, and different layer characteristics
- Value analysis, and value chains and value hierarchy
- Analysis of broadband telecommunications businesses using economic models such as cost–benefit analysis
- Platform mediated networks and broadband telecommunications businesses

Historical Note: Telegraphy

Early telecommunications were made through letters and messages sent by couriers. Very few telecommunications technology advances are recorded until the late 18th century and the development of the visual telegraph by Claude Chappe in 1794 in France. Chappe's telegraph was a relatively complex system and was based on mechanical and optical technologies, resulting in much faster rates of message transmissions compared with traditional means. Development of telescopes in the 17th century extended human visual reach and made it possible to view very distant objects. Visual telegraphy could then transmit messages over distances of many kilometres.

Chappe's invention was a semaphore system with mechanically movable arms constructed over a mast (called a regulator) and used special codes to represent letters and numbers to convey information to a remote station. The arms and regulator had a length of 2 and 4.5 m, respectively, as shown in Figure 2.1, with the whole device constructed on the top of tall buildings such as towers or churches.

The angle of arms with respect to each other and to the regulator was the code which represented a letter of the alphabet or a number as shown in Figure 2.2. Chappe named the system *télégraphe*, a combination of two Greek words meaning 'distance' and 'writer'. To distinguish it from the better known electrical telegraph, Chappe's system is usually referred to as a visual telegraph.

The maximum distance between transmitter and receiver was a function of visual range of the telescopes as well as the local topography. For example, the distance between the stations on the first line which connected Lille (a city in the north-east of France) and Paris varied between 4 km and 15 km. In total 23 relay stations were used to cover the distance of 200 km. A message was relayed from station to station until it reached its final destination.

The semaphore's movable arms were large and a telegraph operator could only change the angles once approximately every 2 min. From this, the transmission speed can be calculated to be

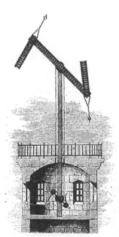

Figure 2.1 A Chappe telegraph transmitter on top of a cathedral [1, 2]. Source: Chappe telegraph. Illustration published in *Les Merveilles de la Science*, Louis Figuier, 1868 [1]

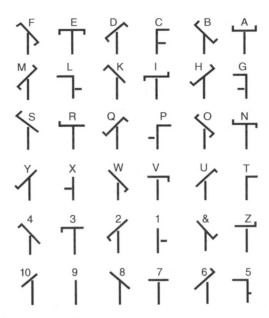

Figure 2.2 Chappe codes. Source: 'Telegraph'. For explanatory text, see entry on 'TELEGRAPH', volume 35 of the Cyclopædia [3]

1/120 = 0.008 character/s. A 50-letter text would therefore take about 100 min to send. Unimaginably slow by today's standard, it was a huge improvement compared with the fastest alternative of a horse-mounted message courier. Moreover, an operator at a relay station could start on transmission after receiving each letter and did not have to wait for the complete message. Therefore the end-to-end transmission delay was a direct function of the number of relay points.

The following example illustrates the Chappe telegraph's superior speed of message transmission. The Lille–Paris link was used to relay the news of the capture of Condé-sur-l'Escaut from the Austrians in a time of less than one hour. A horse-mounted message courier brought the same news some 20 hours later. Clearly the French government understood the national security value of the Chappe system and a network was commissioned throughout France. Figure 2.3 shows the extent of this network by 1850. Wider adaption would have been expected except that a newer, more efficient technology was invented: the electrical telegraph.

Samuel Morse witnessed the operation of visual telegraphs and understood the value of rapid messaging that telegraphy enabled during a trip to Europe. He also learned about the experiments of Ampere and that electricity could be used to send messages. Based on these ideas he built and

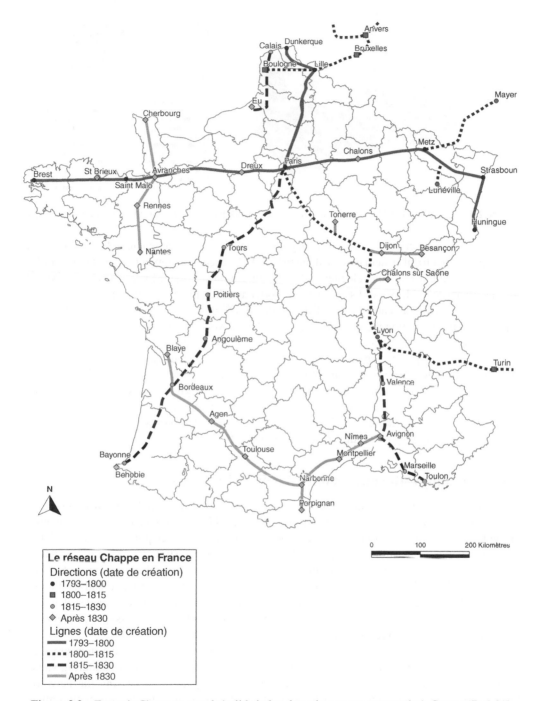

Figure 2.3 France's Chappe network (solid circles show the management nodes). Source: Ref. [4]

Figure 2.4   Morse key and Morse code. Source: Ref. [5], work by Simon A. Eugster, Wikipedia Commons

demonstrated an experimental electrical telegraph in 1837 in the United States. His improvement on Ampere's electrical system extended the range of electrical telegraphy to more than 10 miles (17 km) to rival that of visual telegraphy. Another invention was the 'Morse' alphabet by Alfred Vail, an associate of Morse. Vail's design used shorter and therefore more efficient codes for more frequently occurring alphabet letters. This intuitively efficient approach can be considered an application of information theory, except that this theory would not yet be articulated for almost a century. Figure 2.4 shows a telegraph key (transmitter) and the Morse (Vail) code.

Morse filed for patents, secured funding from the United States government and built an experimental system between Washington and Baltimore – a distance of some 63 km. It was over this link that the famous message 'What hath God wrought' was sent on the 23rd of May 1844, a date regarded as the beginning of modern telecommunications. Morse went on to build a company to operate telegraph lines. The worldwide telegraph system was also generally based on his technology and code, and operated under license [6].

The Technology–Business–Policy Analysis Framework

As presented in Chapter 1, in this book we use a technology–business–policy (TBP) framework to analyse the value of telecommunications systems. The TBP framework enables examination of the relative merits and demerits of a product from these three viewpoints in order to determine its overall value in a marketplace. The framework is illustrated in Figure 2.5, where the three elements of technology, business and policy inter-relate and impact on each other. Each of the three elements depends on a number of factors with quantitative and qualitative impacts. The collective results from these three analyses should then be considered for an overall output. Parameters of importance to each of the technology–business–policy elements are described below with respect to broadband telecommunications.

The complexities of operating in the broadband telecommunications ecosystem are well compensated by the benefits: this is a vital infrastructure for end-users and of enduring value. People need to be connected and are therefore willing to pay for accessing the network. Furthermore, subscribers provide operators in this industry with valuable information about their behaviour, relationships, tastes, hobbies, and so on, information which can be of great value to the players and their industry partners. This ecosystem is one of the most valuable in terms of customer value it creates and customer data it collects.

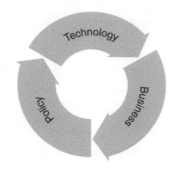

Figure 2.5 Technology–business–policy framework

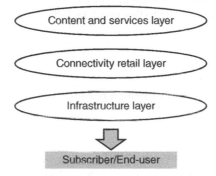

Figure 2.6 Broadband telecommunications industry ecosystem

In this chapter we introduce a number of technology, business and policy issues of concern to management of broadband telecommunications, and demonstrate how these issues are mapped to the ecosystem shown in Figure 2.6.

Technology

Broadband telecommunications infrastructure systems may be classified broadly into two groups of fixed and wireless as shown in Figure 2.7. Three fixed (wired) systems use twisted pair copper wires, optical fibres and coaxial cables as medium for transmissions. Similarly three wireless systems, mobile, fixed and satellite may be identified. One or more of these technologies may be used to provide broadband telecommunications services and indeed Australia's National Broadband Network (NBN) roll-out (see Case Study 2.2) uses all with the possible exception of mobile wireless. Other systems may also be used to provide broadband connectivity, such as balloons, power-lines and Wi-Fi. However, we focus on the systems of Figure 2.7 in discussing broadband telecommunications technologies.

The infrastructure layer technologies are mainly concerned with connecting the end-user to the telephony and internet telecommunications network. In contrast, the retail layer companies, such as NBN retailer layer companies, manage the subscriber and connect them to other subscribers, services and content providers. These companies switch end-user traffic and route them to intended destinations. Furthermore, they manage quality of service (QoS) and

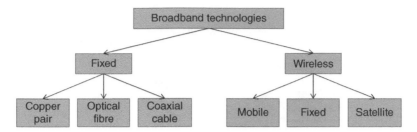

Figure 2.7   Infrastructure of broadband telecommunications technologies

offer value-added services such as filtering, storage, security and customer management. A number of technologies are specific to this layer, including:

- Traffic routing
- Storage and cloud facilities
- Software defined networking
- Information-centric and content distribution networking
- Internetwork Packet Exchange (IPX)
- QoS management

Companies operating at the service and content layer deliver a range of products: video such as YouTube and Netflix; audio streaming such as iTunes and Spotify; news such as NY Times and Yahoo! News; social networking such as Twitter and Facebook; information search such as Google and Bing; and so on. Over-the-top service providers such as Skype and WeChat connect subscribers to each other and enable peer-to-peer content sharing. A large number of technologies are used at this layer, such as source coding, encryption, streaming protocols, data and network analysis, and software programming languages. Most of these technologies are unrelated to the underlying broadband telecommunications systems. This book focuses on information transfer related to broadband telecommunications and does not address the technologies of this layer, except where service quality is impacted by the speed of transmissions at the infrastructure layer. Nevertheless, vertical integration and information collection at these layers leads to many business and policy considerations and therefore many of these companies in this layer are analysed in this book.

Some important parameters for technology analysis are discussed in the following.

Transmission Rate

An important technological parameter in evaluating telecommunications systems is transmission rate or the speed by which messages may be transferred. This depends on the medium used for transmission, signal and noise power as well as the medium 'bandwidth' which determines the amount of data that can be sent at any time interval. For example, a horse-mounted courier's travelling speed and the total number of messages (written on baked clay or paper) a courier can carry are limited. Clearly two couriers can carry twice as many messages: this can be considered as doubling the capacity of the system through doubling the (band) width of the channel.

Table 2.1 Maximum transmission rates for commonly used
telecommunications systems

Telecommunications system	Transmission rate (peak)
Wireless – 3G	10s of Mbps
Wireless – 4G	100s of Mbps
GEO satellite	Several Mbps
LEO satellite	Several Mbps
xDSL	100s of Mbps
Optical fibre	>1 Gbps

Modern telecommunications use electromagnetic waves to transmit signals over a number
of different media including air, copper wires and optical fibres. Electromagnetic waves travel
at, or close to, the speed of light depending on the media. This speed determines the propagation delay, or the time it takes for one symbol of the total message to traverse the medium
between a transmitter and a receiver.

The maximum amount of information that can be carried per second, as will be shown in
Chapter 4, is a function of signal bandwidth or spectrum – in essence the size of the pipe that
carries data. All telecommunications media have a limited bandwidth for transmission because
of physical or regulatory constraints. (For example, a copper wire has characteristics that limit
the bandwidth that may be used for transmission of data; or in the case of wireless communications, mobile operators have access to a limited spectrum.) The wider the pipe is, the more
data that can be sent. Optical fibre systems have the widest bandwidth and therefore can
support fastest transmission rates compared with copper-based and wireless systems. In turn,
fixed and mobile wireless systems in general support faster transmission rates compared with
a satellite system. Message transfer rates for a number of different telecommunications systems are shown in Table 2.1: these systems are further discussed in detail in Chapters 9–11.

Coverage Area

Another parameter of importance for broadband telecommunication systems is coverage area
or the maximum geographic extent within which a central serving node may serve a subscriber
base. In the case of wireless communications, a transmission tower may send messages to
users over a radius of many kilometres. In fact, at times television and radio broadcast antennas
cover entire cities. Mobile communications towers usually have a narrower coverage range of
several hundred metres to several kilometres. Fixed, wired communications systems may also
provide service to users many kilometres away, depending on the medium used for transmission and transmission rates. For example, twisted copper wires may be used for distances
of tens of kilometres for telephony services whereas they can only be used for a maximum of
4 km or so for DSL. Broadband residential/business optical fibre systems (as opposed to backhaul network) have unlimited coverage potential, but cost issues limit their deployment to
hundreds of metres to several kilometres. Wireless systems, both fixed and mobile, can cover
wide geographical areas of radius of tens of kilometres. Satellite systems can cover very large
areas up to one-third of the Earth's surface (Table 2.2). The details of coverage calculation for
satellite, fixed and wireless telecommunications systems are described in Chapters 9–11.

Table 2.2 Coverage area for commonly used telecommunications systems

Telecommunications system	Coverage
Wireless – 3G	~10 km
Wireless – 4G	<1 km
GEO satellite	Continental
LEO satellite	Regional
xDSL	Few km
Optical fibre	Few km

Number of Simultaneous Connections

User capacity or the maximum number of users that can simultaneously access the communications medium is an important parameter for wireless and to a lesser degree fixed broadband telecommunications systems. Fixed systems generally are dedicated: for example, one pair of copper wires connect a DSL subscriber to a telephone exchange at the infrastructure layer. Retail layer networks are shared to serve a large number of subscribers and therefore broadband traffic needs to be managed to ensure high QoS. In contrast, all wireless systems serve subscribers in a coverage area using a common frequency spectrum in the infrastructure layer, and therefore the resources must be shared at both infrastructure and retail layers. For example, the number of simultaneous phone calls that can be made on a mobile telecommunications system is a function of the technologies used and available frequency spectrum. In general, higher possible transmission rates may result in a larger number of users that may simultaneously connect to the network.

Another technology parameter is the total number of users that can remain connected to a network even when they are inactive. This is important for a data communications system where users become active only when they have data to send. Maintaining network connection allows these users to transmit their data quickly without needing to go through the necessary processes of registration with a server.

All broadband telecommunications systems need to carry a significant amount of traffic and therefore user QoS management is an important issue. Again, figures for different broadband telecommunications systems are examined in Chapters 9–11.

Power Consumption

Power requirements differ for different technologies. Generally, the higher the desired transmission rate, the larger the required transmission power. Coverage area also features in determining the transmission power. Fixed telecommunications systems operate at lower transmission power compared with wireless systems. Power consumption features prominently in mobile devices usage and stand-by times and how often recharging is necessary. It also features in operating costs of mobile operators as electricity usage can be quite high.

Power consumption is significant also for power-limited applications. Most battery-powered wireless devices are power-limited, and it follows that technologies which use less power can be advantageous in these conditions. Therefore a system which provides high transmission rate at the expense of high power consumption may be discounted in favour of a low power consumption system with sufficient transmission rate. An example is advanced-antenna systems which can deliver higher transmission rates at the expense of higher power consumption.

Power consumption is also a function of processing complexity. Generally high transmission rates require a large amount of signal processing, and therefore battery consumption. Therefore where high transmission power is not an important factor, such as sensor networks, simpler technologies with lower power consumption are more desirable.

Quality of Service

Another technological parameter of interest is QoS. QoS has several dimensions and includes the probability that a message arrives in error, the delay in receiving a message, as well as the quality of sound or video as perceived by an end-user.

Errors in communications can lead to dropped calls, retransmission requirement, loss of data, and therefore degraded quality. Excessive delays may also contribute to lost data, or temporary disruption of playback and therefore bad quality. Quality of audio and video reproduction can also be impacted by source coding techniques and reproduction fidelity. All of these contribute to QoS and quality of experience and can therefore be important in comparing two broadband telecommunications systems.

Delay is of particular importance to many data communications systems. User-network transmission delay is generally a factor of speed of electromagnetic wave propagation. It is also a function of data frame structure and control mechanism of a technology. End-to-end delay is a significant parameter as it may mean that some services cannot be accommodated. For example, voice communications requires a 200–300 ms end-to-end delay to ensure conversations can be conducted naturally. Some applications such as email and web-browsing may not be so sensitive to delay, but other applications such as computer games require short delays in the order of a few milliseconds.

With the exception of satellite communications, infrastructure layer delay is generally small and of little consequence, although protocol delay as data is packetized can impact delay-sensitive applications (such as high frequency trading or peer-to-peer gaming). On the other hand, retail layer companies relay data from service/content layers and their backhaul links and processing capability can cause delay and reduce QoS.

Another important metric is perceptual quality of sound and video. These are usually functions of transmission rate as well as the probability of lost data. Source coding techniques also contribute to the reproduction quality of streaming media. Different broadband telecommunications systems are impacted equally by these factors, although the transmission rates each technology offers lead to variations in perceptual quality.

Business Issues

A business may be characterised by its cost and revenues. In telecommunications businesses costs are due to network and infrastructure design and build-up – referred to as capital expenditure (CAPEX); and customer acquisition and service, and equipment maintenance – known as operational expenditure (OPEX). Benefits or revenues come from fees paid by end-users/subscribers. A cash flow analysis for a typical business in the broadband telecommunications field is shown in Figure 2.8 [7].

While the cash flow principles equally apply to all three layers of the broadband telecommunications business ecosystem shown in Figure 2.6, the dynamics for businesses across the layers are quite different. This is mainly due to the required initial investment as well as the potential market size and demand.

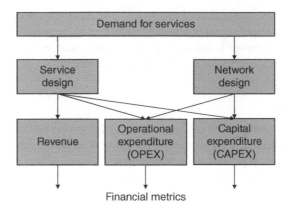

Figure 2.8   Cash flow analysis for a broadband telecommunications business

The infrastructure layer business is characterised by significant fixed-cost CAPEX investment in network equipment and roll-out. Often the private sector is incapable of such large investment and therefore it is left to the government sector to cover the cost: NBN Co. is a very good example.

For efficiency reasons, only a few players may compete at this layer, leading to monopoly or oligopoly business character. As a result the revenue (demand) side can be estimated with high accuracy. In contrast, the retail layer players do not require large initial investments. Their main cost is in OPEX: network operation and maintenance, customer acquisition and customer service. Naturally a company's revenues are a factor of the number of subscribers which it serves as well as the number of competitors in the market.

Similar dynamics may also exist with companies active at the content and service layer. However here network dynamics also apply: there is a tendency for concentration when a small number of players hold a lion's share of the market. Dominant players such as Google, Facebook and YouTube, which have significant share of their respective markets are good examples. Their market leading power can influence retail players to ensure high quality service delivery, whereas smaller players may not be able to enjoy the same priority. This can lead to regulatory oversight requirement.

Another feature of broadband telecommunications business is vertical integration across the layers, where a company is active in two or all three layers. Discriminatory behaviour, where a company slows down traffic of a non-allied company adds to complexities of the ecosystem dynamics and has led to regulatory intervention cases. The prominent example is the discussion around net-neutrality in the United States [8].

Natural Monopolies

Because of economic characteristics of a business at times a company 'naturally' grows to capture the entire market share. This is usually the case with high fixed-cost businesses such as infrastructure and utilities where it is economically more efficient to have a single company serve an entire market. For example, electrical supply lines to homes should not be duplicated as there is little if any value to have two companies compete. Telecommunications operation is also such a business. A large fixed cost is incurred to build telecommunications infrastructure.

Value Analysis

Any product may be analysed based on the value it provides to an end-user. This value can be *measured* by what the end-user, or someone subsidizing him/her, is willing to pay for the product/service. In essence the price at which a product may be sold depends on the value it proposes: the higher the perceived value, the more the price.

Understanding the value of a product is fundamental to business analysis. *Value analysis* enables understanding of the benefits of a product, how these benefits may be enhanced to increase revenue, the contributions business partners make to value provision, and how their role may be managed. Value analysis is therefore an important tool in business analysis.

Value analysis may be divided into two main parts: *value proposition* and *value configuration*. Value proposition focuses on what are the important driver and elements of a product, and what the end-user considers important. In a telecommunications context a value proposition may be simple: secure and swift message delivery between two government centres – as in the Royal Mail (see Historical Note in Chapter 1). Or it may be very complex, as in the provision of broadband telecommunications services to 100% of the Australian population as in the case of the Australian NBN. Value configuration on the other hand focuses on how many organisations and entities work together to deliver a value to an end-user. For example, how government, military, local security and horsemen worked together to realise the Royal Mail of the Persian Empire, or how network designers, manufacturers, retailers, content providers work together to provide the broadband services of the NBN [9].

Value Proposition

A product value proposition, and the price that it commands, can be analysed using a value hierarchy model. This hierarchical model is shown in Figure 2.9 where four levels have been defined [10]. These values start at *Basic* which is the most fundamental aspect a customer expects and without which the product will not sell. For example, in telephony this can be *connectivity* to other subscribers. Next is *Expected* value, which a customer may consider normal for the product and which all players may reasonably be expected to provide. In telephony this can be *good quality* of lines. Following this is *Desired* which includes features that are known to customers but which are not expected as they are not offered as standard. For telephony this may be an *answering service*. At the top of the hierarchy is *Unanticipated* which includes features which go well beyond what a customer knows or expects. For telephony in the 1980s this may have been *mobility*.

This value hierarchy model is useful in setting a product's price: as one moves higher in the value hierarchy, one expects to pay more for receiving these higher values.

It is important to note that value hierarchy is not static. What may be a *Desired* value today is likely to evolve over time to become a *Basic* value. An example is text messaging over mobile networks: it was a *Desired*, even *Unanticipated* value in the 1980s. However, by the 1990s it was a *Basic* service provided by all mobile operators. Similarly, many of the features of early smartphones such as iPhone 3 were *Unanticipated*. Being able to connect easily to the internet or use many applications were highly novel. Many of these features are nowadays *Expected*.

Figure 2.9   Value proposition hierarchy levels

Value Configuration

Value configuration is the list and role of players that contribute to the provision of a product. A valuable contribution is made by each player towards the final product. These players may exist within the same organisation and be fully coordinated. They may also be different companies and form a supply chain to produce and deliver the final product. Value proposition by any individual player may be mixed and matched with a different group towards delivery of different and at times competitive products. While any final product depends on each of these contributions, the reverse is not true: any individual player may opt out of supplying. An example of internal company value configuration is shown in Figure 2.10 where different functional groups within an organisation contribute value towards a final product. A value chain is the collection of all activities that are carried out by a company to produce a value for a customer. It is defined by Michael Porter to consist of [11]:

- Primary activities, such as logistics, manufacturing, marketing
- Support activities, such as human resource management, research and development

Primary activities build on contributions of other groups within the firm whereas support activities contribute to the entire chain. A company makes profit by charging a price for the product which is larger than the sum of costs incurred by all these contributors. The final price charged by the organisation is determined through either cost-based or value-based pricing, which determines the size of the margin.

Porter further defines a value system to be a collection of value chains of independent players who supply each other with valuable products. These value chains where several organisations combine to deliver a value to the customer, are also known as supply chains. An

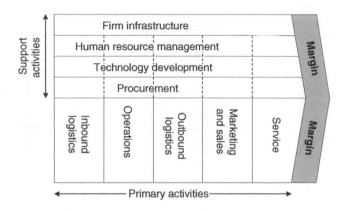

Figure 2.10   Porter's generic value chain

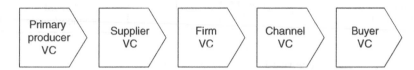

Figure 2.11   A multiple player value chain

example, shown in Figure 2.11, includes the primary producer, supplier, firm, retail channel, and the buyer's value chains. Again each link in the chain adds a value and charges a commensurate price.

Many products have parallel value chain components. A final product depends on services provided by different vendors. An example of video content delivery on a mobile device is shown in Figure 2.12. Here a broadband telecom network operator value chain is supplemented by a video content provision value chain. These two entities work together to provide the combined video-on-demand service to an end-user. Such arrangements between different industry players often exist in telecommunications where one value chain works to provide the infrastructure and another value chain works to provide the content that flows over that infrastructure. In Figure 2.12 an independent value chain may be defined for devices and equipment necessary (e.g. televisions) that facilitate service provision to an end-user. Note that there may be formal or informal linkages between several players (dotted lines) in the business diagram which can lead to policy issues further discussed below.

Pricing

In general the price of a product (goods or a service) is determined from the dynamics of supply and demand. In practice, the price may be set through two different mechanisms. One is value-based pricing, where a customer's willingness to pay is found in order to set a price. In this method, the value of a product to customers is analysed to determine a price which customers are willing to pay with an aim to maximise profits. The method of price setting

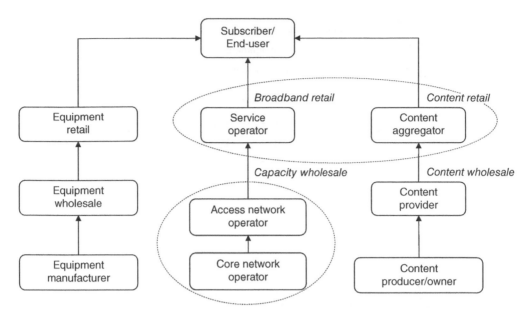

Figure 2.12   A business system diagram

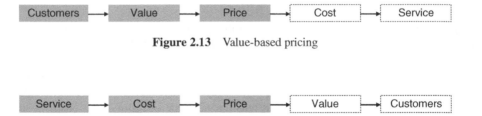

Figure 2.13   Value-based pricing

Figure 2.14   Cost-based pricing

applies mostly to environments where little direct competition exists and where a firm is able to set the price independent of other players. Based on this price point, the cost of product provision and service level may be found as illustrated in Figure 2.13. Such value-based pricing is applied for example to smartphones, particularly when they were first offered to the market in 2007 [12].

Cost-based pricing starts from the opposite side. It analyses a product, calculates the cost for producing it, adds a desired profit margin and arrives at the product price. Based on this price a certain basic value is proposed to customers who purchase the product. This method of price setting is illustrated in Figure 2.14 and generally applies to products which are offered to a very competitive market. In such environments value proposition differentiation between competitors is small, and prices generally fall to be equal to production cost and small margins. A good example is feature mobile phones which differ minimally from vendor to vendor, and are sold at similar prices [12].

The two price-setting mechanisms can be associated with the value analysis model discussed above. Cost-based pricing generally applies to value propositions which are *Basic* or *Expected*. On the other hand, a *Desired* or an *Unanticipated* value proposition can lead to setting the price levels at end-users' willingness to pay using a value-based model.

Cash Flow Analysis

For a business to be sustainable it needs to generate positive cash flow, that is, revenues which are larger than costs. Moreover, this positive cash flow needs to be sufficient to pay off any initial investment for setting the business up in a suitable timeframe. Cash flow analysis is one of the most important processes of a business and is handled at the highest levels of management.

Cash flow analysis is conducted by looking at the revenues and costs associated with a product over a certain period of time, for example a month or a year. This cash flow, combined with forecasts for a number of years in the future determine the expected overall cash flow. These numbers are then analysed with reference to the initial investment needed to set up production. If the overall cash flow over the forward estimates is positive, then a product may get the green light.

Cash flow analysis translates future cash flows by applicable interest rate(s) to show how much value they have in terms of today's money. For example, because of interest $100 in a year's time is worth less than $100 today. In other words, a cash flow analysis adjusts all future cash flows based on the project's interest rate. It yields an important parameter, net present value (NPV) as the sum of all present and future cash flows considering interest rate. It is calculated from the following formula:

$$NPV = \sum_{y=0}^{Y} V_y \cdot \left(\frac{1}{1+i} \right)^y \tag{2.1}$$

where V_y is total cash flow (net value) in year y, and i is the interest rate.

Example 2.1

Expenses and income for a internet service provider over a 10-year period are as follows:

- Initial (year 0) capital cost: $20M
- Maintenance: $0.3M per year, with full system overhaul of $1M in year 5
- Expected number of subscribers in year 1: 20,000
- Growth rate of subscriber: 30% per year over the 10-year period
- Customer acquisition cost: $300 per customer
- Value of capital equipment in year 10: $1M
- Average revenue per user: $250 per year
- Interest rate: 10%

Expenses and revenue are listed and net cash flow per year is drawn in Figure 2.15. What is the NPV for this operator over this period?

Year	Expenses	Revenue	Net cash flow
0	–$ 20 000 000.00	$ —	–$ 20 000 000.00
1	–$ 6 300 000.00	$ 5 000 000.00	–$ 1 300 000.00
2	–$ 2 100 000.00	$ 6 500 000.00	$ 4 400 000.00
3	–$ 2 640 000.00	$ 8 450 000.00	$ 5 810 000.00
4	–$ 3 342 000.00	$ 10 985 000.00	$ 7 643 000.00
5	–$ 4 954 600.00	$ 14 280 500.00	$ 9 325 900.00
6	–$ 5 440 980.00	$ 18 564 650.00	$ 13 123 670.00
7	–$ 6 983 274.00	$ 24 134 045.00	$ 17 150 771.00
8	–$ 8 988 256.20	$ 31 374 258.50	$ 22 386 002.30
9	–$ 11 594 733.06	$ 40 786 536.05	$ 29 191 802.99
10	–$ 14 983 152.98	$ 53 022 496.87	$ 38 039 343.89

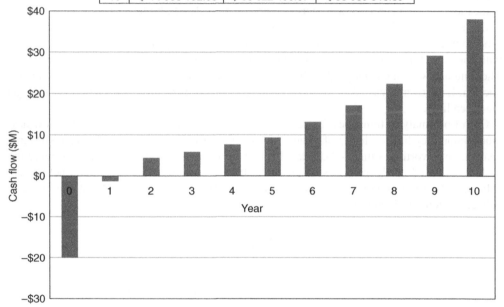

Figure 2.15   Net cash flow over the period of present value analysis

Answer

NPV is calculated from Equation 2.1 to be $46.8M. At the interest rate of 10% this product is profitable.

Closely related with NPV is the Breakeven Point parameter which shows when a product first turns an overall profit considering all expenses-to-date, taking interest rate into account. Adding all cash flows from the table in Figure 2.15 we can see that sometime in year 5 the company breaks even.

Another important parameter measure of product profitability is internal rate of return (IRR). IRR is the interest rate that makes NPV equal to zero. It is in essence the highest possible interest rate that the company may borrow money before it starts losing money. It is denoted by I in the following formula:

$$NPV_{IRR} = \sum_{y=0}^{Y} V_y \cdot \left(\frac{1}{1+I}\right)^y = 0 \qquad (2.2)$$

For Example 2.1, IRR is calculated to be 33%. In effect any $1 invested in this project returns 33¢ a year over the life of the project (10 years). If this is an acceptable investment for the company considering the risks then it will go ahead with the service. Both NPV and IRR are functions within spreadsheet programs such as Microsoft Excel.

Cost–Benefit Analysis

Cost–benefit analysis (CBA) is an important model to decide on whether to proceed with a project. In this method, the pros and cons associated with a project or a policy are listed, given a monetary value, and compared. At times CBA may be carried out for several projects or policies and the outcome compared against other projects. Table 2.3 shows a list of nine steps that a typical CBA should follow [13]. Clearly, a CBA has both qualitative and quantitative concern and parameters, however all parameters need to be translated into monetary terms. This quantitative analysis results in financial metrics, such as NPV and IRR which can be used to evaluate and compare projects. In general, the benefits should be larger than costs for a project to be given the green light.

Accuracy of a CBA depends on a number of factors. These include the reliability of future cost and revenue estimates. Another is how well qualitative parameters are monetised. Moreover CBA cannot by itself indicate whether a project will be effective, especially where the outcomes of different projects cannot be determined with a high degree of certainty. Health and defence policies are good examples where outcomes of different policies cannot be easily monetised: it is difficult to determine the value of human life. Other metrics may need to be included to measure the effectiveness of a project in saving lives for example. Yet another measure may be different projects' ability to result in a fair distribution of benefits to general society. A pure financial CBA generally does not yield how the benefits are distributed.

Table 2.3   Steps in a cost–benefit analysis

1. Specify the set of alternative projects
2. Decide whose benefits and costs count
3. Catalogue the impacts and select measurement indicators
4. Predict the impacts quantitatively over the life of the project
5. Monetise (attach dollar values to) all projects
6. Discount benefits and costs to obtain present values
7. Calculate the net present value of each alternative
8. Perform sensitivity analysis
9. Make a recommendation

This is a major shortcoming in public infrastructure projects such as telecommunications where services' provision to isolated communities is usually cost ineffective: services are generally provided through government subsidy [13].

Externality

A CBA may at times ignore a parameter because it does not incur a cost or a benefit to the project directly. Such a parameter is called an externality, and a good example is environmental pollution such as CO_2: the carbon pollution released into the air had largely been ignored by projects until recently. Clearly, the cost of cleaning the pollution needs to be considered in all projects, which means this externality is now starting to be included in a CBA.

An externality may be negative or positive. A negative externality is an overall *cost* that is not taken into consideration in the production process. A good example is stated above: air pollution. Until recently most factories did not have to pay for the air pollution from their production processes. If such cost is charged through a mechanism such as a regulatory regime, then it is possible that the price of the produced good is increased.

A positive externality occurs when a benefit to the consumer at large cannot be recovered by the producer. For example, a telecommunications operator may charge subscribers for their service usage. However, the social benefits resulting from a robust telecommunications infrastructure, such as security, health, education, and so on are difficult to quantify and be recovered by the infrastructure owner. If such positive externality can be internalised, it may be possible to reduce the subscriber fees.

Network Effects and Platform-mediated Networks

The business dynamics of a three-layered broadband telecommunications ecosystem are shaped by platform-mediated networks, and the degree to which these platforms are open. Telecommunications networks 'connect people' and their value increases with the number of subscribers and the number of possible connections increases.

At times, several seemingly disconnected networks work together to create value. For example a network of competing equipment manufacturers may develop telecommunications devices which are used by service subscribers. These two networks are connected to each other and mutually benefit as growth on one side leads to benefits on the other side. An example is shown in Figure 2.16: a virtuous cycle is created when two networks complement each other. As telephony services gained traction, an increase in the number of subscribers led to an increase in the number of manufacturers entering the market of telephone equipment. More

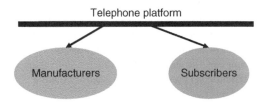

Figure 2.16   A multi-sided platform

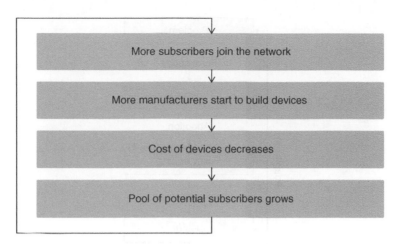

Figure 2.17 A virtuous cycle of platform growth

manufacturers meant more competition and lower prices for network equipment, which in turn meant lower overall cost of telephony and thereby attracting more subscribers, in turn increasing the pool of potential manufacturers. This *virtuous cycle* is illustrated in Figure 2.17 [14].

An analysis of telecommunications business and subscriber networks, especially with respect to services provided by companies in the content layer, is important in the overall value such operations create and how competitive they are.

Policy Issues

A number of public policy issues arise in the provision of a broadband telecommunications service. These are generally related to the role of government in ensuring a robust infrastructure is constructed for social and economic reasons. The government may focus on several issues, such as the cost of development of the infrastructure; the policy framework within which an efficient, competitive broadband telecommunications industry may grow and develop; and ensuring that services are provided to the entire population on a fair and equitable basis.

Public policy development takes place in several steps. It is generally carried out by a government, but private sector may also analyse and contribute to the process in order to arrive at a favourable outcome. The process generally goes through the steps shown in Figure 2.18, namely problem definition, solution proposals and comparison, policy decision, implementation and appraisal. The outcome of a policy setting is a set of government decisions in the form of regulation or intervention [15].

Broadband Telecommunications Public Policy Making

Existence of a case for government intervention depends largely on one's political and philosophical viewpoint. Some consider the role of government is to intervene and regulate many aspects of national economy. Some want the government to remain out of the business of regulation and allow market forces to determine winners and losers.

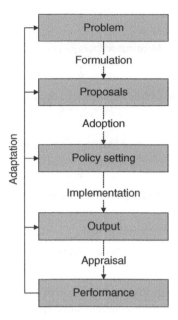

Figure 2.18   Policy cycle

One approach described by Mark Moore justifies government regulation and intervention in two cases [16]. One case is where there is 'market failure' – that is market forces have failed or are incapable of delivering a desirable outcome. Examples may be when private companies create monopolies which need to be regulated, or when no private company is willing or able to build a necessary infrastructure. Moore's second reason for government intervention is when there is a 'claim in fairness'. This is a situation created by failure of market forces to deliver an equitable service to a country's entire population.

Both these cases are applicable to broadband telecommunications. At times the large cost of fixed network build-up and associated risks deter the private sector from investment, which means the government may have to intervene and invest to ensure a robust system is in place. At other times government regulation becomes necessary because of natural monopoly development caused by the high fixed-cost initial investment in this industry. There is also a claim to fairness: because of the high cost of network build-up and relatively low revenue expectation in sparsely populated areas the private sector may not invest to provide universal access. Rural areas may therefore be left poorly connected. Government intervention therefore becomes necessary if basic services are to be provided on a fair basis to all citizens.

We further add a special case to Moore's cases for broadband telecommunications: government intervention is justified when industry needs to comply with various international obligations such as standards or bilateral/multilateral agreements. For example, spectrum allocation for wireless broadband services needs to follow certain international regulations.

Telecommunications has indeed been a highly regulated industry as infrastructure industries with high fixed cost behave as natural monopolies and favour single player and monopoly service provision. Historically telephony services, with very few exceptions, were provided through a single national company around the world. These companies were highly regulated to ensure prices were fair and high quality service was provided to all citizens on a fair basis. In particular, the following important areas have required national policy intervention, regulations, and/or decision making:

- How a present infrastructure (copper wires for telephony) is commonly used by a number of operators. Should the infrastructure be shared by all service providers, or is an incumbent within its rights to withhold access to its competitors? If access is mandated then what price should be charged?
- How present infrastructure is repaired, enhanced and/or replaced: who pays for infrastructure maintenance in a resource-shared regime? Should new investment be made? Is shared access enforceable?
- How to provide telecommunications services to uneconomical remote and sparsely populated areas. If a private company cannot provide service profitably in one region can it be forced to do so regardless?
- How to allow foreign ownership of a nationally strategic asset. What percentage of market share can be owned by a foreign-owned entity?
- Mergers and acquisition of companies of both wireless and fixed services. Can a company dominate the market? Is 50% market share by a single provider acceptable?
- Content provision of service and ensuring net neutrality.
- How to allow foreign companies to supply telecommunications equipment which may incorporate spying devices.
- Whether to harmonise spectrum allocation for wireless telecommunications services and how to allocate frequency spectrum to mobile operators.
- Whether to ratify international standards and what the impact is on possible technologies which may be adopted.
- Negotiations and abiding by international regulations for satellite orbits.

Policy Instruments

When a case can be argued for government intervention, a number of policy options may be considered. The Australian Public Service Commission lists the following as possible policy actions [17]:

- Direct government regulation, where government legislates for a specific course of action
- Self-regulation/co-regulation, where an industry is asked to come up with solutions, and assisted by government action
- Voluntarism, where an industry is trusted/induced to abide by certain regulations as suggested by a government
- Education and information instruments, where market forces are trusted to regulate the industry, helped along by a government information campaign
- Economic instruments, where industry players are incentivised to follow a path of action through economic means (fines or rewards)
- Combinations of policy instruments

Clearly different policy instruments may be used to solve a problem with due consideration to political and philosophical conditions. Different character of the broadband telecommunications industry ecosystem requires a different approach at each of the layers of Figure 2.6.

For example, direct regulation or using economic instruments may be effective at the infrastructure layer, whereas self-regulation or economic instruments may be sufficient at the content and service provision layers.

Policy Dynamics for Broadband Infrastructure

As discussed, broadband telecommunications services business can be defined using three layers: an underlying infrastructure which delivers the 'bits', a retail layer at which customers are served and managed, and a service/content layer which connects the customer to entertainment, news, services and meta-connectivity (such as social networking). Players at these three layers collaborate and compete with each other for a share of revenue from end-users.

Each of these three layers has specific economic characteristics. The infrastructure layer may use optical fibres, copper wires or wireless links to establish connectivity, all of which require substantial initial investment. As described above, such infrastructure businesses behave as natural monopolies. It is therefore economically efficient that a single player provides the service so that the largest possible number of subscribers will use the network. Such a monopoly operation may need to be regulated by government under a policy regime.

Note that natural monopoly tendencies are of different intensity depending on whether the broadband telecommunications infrastructure is wired (i.e. uses optical fibres, copper wires or cables) or wireless. The fixed infrastructure requires major investment in new constructions and upgrading existing infrastructure. It is therefore economically inefficient to construct two parallel sets of fixed infrastructure. It is a similar story in water/electricity/gas infrastructure where only one 'pipe' enters a home. As noted above, such a natural monopoly may need to be regulated.

Wireless infrastructure build-up costs are also significant, but to a lesser degree when compared with fixed infrastructure. Lower cost of infrastructure roll-out means that natural monopoly forces are not very strong and therefore multiple players may compete. In most countries of the world oligopolies of a few major operators have emerged. On the other hand, wireless broadband telecommunications systems require frequency spectrum to operate, which is a national resource, owned and regulated by the government. In these countries government policy decisions are mostly concerned with economically efficient allocation of spectrum and competition maintenance. Here, governments need to also follow an international spectrum allocation regime. While this impinges on a government's national sovereignty, equipment manufacturing considerations mandate a large degree of uniform global spectrum allocation in order to minimise development cost.

Government policies in regards to fixed infrastructure vary in different countries: for example, Australia's NBN is a fully government-owned business; whereas infrastructure ownership in the United States is largely left to the private sector. For fixed infrastructure, policy focus is on how to regulate a monopoly, or how to introduce and maintain market competition. At times geographical division of an infrastructural asset is used as a proxy competition: company A may own and operate infrastructure in one state and company B in another. Policy issues of importance to the infrastructure layer are illustrated in Figure 2.19.

In contrast, the retail layer policy issues concern vertical integration leading to discriminatory behaviour by companies. Furthermore, companies at this layer access and may collect subscriber information and may need to be regulated to ensure such information is kept private and confidential. Monopoly concerns are small as these are generally low fixed-cost businesses

Infrastructure layer

- Natural monopoly
- Service to rural areas
- International regulations

Figure 2.19   Infrastructure layer policy characteristics

Connectivity retail layer

- Vertical integration
- Discriminatory behaviour
- Privacy and confidentiality

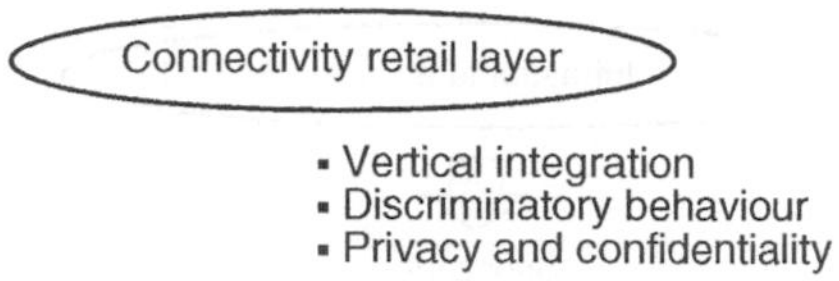

Figure 2.20   Retail layer policy characteristics

Content and services layer

- Vertical integration
- Privacy and confidentiality
- Illegal contents

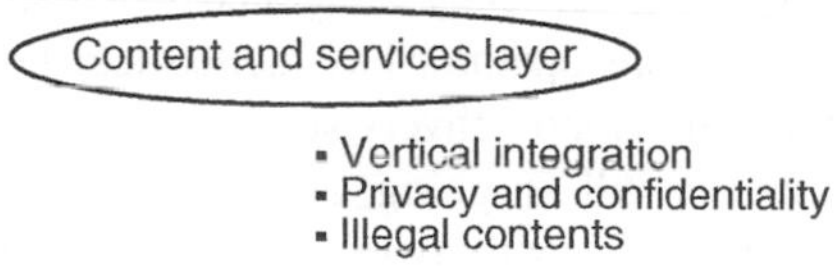

Figure 2.21   Content and services layer policy characteristics

and therefore can be conducted with a high degree of competition and left to market forces to self-regulate. Policy issues of importance to the retail layer are illustrated in Figure 2.20.

Compared with the infrastructure layer, content and retail layers require significantly less initial investment. As a result many internet service providers compete using the underlying infrastructure in many countries. Similarly many content providers and over the top (OTT) service providers, such as Skype or WhatsApp, compete using the retailer and infrastructure layers. Government role in regulating competition in these two layers can be minimal as market forces have led to efficient business systems. Policy issues of importance to the content and services layer are illustrated in Figure 2.21.

Nevertheless, similar to retail layer, the content and services layer may give rise to issues on vertical integration privacy and confidentiality in relation to subscriber information. A vertically integrated player may discriminate delivery of a content of a competitor by slowing it down and reducing its QoS. This behaviour has led to calls for net neutrality, meaning that an infrastructure owner/operator and to a degree a retailer must treat all bits equally. That is, traffic from all content/service providers must be delivered with the same level of QoS. Furthermore, governments may be concerned with the kind of content that may be provided, both in terms of intellectual property rights (piracy) as well as unsuitable content such as pornography, and so on.

Vertical integration across three broadband telecommunications ecosystem layers, as illustrated in Figure 2.22, may lead to discriminatory behaviour. In recent years net neutrality has been a major policy topic in the United States. In essence net neutrality advocates require infrastructure and retail layer companies to be content agnostic and treat all data traffic equally. In particular,

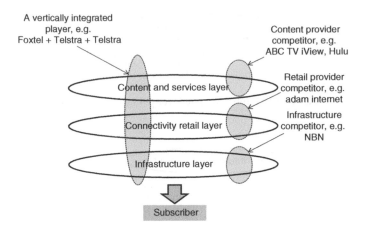

A vertically integrated
player, e.g.
Foxtel + Telstra + Telstra

Content provider
competitor, e.g.
ABC TV iView, Hulu

Content and services layer

Retail provider
competitor, e.g.
adam internet

Connectivity retail layer

Infrastructure
competitor, e.g.
NBN

Infrastructure layer

Subscriber

Figure 2.22   Infrastructure layer characteristics

alleged discriminatory behaviour by a vertically integrated video-streaming service provider + internet service retailer + infrastructure operator has fuelled legislative actions. This company was alleged to have slowed down the video delivery service of a competitor in favour of its own service. Net neutrality discussion is still on-going, although the legislative process appears to have slowed [8]. This demonstrates the capability of infrastructure and retail layer companies to discriminate traffic delivery. Net neutrality and service regulation is being considered in a number of countries around the world. A vertically integrated industry is illustrated in Figure 2.22.

Case Study 2.1: IPMobile

In 2003 the Japanese Ministry of Internal Affairs and Communications (MIC) started the process of selecting a new mobile operator to provide broadband wireless data communications service using the 2010–2025 MHz spectrum. The successful candidate was to acquire the spectrum with no upfront costs based on a 'beauty contest': the ministry's decision of the winning operator was to be based on the strength of the proposed technology and the candidate's business plan viability. This was in stark contrast to contemporary 3G spectrum auctions in many countries around the world where billions of dollars were spent by prospective operators to acquire spectrum rights.

At the time four, mobile telecommunications companies operated in Japan. NTT Docomo was the largest and served some 55% of the market using a home-grown 2G technology (PDC) and a global 3G technology (WCDMA). The second largest was KDDI which had a market share of around 25% and used a 2G/3G technology (cdma-One/CDMA 2000). Third was Vodafone Japan who served some 20% of the market using the same technologies as those of NTT Docomo. Finally, the smallest was DDI Pocket, a subsidiary of KDDI who had a market share of some 10% and used another home-grown technology (PHS).

The main policy goals of the Japanese government appeared to be promoting popular wireless access to internet and reduce the subscription costs which at the time were among the highest in the world. One way to achieve these was to increase competition in the mobile telecommunications markets. As a result the government policy barred the three incumbent players, Docomo, KDDI and Vodafone, from participating in the spectrum acquisition process. As a result, three smaller candidates applied: one was eAccess, a successful ADSL operator with revenues of $270 million and profits of $22 million, who proposed to use a proprietary technology developed by Navini Networks [18]. Another was DDI Pocket, who had revenues of $1.8 billion and profits of $210 million. DDI Pocket proposed to use a proprietary system based on an evolution of its PHS technology. The third was IPMobile, a start-up company consisting of 5 engineers, a few advisers (including the author as the interim CTO), a capitalization of $1 million, no revenue and no operational background. IPMobile proposed to use a 3G global standard called TD-CDMA. A fourth company, Softbank, a major ISP with revenues of $4.5 billion expressed interest to participate in the process with an eye to bid later using a winning technology.

Over the following 2 years a thorough analysis was carried out through a 'shingikai' – deliberation council – under the auspices of the MIC. The interested parties made submissions on the strength and weaknesses of different proposed technologies, and comparisons were made. Major parameters of interest to the ministry were maximum transmission rates, network coverage and maximum possible number of subscribers served. The three candidates were required to provide detailed technology analysis using both theoretical and practical means. This meant that the three companies had to build experimental systems in the Tokyo area and conduct drive tests to verify theoretical results. The parties had to also submit and present reports on how successful commercial and experimental roll-outs of their proposed systems were in other markets. In the case of IPMobile for example, the results of TD-CDMA standard's commercial roll-outs in New Zealand and Germany were analysed and submitted.

After the technology comparison had been made, the companies were asked to provide detailed project and business plans, to demonstrate that they were capable of building and operating the necessary infrastructure for telecommunications services. The information required by the ministry included:

- Network and end-user equipment suppliers
- Network roll-out plans, cell planning companies, the necessary equipment suppliers
- Details of roll-out plan, the number of base stations, place, cost
- Target market and marketing plans
- Investment details, including investors details and conditions

It was highly improbable that IPMobile should emerge as the successful candidate. Certainly the small team at IPMobile considered themselves as rank outsiders. Nevertheless IPMobile was indeed successful. After 2 years of deliberation and much technology, business and policy analysis IPMobile was granted the spectrum licence to build and operate a 3G mobile network based on TD-CDMA technology in November 2005 [18] (Figure 2.23). The business plan submitted by IPMobile to MIC projected a network build up to cover 50% of populated areas in Japan by 2010. This plan called for $1 billion capital to be raised from Japanese and international investors. Ultimately IPMobile was unsuccessful and the company had to fold in October 2007. The 2010–2025 MHz spectrum licence was returned to the ministry.

Figure 2.23 IPMobile President Itsuo Sugimura (right) receiving the spectrum licence from the Minister of Internal Affairs and Communications Heizo Takenaka. Reproduced with permission of Mari Sekine, the communications director of IPMobile at the time

A number of factors were important in the initial success and ultimate failure of IPMobile. Although from technology and policy viewpoints IPMobile had strengths, from a business point of view it had to overcome many obstacles. Many of these had to be overcome early on for IPMobile's project to succeed. Many of these risk factors were not addressed early enough: an early comprehensive business analysis would have improved IPMobile's chances of success significantly.

As discussed, each of the three viewpoints of the TBP framework can be individually analysed and compared based on a number of numerical and qualitative measures. The following is a brief analysis of IPMobile.

Technology Analysis

The three competing telecommunications systems in Case Study 2.1 all used a similar bandwidth. The transmission speeds supported are listed in Table 2.4. As it can be seen, the transmission rates vary widely. This is because the component technologies used to deliver these rates were very different. Nevertheless, there is one technology-based winner through this comparison.

The coverage comparison for eAccess, IPMobile and PHS systems are listed in Table 2.5. Again the difference between the coverage areas is large. Both technological comparisons do not take into account the details of the techniques used by each system. For example, equipment used for each of these systems differed in its bandwidth usage, and therefore by

Table 2.4 Transmission rates for IPMobile, eAccess and PHS proposals

Proposal	Transmission rate in the downlink (Mbps)
IPMobile	2–3
eAccess	5–6
PHS	<1

Table 2.5 Coverage range for IPMobile, eAccess and PHS proposals

Proposal	Coverage range
IPMobile	Several km
eAccess	Several km
PHS	Several hundred m

design delivered different capacity figures. Similarly, they produced different coverage figures as they were originally designed for different applications. In other words, this was not comparing like with like as the technologies mix was different.

Nevertheless, if the comparison is purely on transmission rate and coverage, one technology emerges as winner: eAccess' proposal was superior, the reason being its usage of special antennas. While these special antennas may be used by any technology, the only proposal that did incorporate them was that of eAccess. Regardless, this technological superiority did not determine the final decision: IPMobile won the day in the end as discussed in Case Study 2.1. Therefore while a technological winner using one or two parameters was clear, this by itself was not decisive. In this particular case, business and policy issues were more important in arriving at a final decision.

Business Analysis

A main requirement of the spectrum license grant by the MIC was a thorough business plan where costs of build-up, marketing, customer acquisition, and so on and methods of finance were detailed. The business plan ran to thousands of pages: shown in Figure 2.24 is the entire set of documents on the day they were submitted. The Excel file of the business plan itself was nearly 600 worksheets. The ministry needed to ensure that IPMobile was indeed capable of executing its plan, roll out a network and become a viable business. IPMobile had succeeded in demonstrating the TD-CDMA system's superiority from a technology point of view but that by itself was not sufficient. Business issues were of equally great importance and therefore much effort (and money) went into preparing a high quality business plan.

The ministry officials then conducted a review of IPMobile's internal business analysis on whether the plan was viable. Fundamentally the ministry needed to ensure that necessary capital was in place for network construction and operation until the company could make enough money to survive. IPMobile presented letters of credit from banks and offers of

Figure 2.24 IPMobile business plan set submitted to the Ministry of Internal Affairs and Communications. The folder to the right is the overall summary and all others are reference documents. Reproduced with permission of Mari Sekine, the communications director of IPMobile at the time

support from equity funds and fulfilled the main requirement from the ministry [19]. After the licence was granted, however, many of these investors did not find the IPMobile business model and plan sufficiently compelling and withdrew their support. Ultimately the failure of IPMobile came about because the necessary financial backing could not be secured.

The business analysis carried out by investment funds would likely have looked at financial parameters such as total expected costs of build-up and operation and total expected revenue. They were also likely to have examined IPMobile's business model, including competitive environment, strategies, necessary suppliers and partnership and contributing factors. These and other contributing factors to ensuring a telecommunications technology business is successful are described below.

Policy Analysis

IPMobile's spectrum licence issuance depended on a number of government policy decisions. Many of these decisions had been made well prior to the 'Technology Council' of 2003. These included the participation of MIC in a global harmonisation of frequency spectrum allocation whereby 2010–2025 MHz was set aside for 3G services. International harmonisation of spectrum allocation was in fact a voluntary decision by the Japanese government to partly cede control of its telecommunications policy to ITU, an agency of the United Nations (UN). ITU had approved a set of technologies for its International Mobile Telecommunications for the year 2000 (IMT-2000) frequency spectrum including the 2010–2025 MHz band. This harmonisation placed some constraints on the type of technologies that could be used in this frequency range: in general this meant that non IMT-2000 could not be considered. While as sovereign nation the ministry could choose any technology and disregard an international policy, the frictions it would cause with the UN would be troublesome. This policy was decisive in the choice of TD-CDMA and IPMobile. Other technologies proposed by eAccess and DDI Pocket could not be chosen as neither were part of IMT-2000.

Another government policy helpful to the cause of IPMobile was the government decision to introduce more competition to the Japanese mobile operator market. Two of the incumbent operators, NTT Docomo and Japan Vodafone, would have been likely to choose TD-CDMA to complement their other 3G standard of WCDMA. Had they also been candidates, IPMobile would have had no chance.

The spectrum allocation process itself resulted from a government policy decision. The policy of encouraging usage of broadband wireless communication led to opening of the spectrum and seeking expressions of interest from prospective mobile operators. All these policies factored in the decision by the government to grant the licence to IPMobile.

Private Enterprise Policy Analysis

As noted above, policy analysis can be carried out by the private sector as well as the government. The main goals of a company in policy analysis are two-fold. One is to influence policy decision through lobbying. This can be accomplished by proposing and arguing for alternative policies and courses of action, which are presumably more favourable to the company. The other goal is to understand the consequences of national and international policies and align company strategies to take advantage or avoid conflict.

For example, a policy analysis by the other two candidates for the Japanese 2 GHz spectrum allocation, eAccess and DDI Pocket, could have resulted in abandoning their proposed technologies and opt for TD-CDMA. The Japanese government had made a commitment to ITU, which more-or-less mandated usage of TD-CDMA for this band. However, as far as eAccess and DDI Pocket were concerned, the TD-CDMA alternative was not an option. This was because IPMobile had secured the rights to using TD-CDMA through the only supplier of the technology, IPWireless Inc., a US-based company. The alternative was to try to convince MIC to change its policy; and to achieve this, these companies and their suppliers mounted a push in ITU to ratify alternative technologies for use in the 2 GHz TDD band.

The IPMobile case illustrates several policy issues:

- Market failure: the market was dominated by three players, one of whom had more than 60% of the market share. Resulting prices were higher than most (all?) comparable countries. New competition was needed.
- International obligation allowed only a set of technologies known as IMT-2000. While the government could allow other technologies, and try to ratify these under the IMT-2000 banner, it was a risky and unpredictable path. The government could perhaps go ahead and allow these technologies alongside TD-CDMA, and try to resolve the consequences later on. Neither option was easy. As far as other players, that is eAccess and DDI Pocket, were concerned, their participation and involvement in the Technology Council was therefore nearly futile. Nevertheless, this participation allowed them to be regarded favourably by the government and receive licences in other frequency bands.

- How could spectrum be allocated to entities? The Japanese government had early on decided against spectrum auctions. This decision had already paid dividend in the form of three successful, tax-paying mobile operators. This was in contrast to many operators in a number of European countries who had to write-off assets (and minimise taxes) when they could not gain a return on their investment in acquiring frequency spectrum.
- Increasing competition in the mobile market. In hindsight, the decision of the Japanese government to increase competition through barring Docomo, KDDI and Vodafone from bidding for spectrum has been unsuccessful. As of 2014 there are only three operators in Japan: Docomo, KDDI and Vodafone – now Softbank who acquired them in 2006. DDI Pocket has ceased to be an independent entity, eAccess has been sold off, and the market has been consolidated. This demonstrates the limits of what a government can in fact achieve.

Case Study 2.2: Australia's National Broadband Network

Historically, telephone services in Australia were offered through the Postal organisation. The operation of domestic and international phone services was later transferred to two government commissions: the Australian Telecommunications Commission in 1975 (known as Telecom Australia), and the Overseas Telecommunications Commission in 1946 (known as OTC, where the author worked in 1988–90) [20]. Later on a separate entity was formed to operate domestic satellite services, known as AUSSAT.

In the late 1980s and early 1990s the telecommunications operation was deregulated in order to increase competition, in line with general deregulation policies of the government. First in 1992, the two commissions were merged, named 'Telstra', and gradually privatised in three stages, T1-1997 (33% @ A$3.30/share yielding ~A$14B), T2-1999 (16% @ A$7.40/share yielding ~A$15B) and T3-2006 (34% @ A$3.60/share yielding ~A$16B). AUSSAT was sold in 1992 to Optus, a consortium of Australian and International companies, and competed with Telstra in the long-distance telephony market [21, 22].

Copper lines connecting homes to the domestic telephone network remained the property of privatised Telstra. However the privatisation act mandated that Telstra provide access on this network to all long-distance operators essentially at operating cost. The rationale was that Telstra's network had been developed through national funding by the taxpayers and therefore belonged to all Australians.

The value of this copper network increased significantly at the turn of the century when Asymmetric Digital Subscriber Line (ADSL) technology was introduced. ADSL systems made it possible for subscribers to connect to the internet from their homes/offices at several megabits per second speeds using the old telephone lines. Many telephone operators, including Optus, started offering ADSL services using this Telstra-owned network. The popularity of these services, as well as the need to upgrade and augment the network, caused friction between Telstra and other service providers. Telstra argued that any new copper and fibre cable installation was its own, and therefore it was not subject to the deregulations terms. Therefore it could charge access fees at market rates. The government disagreed, and a stalemate ensued. After a number of attempts to find a private sector solution for the roll out of a fast broadband network, the government decided that a new nationally owned entity was the way forward, and the NBN policy was announced. A new infrastructure monopoly 'NBN Co' was created by this policy, funded by public and private equity. It is intended that NBN Co will be privatised when the national network roll out is completed [23].

The NBN is an Australian government policy to build a broadband telecommunication infrastructure to connect all premises to the internet at very fast transmission rates announced in July 2009. The infrastructure is to be owned by NBN Co, a 100% Australian government-owned Company. NBN roll-out calls for connecting 93% of Australian premises (homes, businesses, schools, etc.) by optical fibre cables to a national grid, with expected transmission rates of 1 gigabit/s (Gbps) or more. The remaining 7% are connected by wireless broadband and broadband satellite technologies with a minimum transmission rate of 12 Mbps. This broadband policy came about after a number of failed attempts by the government to induce the private sector to upgrade and further build a broadband telecommunications infrastructure. The current policy is shaped by the present needs of the country, required transmission rates and therefore telecommunications technology, cost and benefit issues, as well as the policy decisions made by the Australian government when it deregulated the telecommunications industry in the early 1990s.

The birth of NBN can be considered an indirect, unplanned result of deregulation policies of the early 1990s and privatisation of Telstra as a single infrastructure and retail entity. If Telstra had been privatised as two independent entities, then the infrastructure entity could have provided capacity to all retailers, Telstra, Optus and others on an equal basis. Upgrading of this infrastructure would have been then carried out as a matter of course as technology and business circumstances changed, benefiting all retailers equally. There would have been no need for a public-owned infrastructure such as NBN Co.

The current policy has NBN playing the role of national broadband infrastructure owner and operator, as shown in Figure 2.25. It provides the network over which all retail players, including Telstra, can provide service to subscribers on an equal footing.

Figure 2.25 includes two layers of a broadband telecommunications ecosystem, infrastructure and retail. The model is a reflection of the recent development of the telecommunications industry, and its evolution from mainly telephony to internet connectivity. We consider a further layer of content and service providers. Companies active at this layer provide an end-user with information and entertainment content as well as video conferencing, messaging, social network services, and so on. As introduced in the Preface, the ecosystem may be drawn as in Figure 2.6. Each layer uses a number of specific technologies, has specific business dynamics and is subject to different policy regimes. Vertical integration across the layers, as well as competition and collaboration adds to complexities of managing a company in this industry.

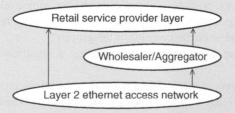

Figure 2.25   NBN and retail service providers [24]

Case Study Questions

- What is the rationale behind the selection of NBN mix of infrastructure technologies?
- What are the business issues associated with the roll-out of NBN?
- How did Australian government policies lead to the broadband telecommunications landscape pre- and post-NBN?
- How may structural separation be conducive to a more efficient telecommunications service provision?
- How do content providers benefit from this structural separation?

Analysis frameworks are one of the most important tools to analysts and managers. For technology businesses, one needs to consider not only the technology but also the associated business issues. For telecommunications systems, policy is another important dimension as governments regulate this business in order to ensure equitable access to services. The TBP framework and its associated models are fundamental to the analysis and management of broadband telecommunications services and content businesses.

Review Questions

1. What are the technology, business and policy issues the Chappe telegraph dealt with?
2. What role did the government of France play in the roll out of the Chappe Telegraph?
3. Assume that a Chappe relay operator can start setting its own transmission after observing the received code. Further assume that it takes on average 2 min to set the semaphore. How long does it take to send a message 'battle is won' from Lille to Paris using the Chappe Telegraph?
4. Why do you think Morse telegraph was a system used by business as well as the government, whereas Chappe's system was mostly used by government?
5. Use the TBP framework and compare the following:
   a. CD and MP3 players
   b. Android vs Windows Mobile
   c. ADSL vs Wireless Broadband
   d. LTE vs WiMAX
   e. Comcast vs Netflix
6. What are the technology, business and policy issues the Morse telegraph dealt with?
7. What role did the US government play in the roll out of Morse Telegraph?
8. Why is Morse believed to be the inventor of telegraphy?
9. Define steps in setting a government policy using a broadband service provision as context.
10. What are the three layers of a broadband telecommunications industry ecosystem?
11. How does service coverage as a technology issue impact on business parameters?
12. What are the major elements of a cash flow analysis? What are special business characteristics for a telecom operator?

13. List two value propositions of a broadband telecommunications service, and described how these values are configured.
14. What were the values offered by a smart phone such as iPhone 3 in 2007? List them in a value hierarchy model.
15. Use examples to describe how value-based and cost-based pricing are different.
16. What were the values offered by an Apple iPad in 2010? Identify a minimum of 10 values and list them in a value hierarchy model. How do you think the iPad price was decided by Apple?
17. When are telecommunications services priced based on value? When are they based on cost?
18. Describe how mobile telephony forms a multi-sided network platform. How can its growth be explained using this model?
19. How can the platform-mediated network concept explain Apple success in popularising MP3 technology?
20. Why does a private sector company participate in a telecom policy deliberation?
21. Give one example of vertical integration in a broadband telecommunications context.
22. Discuss the technology, business and policy implications of using chariots for message delivering on the Royal Highway.
23. Although telecommunications comprises only 3–4% of global Gross Domestic Product, the share of global patent applications in 2013 was nearly 7%. Why?
24. What are the reasons for the rapid growth of mobile communications in the developing world?
25. What are the business drivers for the rapid adoption of broadband services?
26. What technology, business and policy issues does NBN face?

References

[1] Figuier, L. (1868) *Les Merveilles de la Science*.
[2] http://commons.wikimedia.org/wiki/File:Jacques-Auguste_Regnier,_Vue_de_l%27%C3%A9glise_Saint Pierre_de_Montmartre_%C3%A0_Paris,_circa_1820._Coll._part..jpg, accessed 20 August 2015.
[3] http://en.wikipedia.org/wiki/Semaphore_line#mediaviewer/File:Rees%27s_Cyclopaedia_Chappe_telegraph. png, accessed 20 August 2015.
[4] http://en.wikipedia.org/wiki/File:Reseau_chappe77.png, accessed 20 August 2015.
[5] https://commons.wikimedia.org/wiki/File:Swiss_Army_Telegraph_Key.jpeg, accessed 20 August 2015.
[6] Huurdeman, A.A. (2003) *The Worldwide History of Telecommunications*, Wiley Interscience.
[7] Project Optimum – Telenor, (2005) Cited in ECOSYS Report #6, OPEX Models. http://ecosys.optcomm.di.uoa. gr/deliverables/ECOSYS_Del06_v1.0.pdf, accessed 20 August 2015.
[8] Wu, T. (2003) Network neutrality, broadband discrimination. *Journal on Telecommunications and High Technology Law*.
[9] (2004) ECOSYS Report #3, Business Models in Telecommunications. http://ecosys.optcomm.di.uoa.gr/ deliverables/ECOSYS_Del03_v1.0.pdf, accessed 20 August 2015.
[10] Albrecht, K. (2006) *The Northbound Train*. https://www.karlalbrecht.com/books/chapters/NBT00.pdf, accessed 20 August 2015.
[11] Porter, M. (1985) *Competitive Advantage: Creating and Sustaining Superior Performance*, Simon and Schuster.
[12] (2005) ECOSYS Report #5, Tariff Models. http://ecosys.optcomm.di.uoa.gr/deliverables/ECOSYS_Del05_ v1.0.pdf, accessed 20 August 2015.
[13] Boardman, A.E. (2005) *Cost-Benefit Analysis: Concepts and Practice*, 3rd edition, Prentice Hall.
[14] Eisenmann, T.R. (2006) Platform-mediated networks: definitions and core concepts. *Harvard Business Review*.
[15] Dunn, W.N. (2007) *Public Policy Analysis, an Introduction*, 4th edition, Prentice Hall.

[16] Moore, M. (1997) *Creating Public Value Strategic Management in Government*, Harvard University Press.
[17] Australian Public Service Commission. Smarter Policy – Choosing Policy Instruments and Working with Others to Influence Behaviour. http://www.apsc.gov.au/publications-and-media/archive/publications-archive/smarter-policy, accessed 20 August 2015.
[18] http://www.eaccess.net/en/ir/pdf/final_presen_040210.pdf, accessed 20 August 2015.
[19] http://www.jetro.go.jp/en/reports/market/pdf/2006_01_b.pdf, accessed 20 August 2015.
[20] http://www.telstra.com.au/abouttelstra/company-overview/history/telstra-story/, accessed 20 August 2015.
[21] https://www.optus.com.au/aboutoptus/About+Optus/Satellite/About+Optus+Satellite/Satellite+Key+Milestones, accessed 20 August 2015.
[22] http://trove.nla.gov.au/people/602024?c=people, accessed 20 August 2015.
[23] http://www.nbnco.com.au/assets/documents/nbn-co-corporate-plan-6-aug-2012.pdf, accessed 20 August 2015.
[24] Bayley, R. (2011) Network Operations. http://www.nbnco.com.au/content/dam/nbnco/documents/operations-breakout-session.pdf, accessed 20 August 2015.

3

Voice Communications

Preview Questions

- How is voice communicated electrically?
- What is the bandwidth of human voice? Why does it matter?
- How are multiple telephone calls carried over the same physical link?
- How can telephone networks be dimensioned?
- How did telephone operator businesses monopolies emerge and how was competition introduced?

Learning Objectives

- History of voice communications and emergence of telephone operator businesses
- Principles of voice communications
- Analogue modulation and signal multiplexing
- Statistical multiplexing and network dimensioning
- Regulation and deregulation of telecom operation industry

Historical Note

The word telephone was coined by Gottfried Huth in 1796 to describe his invention of a network of mouth trumpets or speaking tubes that could 'pass messages from tower to tower' [1]. The distance over which such *telephony* could be carried out was clearly limited by the power

Broadband Telecommunications Technologies and Management, First Edition. Riaz Esmailzadeh.
© 2016 Riaz Esmailzadeh. Published 2016 by John Wiley & Sons, Ltd.
Companion Website: www.wiley.com/go/BTTM

of the 'loudspeaker' trumpets. However this was only 2 years after the first official Chappe telegraph message, and before the widespread use of electricity and understanding of electromagnetic waves. The mere fact that *telephony* was considered is noteworthy.

A system (pictured)[1] for transmitting sound using electrical signals was first demonstrated by Johann Philipp Reis, a German scientist, in 1861. His invention used an animal membrane stretched over a wooden cavity to convert sound energy to an electrical current. The receiver used a coil which magnetised a needle as the current passed through it, and the movement of the needle reproduced sound. Fundamentally, the principle of converting sound energy into electrical current and transmitting it to a remote location is what is used in analogue telephones to this day. Reis, however, was unsuccessful in selling his invention to Francis Joseph I, the emperor of Austria, and died penniless in 1873 [1]. This is in contrast to the story of Alexander Graham Bell (pictured)[2] who succeeded in building a phenomenally successful business out of his telephone invention, and managed to popularise telephony service. The fact that he is widely credited as the inventor of the telephone may owe much to his business success.

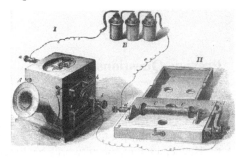

This is not to discount Bell's technological contributions. Most importantly he improved on the quality of the telephone microphone and speaker, and also the wires that could carry the electrical signal over very long distances. The fidelity of sound was important in the marketing of the device. Better transmission medium and technology facilitated a network that could be rolled out economically. Both of these technological improvements impacted greatly on the business success.

One reason often cited for Bell's success is his patenting of the device he invented. His patent application was filed at 2 pm on 14 February 1876, a mere 2 h earlier than that of a competitor, thereby giving Bell monopoly rights on the technology in the United States. However, the success of the Bell Telephone Company (BTC) owes much to a well-managed operation led by Theodor Vail as the managing director. Vail was an experienced manager and the general superintendent of the US Mail service prior to joining BTC. Vail led BTC during 1885–1889 and again between 1907 and 1919, and was instrumental in popularising the telephony service as a necessary infrastructure, and led BTC growth in the face of competition and protracted litigation by a large number of claimants. He also managed BTC relations with government to keep the threat of anti-trust litigation at bay [1].

---

[1] Johann Philipp Reis telephone: https://commons.wikimedia.org/wiki/File%3AJohann_Philipp_Reis_telephone.jpg.
[2] Alexander Graham Bell: Moffett Studio (Library and Archives Canada / C-017335) https://commons.wikimedia.org/wiki/File%3AAlexander_Graham_Bell.jpg.

Soon after establishing BTC, Bell travelled to Europe on his honeymoon. He took his telephone, demonstrated it to government and business officers and applied for patents wherever he could. Two countries where he did not manage to file for patents were Germany and Sweden. Two companies based in these countries, Siemens and Ericsson, rose to become major manufacturers of telephony equipment, and have remained major players in the industry. It is interesting to ponder whether these companies would have become so successful had Bell managed to patent his telephone in Germany and Sweden.

Although telegraphy had enabled fast message transfer, it was not intended to facilitate a real-time conversation. At its core, telegraphy was a one-way messaging system and not too dissimilar from the courier systems that had preceded it over millennia. In contrast, telephony technology was the first time a real-time conversation could be held between two remotely located parties. To converse with others, even if they are not physically nearby, is a basic human need. The invention of telephony clearly had the potential to become a great business success [1].

This chapter is an introduction to telephony, a complete telecommunications system with its specific technologies, business dynamics and policy issues. Moreover, most present telecommunications operators and manufacturers trace their roots to telephony services and equipment, and many of the technologies and processes developed for telephony systems are still in use. Therefore an examination of the telephony system can greatly inform our study of present day broadband telecommunications.

Many technologies have been developed to enable and enhance telephony systems. Initially technologies were focused on sound-to-electricity-to-sound conversion. Research and development into call initiation, called party information, switching, and call termination were also important. Later on signal modulation, multiplexing, network dimensioning as well as automatic switching improved telephony operations. Most of these technologies are important to provision of wireless and fixed broadband telecommunications services. An introduction to these early telephony technologies will be given in this chapter and will be further expanded in later chapters.

Sound and Electrical Signals

A telephone comprises a microphone which converts sound(human voice) into an electrical signal, and a speaker which converts the electrical signal back to sound, as shown in an early illustration of Bell's telephone in Figure 3.1. As noted above, Bell improved on technologies that made these two conversions possible and facilitated higher quality voice communications compared with earlier devices.

The generated electrical signal is transported to a receiver using a physical medium – commonly a twisted pair of copper wires, a medium used since the days of Bell. Copper wires have good electrical conductivity, and are robust, flexible and relatively inexpensive. Furthermore, they can transport the electrical signals representing voice for very long distances with small loss.

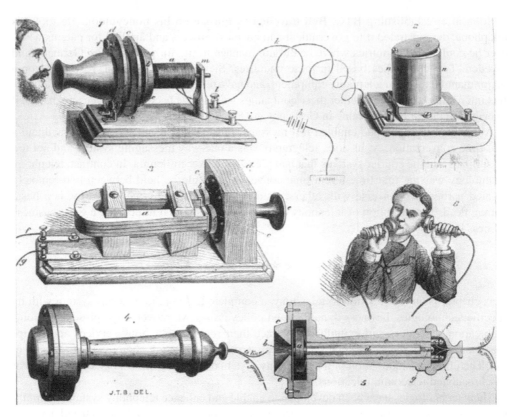

Figure 3.1 Bell's telephone [2]. Source: Illustrated London News – published 1872 (cited: http://hwdp1.blogspot.com.au/)

As noted above, early telephony technological enhancements focused on the development of better wires as well as better microphones and speakers in order that voice signals may be transported over long distances and with good quality. Another major research focus was the development of switches, through which many subscribers could be connected to each other at low infrastructure costs.

Switching

Two telephones need to be connected by wires (circuits) which carry the electrical signals between them. Although the connection can be direct, it is generally made through one or more intermediary switches. These switches are located in a telephone exchange, as illustrated in Figure 3.2. Switching was an important initial step in the development of telephony services as it made the roll out of large networks economically feasible.

The importance of switches was already well known from telegraph roll-outs as messages between cities were relayed through major hubs: direct connections between every two cities with dedicated lines necessitated a very large network and was impractical. The difference is

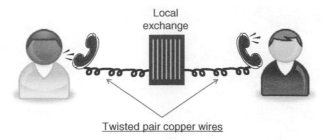

Figure 3.2   A telephone call connected through a local exchange

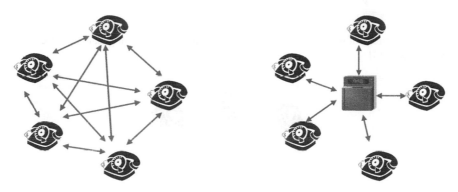

Figure 3.3   Switching efficiency

illustrated in Figure 3.3: a central hub enables 5 nodes to be connected to each other using 5 links, whereas one-to-one direct connections require a total of 10 lines. Furthermore, the length of individual connections can be prohibitive. A central switch reduces the number of required links, as well as the total length, to be proportional to the number of subscribers (N) whereas full connectivity requires links proportional to N^2 [precisely it requires $N*(N-1)/2$ links]. Clearly a directly connected network is not practical.

Initially telephony switchboards were manually operated. A caller told the name, and later on the number, of a desired party to a switchboard operator (illustrated in Figure 3.4) who connected the parties and disconnected when the call ended.

Switchboard operators were mainly young women, as it became clear that they could concentrate and remember subscriber details better than men. Female operators however usually lost their jobs when they married, as marriage was considered a distraction and a valid reason for dismissal. This resulted in a rather limited workforce size. Furthermore, training costs and salaries in general meant that switching was a major operational cost. These motivated development of automatic switching in the early 20th century. Indeed, a number of early telephone operators relied on automatic switching as a competitive advantage vis-à-vis the BTC [1].

With the growth of subscriber numbers it became necessary to use multiple switches in a hierarchical configuration, as shown in Figure 3.5, to reduce the cost of wiring. While each subscriber still had a dedicated connection, the length of connection to the switch could be significantly reduced. Furthermore, hierarchical switching also allowed for a larger number of

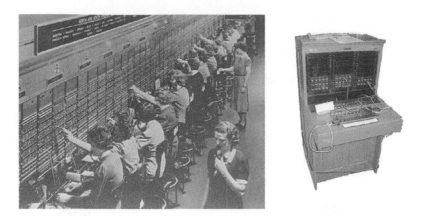

Figure 3.4   Photograph of women working at a Bell system telephone switchboard and an early switchboard. Source: Ref. [3]

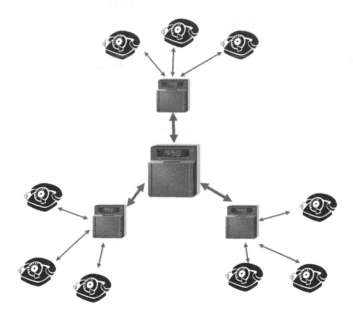

Figure 3.5   Hierarchical switching

subscribers to be served as the capacity of a switchboard was limited. Due to the hierarchical switching configuration, long-distance connections needed to be set up through operators in advance. Automatic switching allowed for a faster and less expensive call set up.

Nowadays, a typical telephone call goes through a number of automatic exchanges, which switch the call to its destination, as shown in Figure 3.6.

The connection between a customer's premise and the local exchange has remained largely the same to this day and is made of twisted pair copper wires. As noted, these wires are inexpensive and have excellent characteristics for transmission of voice signal for long distances.

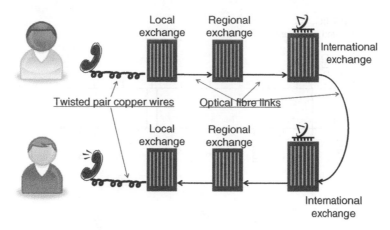

Figure 3.6 A typical telephone call connected through a telecommunications network

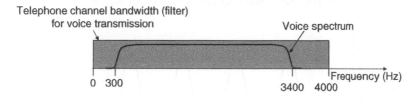

Figure 3.7 The voice signal frequency range

This is because human voice frequency components are mostly limited to within a range of 300–3400 Hz (Figure 3.7), and the twisted copper wires can transmit this range of frequencies for tens of kilometres with little signal loss. Voice signal is filtered in the telephone so the transmitted signal is confined to within 4 kHz to minimise the amount of added noise at the receiver. Usually each subscriber has a dedicated link to the local exchange, which is used to carry one voice call at a time.

Frequency and Bandwidth

To understand telecommunications better let us first describe signal frequency and bandwidth concepts. Modern telecommunications are carried out using electromagnetic waves. These waves are natural phenomena and are infinite in their range: from very low frequencies to very high frequencies. Although the range of frequency spectrum is infinite, only certain frequency ranges can be used for telecommunications as shown in Figure 3.8.

The concepts of frequency and bandwidth are important in understanding how information signals are generated and communicated. Frequency is defined as the number of cycles per second: it is the number of times a waveform ebbs and flows each second. For example, Figure 3.9 shows a sine waveform with five cycles a second. The unit of frequency is hertz (Hz) in honour of Heinrich Hertz (1857–1894), who first experimented with the transmission of electromagnetic waves. Frequency is the standard metric for measuring the variability of a signal with respect to time. For example, the signal represented in Figure 3.9 is said to have a

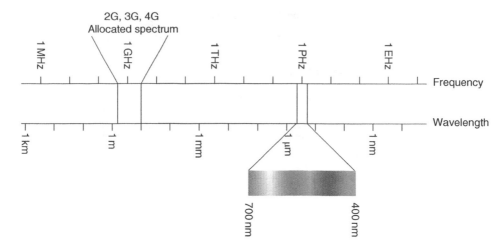

Figure 3.8   Frequency of electromagnetic waves

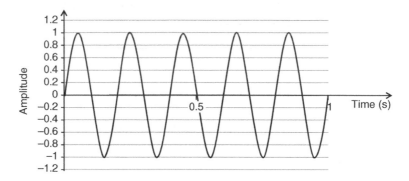

Figure 3.9   A waveform with a frequency of 5 Hz

frequency of 5 Hz. The maximum amplitude of the sine waveform here is 1 volt (V), named in honour of Alessandro Volta (1745–1827), the inventor of the battery.

Similarly Figure 3.10 shows three waveforms (a, b and c) with frequencies of 5, 15 and 25 Hz, with maximum amplitudes of 1, 0.33 and 0.2 V. Figure 3.10 also shows the sum of these signals in (d), a waveform which has nearly a square wave shape. As Figure 3.10(d) is the sum of the three signals a, b and c, it has exactly three frequency components, 5, 15 and 25 Hz. The *signal bandwidth* is defined to be the maximum frequency component minus the minimum frequency component. Therefore, the signal of Figure 3.10(d) has a bandwidth of 20 Hz.

Most natural signals contain many frequency components. An example is the human voice. Our voice is generated by vocal cords which resonate at a particular frequency which is then modulated into a range of frequencies and harmonics by our throat, mouth, teeth and lips. The range of human voice frequencies, as shown in Figure 3.7, is generally measured to be 300–3400 Hz. This means that the human voice frequency bandwidth is 3100 Hz. Figure 3.11 shows a sound clip in the time domain (top) and in the frequency domain (bottom). The lowest pale strip in the frequency domain shows the fundamental resonance frequency and the lines above, its multiple harmonics.

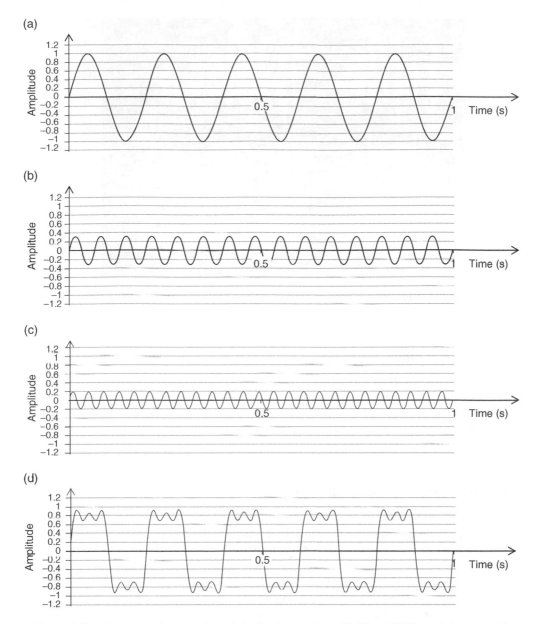

Figure 3.10 Three waveforms (a, b, and c) with frequencies of 5, 15 and 25 Hz, and their sum (d)

The bandwidth of a signal is measured in the baseband, which is the frequency components naturally produced by a source. Modulation and signal processing may increase a signal's bandwidth before it is transmitted.

Generally the more information contained within a signal, the more the bandwidth. For example, an audio signal (such as music) may have a bandwidth of up to 20 kHz and a TV video signal a bandwidth of 7 MHz.

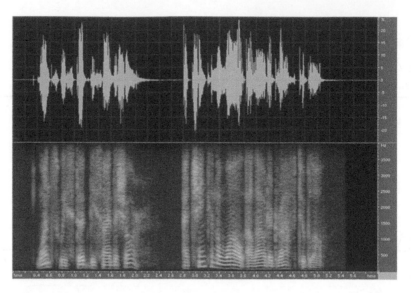

Figure 3.11 A sound clip in the time (top) and frequency (bottom) domains

Circuit Switching

The term circuit switching was coined to describe telecommunications systems which can only start after a link has been established. The switching of intermediate links between a caller and a receiver through a switchboard or an automatic exchange establishes such an end-to-end connection which generally remains until the call ends.

Inter-exchange Telecommunications

Because of the good conductivity properties of twisted copper wire in the 300–3400 Hz frequency range, the required power for transmission of voice electrical signal is quite low. These links have been traditionally used to connect a subscriber's premises to the local exchange. The necessary power to transmit and receive voice signal is generally supplied through the local exchange and no local power source is necessary at the customer's premises, which is why the telephones work even when electricity is out.

Commonly many subscribers served by a local exchange need to connect to subscribers of other local exchanges, and therefore the links connecting the exchanges need to carry multiple calls simultaneously. The twisted pair copper wires were generally set up to carry one telephone call. Therefore, multiple pairs of wires or media capable of carrying multiple calls are needed to connect the exchanges together. The latter solution uses a technology known as multiplexing, where two or more connections are carried over the same physical medium.

To understand signal multiplexing we first define modulation and demodulation technologies. These technologies are fundamental to telecommunications and were among the early developments of this field and are in use in present broadband systems.

Modulation

Modulation is the process of carrying desired data contained within a signal through varying a carrier signal. The data can be embedded within the signal amplitude as shown in Figure 3.12. Here an information signal $x(t)$ modulates a carrier signal, in this case a cosine waveform, $\cos(2\pi ft+\theta)$, where f represents the frequency of the carrier signal. Since the amplitude of the carrier signal is modulated by the information signal, this method is known as amplitude modulation (AM). In AM systems, a frequency waveform's amplitude is changed relative to a data signal, and as a result the envelope of the modulated signal carries the data.

Data can also be embedded in the frequency of a carrier signal. This is known as frequency modulation (FM). In FM systems the frequency of the carrier wave is changed relative to the information signal's amplitude: the higher the amplitude of the information signal, the higher is the frequency of the modulated signal. Conversely, the lower the amplitude of the modulating signal, the lower is the frequency of the modulated signal. This is illustrated in Figure 3.13. AM and FM technologies are well known in radio broadcasting.

Both AM and FM technologies are commonly used for multiplexing in telephony systems. In general FM is more robust against noise and interference and is used in wireless telecommunications. For example, the first generation of mobile phones used FM. AM radio is less robust and is therefore used in applications where noise and interference levels are relatively low, such as in fixed line communications. A variation of AM technology was used to multiplex calls on inter-exchange links.

Amplitude modulation can be accomplished using a simple signal multiplication. Assume the information signal is $x(t)$ and the carrier signal has a frequency of f_c and a signal amplitude of A written as:

$$A\cos\left(2\pi f_c t\right)$$

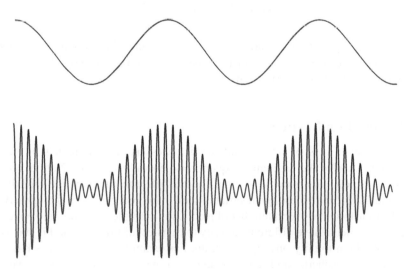

Figure 3.12   Amplitude modulation

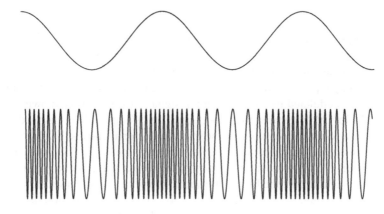

Figure 3.13 Frequency modulation

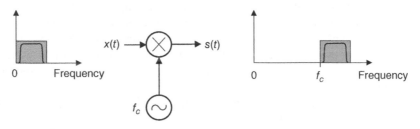

Figure 3.14 Amplitude modulation process

The modulated signal $s(t)$ is simply the product of the information signal and the carrier signal and is written as:

$$s(t) = Ax(t)\cos(2\pi f_c t)$$

The process of amplitude modulation can be implemented in hardware as illustrated in Figure 3.14. The modulation process transfers a baseband signal (e.g. a 0–4 kHz voice signal) to higher frequency band of a carrier signal.

Why is Modulation Needed?

There are a number of reasons why modulation is necessary. The frequency of the original information signal, called the baseband frequency, is generally very low. Natural phenomena frequencies and bandwidth are not very high. For example, the human voice's maximum frequency is some 4 kHz, and video signal variations are in the order of several megahertz. These signals can be transmittable over a number of media, for example air, space and optical fibres. However different media have different capacity to transfer information, and it may be inefficient to dedicate them to only one signal stream. For example, coaxial cables have a bandwidth of several megahertz and are capable of transferring hundreds of voice calls simultaneously. Should transmissions be carried out at baseband, only one voice signal can be accommodated over the channel. This is because two voice signals transmitted simultaneously at baseband over the same medium would mix and interfere with each other. Modulation to different carrier frequencies allows for these two calls to be carried simultaneously over the same medium. This process is called call multiplexing, and is further explained below.

Another reason for modulation is that transmission of signals may not be possible at its baseband frequency over a medium. For example, transmission of voice signals over optical fibres is not possible in the baseband. The signals need to be modulated to higher frequencies – such as those of the visible and invisible light range – for them to be transmittable over optical fibre links.

Yet another reason for modulation is the range of transmission. Voice waveforms, in their baseband frequency, can be carried by air and be detected by human ear over a distance measured in tens of metres. The same voice signal may be modulated using amplitude modulation and transmitted to a distance of tens of kilometres using modulating frequency in the AM radio range (500–1300 kHz) over the air. The same voice signal may be modulated in the short wave radio range (35–100 MHz) and transmitted halfway around the earth, since signals transmitted at this range use the earth's stratosphere as a communication channel. Modulation of signals using different carrier frequencies enables different range and quality of transmission.

Demodulation

At the receiver a signal must be transferred back to the baseband so that its information is extracted. For example, a sound signal can only be heard by human ear if it is played back at baseband. Principally AM signals are demodulated through multiplication by the carrier signal followed by baseband filtering. This is illustrated in Figure 3.15.

Mathematically this process is shown below, where $r(t)$ represents the signal after the received signal $s(t)$ is multiplied by the carrier signal:

$$r(t) = s(t) A \cos(2\pi f_c t)$$

$$r(t) = x(t) A^2 \cos^2(2\pi f_c t)$$

$$r(t) = 0.5 x(t) A^2 [1 + \cos(2^* 2\pi f_c t)]$$

A low pass filter suppresses the signal with $2f_c$ frequency and the baseband signal $x(t)$ multiplied by a constant factor is obtained.

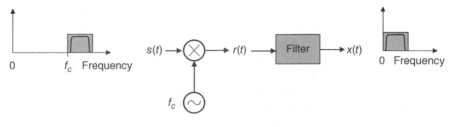

Figure 3.15   AM demodulation process

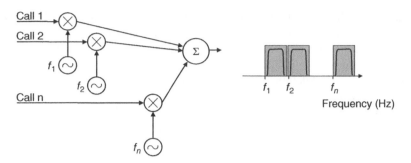

Figure 3.16   Frequency division multiplexing process

Multiplexing

The links that connect exchanges together need to carry multiple calls as discussed above. As shown in Figure 3.6 these links are nowadays mainly made of optical fibres which can carry a large number of calls simultaneously. Before optical fibres these links were made of coaxial cables which were capable of carrying tens to hundreds of calls simultaneously. Each voice call needs its own dedicated channel over the coaxial cable, or optical fibre link, without interfering with other calls. This process is known as multiplexing, and refers to the process whereby several signals are simultaneously transmitted together using a shared medium. Several types of multiplexing exist. Frequency division multiplexing (FDM) is when user signals are modulated to different frequency bands. The frequency bands are distinct and as a result signals do not interfere with each other. Another method is time division multiplexing (TDM) where multiple signals are each sent at a different time slot.

FDM technology is widely used for transmission of voice signals over high capacity coaxial cables between exchanges. The FDM process is illustrated in Figure 3.16, where three modulators operating at three different carrier frequencies are shown. The resulting multiplexed output depicts the voice signal spectrum side-by-side. The FDM process may be likened to a freight train that loads cargo containers, and carries them side by side between major train stations. At the destination train station the container is downloaded and carried to its final destination individually. Similarly on the receiver local exchange, each modulated signal is transferred back to its baseband signal and delivered to the intended recipient.

Teletraffic Engineering

High capacity media which connect exchanges need to be dimensioned in the number of calls they may carry. An important parameter in inter-exchange capacity is the necessary number of lines to ensure a high degree of connectivity. For example, if an exchange serves 1000 local subscribers, the number of links that connect the exchange to other parts of the network is an important technological, business and indeed policy parameter. One factor of interest is the probability that a subscriber will not be able to make a call because all the inter-exchange links are in use by other subscribers, and is therefore blocked. Under-dimensioning, or too few links, requires small infrastructure investment. However, it will result in blocking probability which is too high. Conversely, too many links requires too high an infrastructure cost, while the blocking probability will be quite low – perhaps even less than required from business or policy constraints.

As noted above, a capital expenditure point of view mandates an operator to keep the number of lines to a minimum as each extra link adds to the cost of network roll-out. The question is then how many multiplexed channels are needed between two exchanges considering the *likely* number of calls that may be active at any one time. Assuming each call requires a dedicated circuit for the duration it is active, and that all circuits are busy, then any new call attempts will not find a channel onto which they may be multiplexed. Such a call is blocked and the subscriber will be given a busy signal. On the other hand, dimensioning the cable to have extra lines increase both capital and operational expenditures and therefore it is desirable to avoid unnecessary extra lines. A field of telecommunications technology, teletraffic engineering, was developed with focus on calculating the necessary number of lines for a certain probability of call blocking.

Teletraffic engineering techniques calculate the number of necessary lines for carrying voice (and later data) traffic over a finite number of telephone lines. These inform a telecommunications system designer on the number of links that are necessary to connect an exchange to the rest of the network. This number is found based on the number of subscribers, and the likely amount of traffic that at any time may flow from the exchange. Through teletraffic engineering, the size of the inter-exchange links is calculated based on the expected number of calls that would be made between users in different localities, and the required probability of blocking.

Call traffic per user is defined as the portion of the time a subscriber is active. For example, if a subscriber is active 3 min on average in a 1-h period, the amount of activity factor for the subscriber is $3/60 = 5\%$. Call traffic is measured in units of erlang (E), named after Agner Erlang (1878–1929) a Danish mathematician who founded the field of teletraffic engineering. One erlang of traffic is defined as the amount of traffic that occupies one telephone link. In the above example the subscriber is generating 0.05 E of traffic [4].

The traffic amount depends on two factors. One is the length of a phone call, known as the *holding time*. This is a random process, and is defined by the average length of a call (usually denoted using μ) and its probability distribution. The other is the time between two call attempts, known as call inter-arrival. This also is a random process defined by its average (usually denoted using λ) and probability distribution. The amount of traffic (ρ) is calculated by dividing average holding time by average inter-arrival time: $\rho = \mu/\lambda$. The holding time and inter-arrival processes are illustrated in Figure 3.17, where different μ's show the amount of time a call occupies a multiplex channel and λ's show the time between the arrival of two phone calls as seen by the exchange. Figure 3.17 shows that at least three circuits are required for this system if no call is to be blocked.

Blocking probability is of great importance in telephony as it is a measure of the quality of the telecommunications network. Network subscribers will not be satisfied with the service if too many calls are blocked, and may migrate to a different operator, which clearly has business implications. Blocking probability P is defined as the probability that an attempted call cannot find a free multiplexed channel and is therefore blocked. It is calculated from the following Erlang-B formula:

$$P = \frac{\dfrac{\rho^n}{n!}}{\displaystyle\sum_{x=0}^{n} \dfrac{\rho^x}{x!}}$$

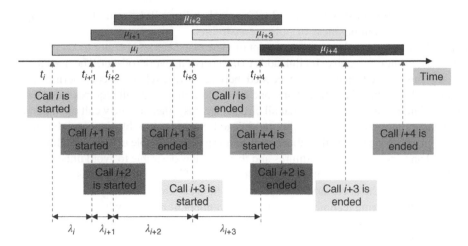

Figure 3.17   Holding time and inter-arrival processes

where n is the number of voice channels and ρ is the amount of offered traffic. Blocking probability is also known as the grade of service (GoS). GoS has traditionally been a parameter of great importance in network design and has also been used in evaluating the competitive advantage of different long-distance telephone operators. GoS levels have also been regulated by governments in many countries as a minimum measure of performance for monopoly operators. In particular, in countries where the incumbent telephone company was government-owned, quality of service provision could be a political issue, and therefore the regulator used GoS to ensure high quality service provision.

The Erlang-B formula is complex to calculate manually. It can easily be calculated however using a simple computer program and is found on numerous websites. The Westbay Engineers site [5] has a simple interface for calculation of GoS based on offered traffic amount and available voice circuits. Alternatively it can calculate the number of necessary lines for a desired level of GoS.

Example 3.1

One thousand users each generate on average one call every 2000 s, or on average, one call is generated every 2 s. Each call goes on for an average of 100 s. What is the amount of offered traffic? What is the GoS for 50 lines and 64 lines?

Answer

- $\mu = 100$ s
- $\lambda = 2$ s
- $\rho = 100/2 = 50$ E
- For 50 circuits: GoS = 11%
- For 64 circuits: GoS = 1%

Case Study 3.1: Transatlantic Cable 1 (TAT-1)

The first transatlantic submarine telegraph cable was laid in 1858, only 14 years after Morse's first official message was sent. The line was very slow (it took 16.5 h to send a 96-word message from Queen Victoria to President Buchanan [6]). With the invention of telephony, transatlantic telephony was considered, but clearly the required capacity was not available. The main problem was signal dissipation through the coaxial cable, which meant that a transmitted voice signal would be received well below noise level. With no fixed line telephony, transatlantic voice communications was carried out using wireless technologies. These were unreliable as short wave radio signals were susceptible to signal fading and noise.

To increase received signal power from coaxial cables repeater devices needed to be installed at intervals of some 70 km. These required power which had to be supplied from both ends. The technology for such repeaters needed for submarine cable telephony was not developed until the 1950s. The first transatlantic cable (TAT-1), connecting the UK, Canada and the US was commissioned in 1954 and started operation in 1956.

TAT-1's useful bandwidth was 144 kHz and could provide 3 groups of 12 voice channels (36 in total). The cost of manufacturing and laying the cable down was $36 million in 1954, or the equivalent of some $320 million today (2014), which averaged over its 22 years of operation cost some $34 000 per channel per year. The cost of a phone call needed to be set in the order of $3 per minute to pay back the initial investment and operating costs [or $27 today (2014)]. This was still a major improvement over the radio calls which had inferior voice quality, and cost some $9 per minute [7].

The transatlantic telephony link became very popular and traffic grew at an annual rate of 20%. Soon extra capacity was needed, and therefore link capacity had to be increased by reducing the bandwidth required for voice communication from 4 to 3 kHz. This increased the number of circuits by 33% to 48 channels at a reduced voice quality. Even at this quality, 48 circuits could only support 36 E of traffic. To maximise the links' usage transatlantic calls were handled by booking through operators.

With the popularity of the link, more submarine cable systems were commissioned and laid around the world. Seven generations of coaxial cable-based transatlantic cables were laid before optical fibre technology was developed. Table 3.1 shows these seven transatlantic cables and their respective capacities, and cost per speech channel per year.

Table 3.1 Transatlantic cables, number of channels, and their cost [1]

Cable name	Years in service	No. of channels	Cost per channel ($)
TAT-1	1956–1978	48	34 000
TAT-2	1959–1982	48	15 000
TAT-3	1963–1986	138	10 000
TAT-4	1965–1987	138	10 000
TAT-5	1970–1993	845	3000
TAT-6	1976–1994	4000	1500
TAT-7	1978–1994	4000	900

With higher capacity, statistical multiplexing efficiencies grew reducing the need for call bookings. Subscribers could place international calls directly and without the need to go through operators. The cost of calling also dropped to \$1–2 per minute and was within the reach of many subscribers. As a result, calling relatives and friends in other countries, especially on special occasions such as at Christmas and New Year, became common. After the mid-1960s geostationary telecommunications satellites also added capacity to the international telephony market, further reducing the cost of the long-distance phone calls.

Case Study Questions

- How important was an understanding of teletraffic engineering to the designer of transoceanic international telephony cables such as TAT-1?
- Why was TAT-1 rolled out so long after the invention of telephony?
- What were the business issues and why did the price per call drop so dramatically?
- Who were the initial TAT-1 customers? Why did this change?
- What were the roles of governments in rolling out of transoceanic telephony cables?

Telephone Operator Business

Alexander Graham Bell envisioned that telephony would become an infrastructural utility. As he wrote to his father on the evening after his successful experiment: 'I feel that I have at last found the solution of a great problem, and the day is coming when telegraph wires will be laid on to houses just like water or gas is, and friends will converse with each other without leaving homes' [8]. As Bell predicted, the telephony business soon became a necessary utility, and behaved as a natural monopoly business similar to gas and water. In fact telephony had a higher propensity to become a natural monopoly compared with gas and water. The value of these utilities was inherent in themselves and largely independent of whether one's friend or neighbour also had access to them. In contrast, the value of telephony services largely depended on whether one's friend and family had a telephone subscription: the value indeed increased as more subscribers joined the network. From a natural monopoly point of view, an operator with the largest number of subscribers was likely to attract more subscribers since most of one's network were also likely to be subscribers.

On the other hand, similar to all large fixed-cost utility businesses, telephony service provision duplication was inefficient. No house is served by two set of water or gas pipes or electricity wires. These factors meant that telephony companies in each local, and soon national, market became natural monopolies. While in the United States, BTC remained a private company, in almost all other countries telephony operation was conducted through government-owned companies. These companies remained monopolies until the late 20th century when they were largely sold-off following a global trend in deregulation and privatisation.

These operator companies owned the infrastructure, generally the twisted pair copper wires connecting a subscriber's premises to an exchange, the exchanges and switches and cables connecting the exchanges together. They also were the retailer of the telephony service: managing the subscriber and ensuring a mandated GoS was maintained. Furthermore, they controlled and provided most of the content provided over telephony networks: contents such

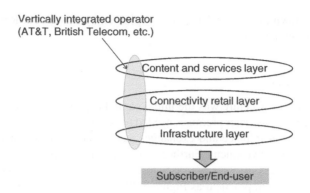

Figure 3.18 Vertically integrated operators of the first century of telephony services

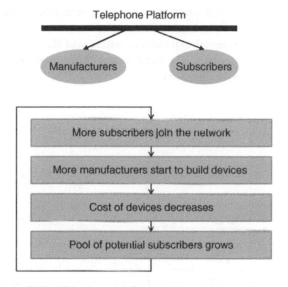

Figure 3.19 Telephony platform and its virtuous cycle growth

as directory assistance, information on weather, stock, horoscope, and so on were provided mostly through the incumbent operator. Anyhow, the content and services constituted a very small portion of an operator's revenue. In summary, the telephone operator was generally a vertically integrated company providing all the services a subscriber needed for over a century after the invention of telephony (Figure 3.18).

As noted in Chapter 2, telephony operators also created and controlled a platform-mediated network, which grew as the number of subscribers increased and more manufacturers entered the market. The growth in the number of subscribers reduced the per subscriber fixed cost of network roll-out. It further helped attract more third-party manufacturers and investment in research and development (R&D) to enhance the efficiency and quality of the networks, as illustrated in Figure 3.19. The increase in the number of manufacturers had the added impact of extra investment in R&D of new technologies and products. The reduction in cost per circuit shown in Table 3.1 is partly explained by this global R&D effort by both the operator and manufacturers.

Capital Expenditure

Most of the operator cost structure was the initial capital expenditure associated with the roll out of the infrastructure. Customer premises needed to be wired and connected to an exchange, switches needed to be put in place and exchanges need to be connected with each other. All these required significant investment. Furthermore, early networks used telegraph poles and overhead wires to connect customer premises to the local exchange. As the number of subscribers grew this became impractical as Figure 3.20 illustrates. Telephone companies had to build underground ducts to house these copper wires as a significant capital outlay. Further capital expenditure was required for automatic network switching equipment and high-capacity, long-distance links roll-out.

Historically, most operators grew gradually, initially connecting the central part of a city, and then connecting the outer suburbs. Similarly, telephony services were offered in large cities before small cities. The capital expenditure outlay was therefore made from the cash flow from serving present subscribers. The large capital expenditure played into the hands of an incumbent operator as a major entry barrier.

Figure 3.20   Overhead telephone and telegraph wires in Broadway, 1890. Source: Ref. [9]

Operational Expenditure

The telephony network roll-out costs were very large. Thereafter however, operational expenditure was minimal. These included the cost of human resources (switchboard operators) and later on the cost of electricity for switching the calls. Furthermore, there existed on-going costs of network maintenance and customer service. The costs of customer acquisition remained small as telephony services became a necessity. Furthermore, most new customers paid for having their premise connected to the nearest exchange.

Many telephony operators maintained an R&D department to enhance telecommunications equipment and services. The most prominent of these was the Bell Laboratories which conducted a wide range of research and earned seven Nobel Prizes for a number of inventions such as transistors and discoveries such as cosmic microwave background radiation. While many of these innovations helped the operator business, some did not contribute to the main business of the operator.

Revenues

Telephone operator revenues were generally from subscription fees paid for services. These varied from operator to operator. Some operators charged a fixed monthly fee and allowed unlimited local calls while charging extra per minute for long-distance calls. Some other operators charged only for the calls a subscriber made, with a fixed rate for local calls and per-minute charges for long-distance calls. Some operators even charged local calls per minute. A combination of these models also existed. This fee-based approach, where a customer is charged for usage of the telecommunications link is also known as charging for the 'pipe'. This means that an operator owns a telecommunications pipe which it rents to a subscriber for a certain fee.

Early telephone operators also raised some revenue through provision of services and contents. These included directory assistance, stock and weather information, horoscope, and later on an answering service and call transfer. The revenue raised from these value-added services was small however compared with the revenue raised from operating a 'pipe'. Pipe and content businesses are illustrated in Figure 3.21.

Policy

The natural monopoly character of telephony operators in many countries run against governing laws on anti-trusts and monopolies. The nationalisation of most operators worldwide was one response to ensure service provision was well regulated and controlled. Another reason for nationalisation of the industry in many countries was the fact that the initial capital expenditure was beyond the means of the local private sector, and therefore the government had to step in. The case of the BTC in the United States was different as it remained a private sector company. Nevertheless it had to battle the Department of Justice as it established its monopoly of service provision to ensure a balanced regulatory regime.

Another major policy concern was service provision equity. A publicly owned entity afforded governments a strong role in ensuring sparsely populated areas had equal access to telephony services. Such service provision necessitated subsidisation as revenues did not cover costs. Nevertheless, the incumbent national operators were generally profitable on the whole.

Government departments were also heavily involved in negotiating and funding transoceanic and international telephony cables. These infrastructures were funded and rolled out on a bilateral/multilateral basis. Telecommunications satellites were a global undertaking as they

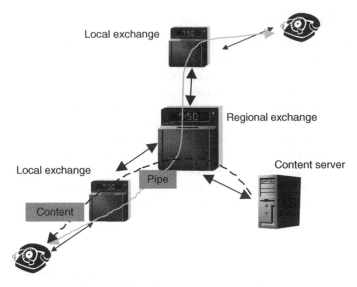

Figure 3.21   Pipe and content provision

served entire continents. These infrastructure projects were negotiated under the auspices of the United Nations and through the International Telecommunications Union (ITU).

Global Standardisation

National telephone companies followed global standards in various aspects of telephony systems. Analogue telephony transmitted in the baseband can easily be converted to sound. However, signalling systems needed to be standardised, especially after automatic switching was introduced. The called party number needed to follow standard specification in order that the receiving exchange could route the call to its destination.

The telecommunications government department in most countries was represented in standardisation bodies – at times through the incumbent operator, and worked together with manufacturers in ratifying these standards. They then ensured that national operators worked within these standards.

Case Study 3.2: AT&T

The case of the BTC and its successor the American Telegraph and Telephone Company (AT&T) illustrates the business and policy issues associated with telephony service provision. Bell's contributions as discussed in the Historical Note included perfecting the technologies necessary for transmission of voice using electrical wires. However, these by themselves were not enough to ensure the success of telephony service provision. Bell also created a company which in time would provide a universal telephony service to almost the entire North American market. The BTC (and its successor, AT&T) was greatly successful and grew to become one of the most valuable companies on the New York Stock Exchange.

The success of the BTC was based initially on its utilisation of Bell's patented technology, licensing fees and ensuring competitors remained at a disadvantage. In the process BTC successfully defended its patents in courts over a period of some 20 years [1]. As it grew, BTC's cost advantage allowed it to outperform and acquire competitors one after another: eventually becoming almost the sole provider of telephony services in the United States.

Telegraphy and telephony were nascent industries: BTC established an R&D entity, and equipment manufacturing subsidiaries to provide it with the necessary technologies and devices as well as leadership in the field. In particular, Bell Laboratories – the research arm of the company – was the place which attracted many of the brightest mathematicians and scientists of the day. These researchers discovered and formulated principles not only associated with telecommunications but also many other fields as diverse as astronomy and quantum physics. In time Bell Laboratories became a world leading research institute and home to seven Nobel Prize laureates as well many other prestigious award winners. By any measure BTC's technological endeavour was a great success.

BTC's business dynamics were characterised by the high infrastructure roll-out cost, and network effects. It was soon realised that no market could afford to have more than one operator as it was inefficient to duplicate wiring, and difficult to connect circuits between two operators. This led to the establishment of monopoly operators in different geographic regions, who were then acquired one after another by BTC, resulting in a near-monopoly operation in North America. Indeed, a theory was developed by AT&T president Theodore Vail in 1907, that the telephone, 'by nature of its technology, would operate most efficiently as a monopoly providing universal service' [10].

This monopoly control of the market was in conflict with anti-trust laws of the United States, and therefore BTC had to negotiate with the US government to allow it to run a monopoly business despite existing strong anti-trust laws. Over the years several anti-trust suits were launched but BTC continued to operate successfully for over 100 years.

Despite all its advantages, BTC–AT&T fell into decline in the mid-1980s and was finally acquired by competitors, a fall which can be attributed to all three technological, business and policy factors.

First, on the policy side, monopoly operation and lack of competition may have had a role in AT&T's reputation for inadequate customer service. In 1974 another anti-trust suit was launched by the US government which resulted in a settlement which broke the monopoly operation that BTC had enjoyed. The company was divided into seven regional telephone operators known as Baby Bells. An eighth company connected these regional companies as a long-distance operator under the name AT&T. While BTC–AT&T had remained successful from the technological and business viewpoints, it had failed to stay with the policy shift and adjust accordingly.

AT&T continued to operate profitably as a long-distance operator, carrying calls between regional Baby Bell telephone companies, and later as a mobile telephone operator. However, technological developments in whose development AT&T had played a prominent role led to its own demise.

Bell Laboratories had been one of the major contributors to the development of optical fibre systems. These cables have very large capacities and can carry a very large number of simultaneous telephone calls. With the deployment of optical fibre cables in place of copper cables, and the deregulation of operator business, the cost of long-distance telephony dropped, and AT&T's business as a long-distance operator suffered.

Mobile telephony was the other technological development which Bell Laboratories had pioneered. The growth of mobile operators and the competitive pressures on AT&T mobile division, in addition to its dubious strategic moves into content provision, finally led to high debt burdens and eventual fall. In effect, the technological break-throughs pioneered by AT&T had been used by its competitors in a more effective way.

The 'AT&T' brand however was still very valuable and the acquiring companies adopted the name. The policy–business–technology developments of the last decades in the 20th century transformed the industry landscape and led to emergence of new companies. It also signalled the end of arguably the most iconic telecommunications company [11].

Many telephone operators went through the same experience as AT&T. The majority were government-owned, and many have stayed afloat after deregulation. The emergence of mobile and broadband telecommunications has created new revenue streams as the case of Australian operator Telstra demonstrated (Chapter 2). Nevertheless, the dynamics of technology, business and policy have been in play in every market, and have been instrumental in the success or failure of operator companies.

Case Study Questions

* How did AT&T establish a monopoly?
* How did AT&T manage the technology–business–policy framework?
* Was AT&T successful in R&D?
* Why was AT&T broken up?
* In hindsight, is there anything AT&T could have done differently?

Private Branch Exchange and Leased Line Business

We finish this chapter with a wholesale business model offered by telephone operators in the second half of the 20th century. This wholesale business appeared because of the need for internal telephony systems by large organisations. A number of organisations such as large companies and governments have offices in remote places that need to be connected on a regular basis. Officers in these organisations make many calls over the course of a working day to remote offices. Until the mid-1990s the cost of long-distance telephony was quite high and therefore it made business sense for these organisations to enter into special arrangements with telephone service providers such as Telstra or AT&T. In essence these companies managed to buy telephony services on a 'wholesale' rather than a 'retail' basis. To cater to the needs of these customers, a new line of wholesale business emerged. Telephone operators started leasing a dedicated bundle of telephone lines to a customer for a fixed monthly charge. These large organisations could then use leased lines to link their geographically separated buildings, connecting them through a single internal network. Their internal network used a private switch, known as a Private Branch Exchange or PBX. Such a network structure is illustrated for the Carnegie Mellon University (CMU) campuses in Pittsburgh and Adelaide in Figure 3.22.

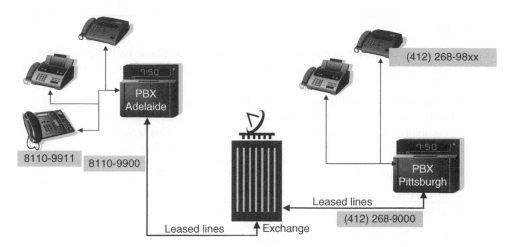

Figure 3.22 Leased telephone lines to connect the CMU campuses in Adelaide and Pittsburgh

Private switches were introduced into the telephony landscape early on. With the spread of telephones, large organisations started to install a local network and switchboards for internal communications. These included hotels, hospitals, government offices and large companies. These private switchboards evolved as automatic switches were introduced.

The usage of PBXs became widespread in the second half of the 20th century in conjunction with leased line connections between remote offices of companies. The capabilities of PBXs grew with electronic switches, and functions such as call forwarding, voice mail, and conference calls were developed.

Telephone operators leased telephone lines to these large organisations as dedicated links of a fixed quantity, for example 20, 50 or 100 telephone circuits, between any two or more remote offices of the organisation. The telephone company charged for these lines on a monthly basis, hence the name 'leased lines'. Leased lines allowed the operator to lock in a customer for a fixed period and fixed revenue. The leasing organisation also benefited by having the cost of telephone communications between the offices fixed, and usually at less than long-distance charges. The lease charges depended on the number of lines and the distance between the offices.

Initial leased lines were physically dedicated. That is, a number of physical links were set aside for the customer. Since no switching to a local number was possible, a local switch needed to be installed at each of the customer's local offices. These local switches were common for internal office communications, and the leased line simply connected into these. The communications between offices could now be configured as internal calls. The internal switchboard and other communications expertise needed necessitated the formation of a technical group which in time grew to become the information technology (IT) divisions of today.

The lease size calculation followed the Erlang-B formulas discussed earlier in this chapter. Indeed the cost–benefit analysis of whether to lease or not depended on the amount of inter-office traffic and its cost compared with lease set-up, monthly lease and internal IT operational costs.

Example 3.2

A company's 150 health insurance agents on average make one phone call every 10 min, and each call lasts on average 5 min. Assume each agent works 250 days a year. Assume a GoS of 1% is necessary.

(a) What is the total phone traffic generated by this company in erlangs?
(b) If each call costs 15 ¢ on average, what is the daily (8 h) telephone bill?
(c) As the company telecom manager you have found the following lease options. Which one will you choose and why? Clearly show your assumptions and work.

Option 1: Company X		Option 2: Company Y	
One-time set-up cost ($)	35 000	One-time set-up cost ($)	75 000
First 50 lines/year ($)	130 000	First 40 lines/year ($)	100 000
Each extra 25 lines/year ($)	60 000	Each extra 20 lines/year ($)	40 000

Answer

(a) 75 E
(b) $150*6*8*0.15 = \$1080$
(c) Yearly cost as is $= 250*\$1080 = \$270\,000$.
    Required number of circuits for GoS of 1% is 91 circuits.
    Option 1 $= \$35k + \$130k + 2*\$60k = \$285k$, first year, and $250k/year afterwards.
    Option 2 $= \$75k + \$100k + 3*\$40k = \$295k$, first year, and $220k/year afterwards.
    If a multi-year lease option is considered, then option 2 is more economical.

Virtual Network Operators

Deregulation of the telecommunications industry, added to the long-established leased line practice, opened the door to a new type of operator: those without a network of their own. These became known as virtual network operators or VNOs.

VNOs lease capacity at wholesale price from an incumbent operator and then resell it to their customers at a lower retail price compared with the incumbent. Lease prices are usually set by the regulator and therefore are low enough that a VNO with low fixed and operating cost can make a profit. In essence these players were active at the retail layer of the telecommunications multi-layer ecosystem as shown in Figure 3.23.

A number of VNOs appeared in the Australian market after the telecommunications industry deregulation of the early 1990s. Several such operators were started in 1990s in Australia, for example AAPT, who offered long distance telephony and later on data communications services [12]. Capacity was initially leased from Telstra over which long-distance telephony was provided to business and individual subscribers.

A VNO business model is generally constructed around low fixed cost of set up and variable costs of serving customers. For example, a VNO may lease 20 circuits for voice traffic between two cities. It will then need to find enough customers to utilise these 20 lines. If grade of service is 1%, then some 12 E of traffic may be accommodated, for a usage efficiency of 60%. The cost structure is the fixed costs of network set-up, and variable costs of lease, marketing, advertising

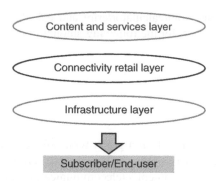

Figure 3.23 VNOs' business layer

and customer service. As the number of customers increases, more lines may be leased. With 40 leased lines, some 20 E of traffic may be supported, for a usage efficiency of 73%. Clearly, a larger customer base increases the efficiency of the business model.

Many VNO businesses spent heavily to increase their customer base, through offering very competitive rates to the incumbent operator. Some other operators stayed small, but increased their usage efficiency through inferior grade of service (e.g. 90% usage efficiency is possible over 20 circuits if a blocking probability of 11% may be tolerated.) These VNOs offered inexpensive long distance calling for users who did not mind the inconvenience of lower quality. With the emergence of mobile telephony, the concept of mobile VNO (MVNO) has gained currency, with many MVNO companies re-selling voice and data services of incumbent network operators.

VNO businesses are policy sensitive. Many countries have an incumbent infrastructure company, which also operates as a retail provider and can compete with the VNOs. An example is Telstra in Australia. Should the lease price be variable, or subject to undue risks, a VNO could soon find itself at the mercy of the incumbent infrastructure owner. Regulators in most countries have strict guidelines on how much an infrastructure owner may charge VNOs for accessing their network. In many places these charges are little more than the costs of operating the infrastructure.

As the first real-time telecommunications system, telephony soon became a global success. The initial technological, business and policy issues and solutions were complex. These have continued to influence how these services are provided over nearly one and half centuries since the development of these systems. An overview of how telephony technologies, and associated services and businesses developed is informative on how nascent broadband telecommunications services and associated ecosystems evolve and how they may be analysed and managed.

Review Questions

1. What value do switches provide in telephony?
2. What is multiplexing and why is it necessary?
3. How does a telecom engineer decide on the number of lines necessary for connecting two local exchanges?

4. Access http://www.erlang.com/calculator/erlb/ and calculate the number of necessary circuits to provide a 1% GoS for the following amount of traffic:
   (a) 1 E
   (b) 5 E
   (c) 10 E
   (d) 30 E
   (e) 120 E
   (f) 180 E

5. Draw the result from question 4 on a graph with offered traffic on the horizontal and required links on the vertical axis. What do you observe? Is there a business consideration?

6. How does GoS inform a government's telecommunications policy?

7. Why is telephone infrastructure provision a natural monopoly?

8. A call centrr has 60 employees on-line, each of whom make one phone call every 3 min on average, and talk on average for 2 min. Each call costs 15 ¢.
   (a) What is the total traffic in erlangs?
   (b) How many circuits are required for a GoS of 1%?
   (c) How much should the yearly (250 working days, 8 working hours a day) lease price be for the company to make a 50% saving in its telephone bills over the first 3 years? Assume the lease set-up fee of $30 000 should be paid over a period of 3 years.

9. What factors contributed to the decrease in cost of transatlantic telephony?

10. How did satellite communication impact the undersea cable telephony business? Was it complementary or competitive?

11. Undersea cables were a major infrastructural investment. How do you think these were repaired when a technical problem occurred?

12. How vertically integrated were the early telephone operators? Why?

13. What portion of early telephone operators' revenue came from provision of a 'pipe' and what portion from provision of 'content'? How do you think this has changed in recent years?

14. What was the importance of AT&T's technological leadership in its early history? How did it impact the business in later years? Was there anything AT&T could have done?

15. How successful do you think the Bell break-up model was? Can this have some implications for the eventual privatisation of NBN Co?

16. A company's 100 salespeople on average make one phone call every 5 min, and each call lasts on average 3 min:
    (a) What is the total phone traffic generated by this company in erlangs?
    (b) If each call costs 10 ¢ on average, what is the daily (8 h) telephone bill?
    (c) As the company telecom manager you have found the following two lease options. Which one of the two will you choose and why? Clearly show your assumptions and work.

Option 1		Option 2	
Set-up cost ($)	10 000	Set-up cost ($)	25 000
20 lines/year ($)	75 000	25 lines/year ($)	80 000
40 lines/year ($)	135 000	50 lines/year ($)	140 000
Each extra 20 lines/year ($)	40 000	Each extra 25 lines/year ($)	50 000

17. How did the leasing business change the telecommunication operator business ecosystem?

18. What are the strengths and weaknesses of an operator with its own infrastructure compared with a virtual network operator?
19. Perform a value analysis on a leased phone service.
20. What do you see as policy implications in forcibly breaking up telephone monopolies such as that of AT&T in the United States?

References

[1] Huurdeman, A.A. (2003) *The Worldwide History of Telecommunications*, Wiley Interscience.

[2] Illustrated London News (1872) Cited in http://www.educationscotland.gov.uk/scotlandshistory/makingindustrialurban/alexandergrahambell/index.asp, accessed 23 October 2015.

[3] The US National Archives. http://commons.wikimedia.org/wiki/File:Photograph_of_Women_Working_at_a_Bell_System_Telephone_Switchboard_(3660047829).jpg#mediaviewer/File:Photograph_of_Women_Working_at_a_Bell_System_Telephone_Switchboard_(3660047829).jpg, accessed 20 August 2015.

[4] Angus, I. (2001) An Introduction to Erlang-B and Erlang C. *Telemanagement* July–August.

[5] http://www.erlang.com/calculator/erlb/, accessed 23 October 2015.

[6] http://atlantic-cable.com/Books/Whitehouse/DDC/index.htm, accessed 20 August 2015.

[7] http://news.bbc.co.uk/2/hi/uk_news/scotland/glasgow_and_west/5375796.stm, accessed 20 August 2015.

[8] Froehlich, F. E. and Kent, A. (1997) *Encyclopedia of Telecommunications*, Dekker.

[9] http://www.vny.cuny.edu/blizzard/building/building_fr_set.html, accessed 20 August 2015.

[10] http://www.u-s-history.com/pages/h1803.html, accessed 20 August 2015.

[11] http://www.marketwatch.com/story/sbc-to-acquire-att-in-16-billion-deal works, accessed 20 August 2015.

[12] https://www.aapt.com.au/aapt/about-aapt, accessed 20 August 2015.

4

Information Theory

Preview Questions

- Is a picture really worth a thousand words? How can one calculate the information content of a picture/a text?
- How long does it take for the Mars Discovery probe to transmit a photo of Martian surface to Earth?
- How can information generated by a source be converted to a binary sequence?
- How does the power of a message signal matter in its accurate transmission and reception?
- How do you calculate the maximum possible transmission rate of a communication system?

Learning Objectives

- Information content measurement
- Source information generation rate and entropy
- Principles of source coding
- Link budget calculation
- Channel capacity calculation

Broadband Telecommunications Technologies and Management, First Edition. Riaz Esmailzadeh.
© 2016 Riaz Esmailzadeh. Published 2016 by John Wiley & Sons, Ltd.
Companion Website: www.wiley.com/go/BTTM

Historical Note

Claude Shannon's (pictured)[1] [1] seminal 1948 paper, 'A mathematical theory of communication' [2] is widely considered to be the foundation upon which the fields of information theory and digital communications have been built. The paper is the theoretical foundation which determines how we encode, communicate, and store information in discrete digital form.

Shannon was a researcher at Bell Laboratories, the research arm of the Bell Telephone Company in the middle of the 20th century. At the time a number of communication technologies were known and used: among them were telegraph (Morse, 1844), telephone (Bell, 1876), wireless telegraph (Marconi, 1894), amplitude modulation (AM) radio (Fessenden, 1906), frequency modulation (FM) radio (Armstrong, 1936), analogue voice to digital conversion and pulse-code modulation(PCM) (Reeves, 1937), voice spectrum analysis and synthesis coding (Vocoder) (Dudley, 1937), and spread spectrum (Antheil and Lamarr, 1942).

These technologies all impacted on how information in a discrete form may be transmitted. Morse code had already demonstrated that it is more efficient to use unequal code lengths to represent a letter of the alphabet as a function of its frequency of occurrence. Telephony and AM radio used a bandwidth proportional to that of the baseband signal, showing the minimum required capacity for analogue transmission. FM and spread spectrum techniques had shown that the quality/reliability of reception was a function of utilised bandwidth and therefore there existed a trade-off between frequency resource utilisation and reception quality. (The reason FM quality of sound is better is because a 3–10 times wider bandwidth compared with AM is used for the transmission of the same sounds.)

Representation of voice from its analogue form to digital heralded an alternative of information transmission and 'voice coding' technologies demonstrated that the utilised bandwidth may be reduced if quality impairment can be tolerated. This led to two questions which were addressed by Shannon's paper: (1) What is the minimum number of discrete symbols (such as Morse code's dots and dashes) that are required to faithfully represent information output from a source? (2) How much discrete information may be transmitted on a band-limited channel in the presence of noise?

Shannon mathematically answers these two questions using a new field of science known as 'information theory'. One equation calculates *entropy* as the average amount of information that a source generates. Source entropy is useful in that it can be used to determine the minimum required transmission rate of a link for communicating the information. Source entropy is also important in designing efficient loss-less source coding. Business implications of source entropy are in the design of audio and video source coders used in mobile phones for example and in data compression applications (such as 'winzip').

Shannon's other equation defines the maximum amount of information that may be transmitted over a band-limited channel. This is called *channel capacity* and was initially used by

---

[1] Claude Shannon: www.flickr.com/photos/tekniskamuseet/6832884236.

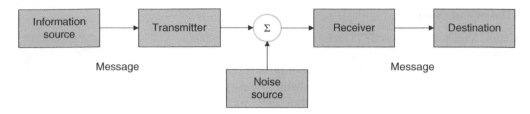

Figure 4.1   Communications system block diagram from Shannon's landmark paper

AT&T to determine the maximum possible rate of transmission of telegraph messages over their network in the presence of noise. At the time it was well understood that channel capacity was directly related to the available bandwidth: that is the larger the size of the pipe, the more information it can carry. However, the effect of noise was not known. Figure 4.1 shows a block diagram of a telecommunications network, including signal power and additive noise as drawn in Shannon's paper [2]. Shannon's equation calculates channel capacity in the presence of such random noise. This equation is fundamental to the design of all digital telecommunications networks, from space probes now travelling, and communicating with earth at the edge of the Solar System, as well as digital mobile systems we use every day. The equation is known as Shannon's theorem, and is widely used to design, dimension and compare communication systems and technologies. Its business implications are evident in the design of all digital communications systems we use today. These contributions have cemented Shannon's place as the father of information theory.

Information Theory

Information theory can be defined as a field of science and technology which quantifies the amount of information included in a message. The following example may serve to better explain the theory. The following two statements convey information on winter weather in two cities where Carnegie Mellon University has a campus. Which of the two statements has more information?

- Last winter it snowed in Adelaide, Australia.
- Last winter it snowed in Pittsburgh, USA.

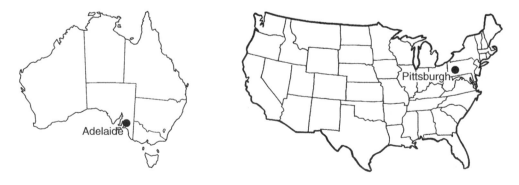

It is difficult to know which of the two statements conveys more information without knowing about the general probability of snowing in each city. Therefore, let us assume that over the past 100 winters there has been only one snowfall in Adelaide, whereas there has been only one winter when it has not snowed in Pittsburgh. In other words, the chance (probability) of snowfall during an Adelaide winter is 1% whereas it is 99% for Pittsburgh. Given this information which statement gives a recipient of the message more information? Which one is more newsworthy? It is reasonable to say that the first statement is more likely to make it to the news as it occurs less frequently (which in fact does – a mere snow flake or two on the hills near Adelaide is sure to make the headline news on all TV stations!)

Given the close relation between information content of a message and the probability of its occurrence, information theory quantifies the information content of a message, $I(m)$, by:

$$I(m) = -\log_2 p(m) \tag{4.1}$$

where $p(m)$ is the probability that message m is generated. The unit for information content is a 'bit'. For example, the information content of the 'Last winter it snowed in Adelaide, Australia' message is $-\log_2 (0.01) = 6.64$ bits, whereas the information content of the 'Last winter it snowed in Pittsburgh, USA' message is $-\log_2 (0.99) = 0.01$ bits [note, $\log_2(x) \approx \log_{10}(x)/0.3$]. This quantifies how informative these two statements are, and makes a comparison possible. We can now say quantitatively that 'Last winter it snowed in Adelaide, Australia' has more information than 'Last winter it snowed in Pittsburgh, USA'.

The quantised information content as defined earlier allows for:

- If the probability of an event is 1, then its information content is 0 ('the sun rose from the east today' conveys no information).
- The information content is never negative: if there is a non-zero probability of an event occurring, there exists positive information content.
- If the probability of an event is 0, then it has no information content, or its information content is not defined ('the sun rose from the west today' conveys no information as it is impossible).
- The lower the likelihood of an event, the higher the information content.
- The joint probability of two independent events occurring, for example 'it rained in Adelaide today' and 'temperature on Mars was −60 °C yesterday' is the product of their individual probabilities. The total information content is the sum of individual information contents.
- Usage of base-2 logarithm fits with the binary character of digital communication of information. In general the occurrence probability of a '1' is equal to probability of '0' and equal to 50%. The information content is one bit, and can be represented by one binary digit or 'bit'.

Example 4.1

A source outputs one of four possible messages, m_1, m_2, m_3 and m_4, with the following probabilities:

$$p_1 = 1/2; \; p_2 = 1/4; \; p_3 = 1/8; \; p_4 = 1/8$$

The information content of each message is calculated as:

$$I(m_1) = -\log_2(1/2) = 1\,\text{bit}$$

$$I(m_2) = -\log_2(1/4) = 2\,\text{bits}$$

$$I(m_3) = I(m_4) = -\log_2(1/8) = 3\,\text{bits}$$

An important application of information quantisation is that it can be used to calculate the length of a code needed to represent a message. For example, if a message is generated frequently it should be assigned a shorter code and if it is infrequent then a longer code. This is an intuitive principle and was used in the Morse code design: for example 'E' a frequently occurring letter is represented by '•', whereas 'Z' an infrequently letter is represented by '––••'. Information theory provides the theoretical background to calculate the optimal code length for representing a message. This design is efficient in that it allows for faster message transmission and therefore lower average operational cost per message.

Source Entropy

As discussed earlier, knowledge of a message's information size is needed in order to design and dimension a network to connect a source to a sink. At times a source generates a range of messages with different probabilities and therefore information size (such as different letters of the alphabet). To dimension the necessary network, the designer needs to know the *average* information size. To calculate this, all possible messages and their corresponding probabilities are considered to calculate average source information. This average source information is called 'source entropy' and is the first of the contributions made by Shannon.

Source entropy is denoted by H, and has a unit of bits/message. It is the sum of individual messages' information content weighted by their corresponding probability. For a source that produces M different messages, entropy is calculated from the following equation:

$$H = -\sum_{i=1}^{M} p_i \log_2(p_i) \tag{4.2}$$

where p_i is the probability message i is generated. A related definition of importance to telecommunications dimensioning is the average source information rate, or entropy rate, which is defined as the average amount of information generated per second. This figure is important to an operator as it informs the necessary size of the telecommunications network.

Example 4.2

The entropy for the source of Example 4.1 can be calculated as:

$$H = -\big[(1/2) * \log_2(1/2)\big] - \big[(1/4) * \log_2(1/4)\big] -$$

$$\big[(1/8) * \log_2(1/8)\big] - \big[(1/8) * \log_2(1/8)\big]$$

$$H = 1.75 \text{ bits/message}$$

If this source generates one message every second then the source information rate is 1.75 bits per second (bps). Theoretically, a telecommunications link with a capacity of 1.75 bps will be sufficient to connect this source to a receiver of information.

Example 4.3

Table 4.1 shows the probability of an English letter appearing in the text. What is the information rate for a source that produces 100 letters every second?

Table 4.1   Typical probability of occurrence of individual English letters

A	8.17%	J	0.15%	S	6.33%
B	1.49%	K	0.77%	T	9.06%
C	2.78%	L	4.03%	U	2.76%
D	4.25%	M	2.41%	V	0.98%
E	12.70%	N	6.75%	W	2.35%
F	2.23%	O	7.51%	X	0.15%
G	2.02%	P	1.93%	Y	1.97%
H	6.09%	Q	0.10%	Z	0.07%
I	6.97%	R	5.98%		

Source entropy can be calculated to be 4.176 bits/letter. Source information rate is therefore 418 bps. Again a telecommunication network with a capacity of 418 bps will be sufficient to connect this source to a receiver of information.

Source Coding

Digital communications require representing of information from a source, with a series of binary bits, that is 0's and 1's.

Consider a source that produces two messages m_1 and m_2 with equal probability of 50%. The source entropy can be calculated from Equation 4.2 to be 1 bit/message. A binary representation, or source coding, for the messages m_1 and m_2 this source generates can be using '0' for m_1 and '1' for m_2. The average code length, A, a measure of how many bits are used on average for coding this source, is calculated from the following formula. This considers the probability of a message and the number of bits used to represent it:

$$A = \sum_{i=1}^{M} p_i n_i \tag{4.3}$$

where p_i is the probability and n_i is number of binary bits used to represent message i. The average code length can also be calculated to be equal to 1 bit/message ($50\% \times 1 + 50\% \times 1 = 1$). Source coding efficiency, defined as the ratio of entropy (H) to average code length (A), is equal to 100%. This is an optimal source coding scheme.

Now consider the source from Example 4.1 which generates four messages, m_1, m_2, m_3 and m_4. The source entropy was calculated to be equal to 1.75 bits/message. One binary

	p_i		n_i	$p_i n_i$
m_1	0.5	00	2	1.0
m_2	0.25	01	2	0.5
m_3	0.125	10	2	0.25
m_4	0.125	11	2	0.25
Average code length (A)				2.0

Figure 4.2   One possible coding method

	p_i		n_i	$p_i n_i$
m_1	0.5	1	1	0.5
m_2	0.25	01	2	0.5
m_3	0.125	001	3	0.375
m_4	0.125	000	3	0.375
Average code length (A)				1.75

Figure 4.3   An alternative coding method

representation for this source is shown in Figure 4.2. The average code length is calculated to be 2 bits/message, which is slightly larger than the entropy of 1.75 bits/message. Source coding efficiency is $1.75/2 = 87.5\%$. By definition this is a suboptimal source coding scheme.

A different binary representation is shown in Figure 4.3. The average code length is 1.75 bits/message now and is equal to the source entropy, resulting in a 100% efficient code. Knowledge of entropy informs the designer on the limit of source coding methods.

This method of coding is called entropy coding. These codes are also called prefix-free codes, which means that no code is a prefix to another code. We will discuss source coding in more depth in Chapter 5.

Another example is usage of ASCII 7-bit length binary for source coding the English alphabet as shown in Figure 4.4. The coding efficiency of 7 bits/message is significantly larger than the source entropy for English text as found in Example 4.3 as 4.176 bits/message. The coding efficiency is calculated to be $4.176/7.0 = 59.7\%$, which is quite low.

In practice it is possible to design a source code with near entropy efficiency, as shown in Figure 4.5. This code is also prefix-free and is found using a method known as Huffman coding [3]. The average code length here is 4.205, resulting in a coding efficiency of

A	100 0001	J	100 1010	S	101 0011
B	100 0010	K	100 1011	T	101 0100
C	100 0011	L	100 1100	U	101 0101
D	100 0100	M	100 1101	V	101 0110
E	100 0101	N	100 1110	W	101 0111
F	100 0110	O	100 1111	X	101 1000
G	100 0111	P	101 0000	Y	101 1001
H	100 1000	Q	101 0001	Z	101 1010
I	100 1001	R	101 0010		

Figure 4.4   ASCII coding for the English alphabet

A	1111	J	001111101	S	1010
B	101110	K	0011110	T	000
C	10010	L	11010	U	01110
D	11011	M	01111	V	001110
E	010	N	1000	W	10110
F	00110	O	1110	X	001111100
G	100110	P	101111	Y	100111
H	0110	Q	001111110	Z	001111111
I	1100	R	0010		

Figure 4.5   A near entropy efficiency code for the English alphabet

4.176/4.205 = 99.3%. This is a good illustration of Shannon source coding techniques and how highly efficient source codes may be found. A number of compression coding techniques for data storage are based on Huffman's work [4, 5].

One result of Shannon's paper is that for a binary coding method (i.e. using 1's and 0's) the code length on average is equal to $-\log_2(p_i)$, where p_i is the probability of message i. This is an important definition as it determines how messages from a source may be efficiently represented using binary 'bits'.

Clearly significant improvement in coding efficiency can be obtained with entropy coding. However, this efficiency gain comes at the expense of complexity and error propagation as Example 4.4 demonstrates.

Example 4.4

Consider the source of Example 4.1 which produces four messages, m_1, m_2, m_3 and m_4. Two source coding methods were given in Figure 4.2 and Figure 4.3. The sequence '$m_2 m_1 m_1 m_3 m_1 m_4 m_2 m_2 m_4 m_1 m_1 m_1 m_3 m_1 m_1 m_2$' will be source coded:

With the first source encoder as:
01000010001101011100000010000001

With the second source encoder as:
0111001100001010001110011101

The second source coding method uses 4 bits fewer bits (or 87.5%) and is more efficient. You can verify that both sequences are decoded to produce the original message sequence using their respective coding methods.

However, the first coding method is more robust. Assume 1 bit arrives in error (shown in bold) because of transmission imperfection: '01000010001101011**1**01000010000001'. A decoder using the first source coding method produces only one decoding error: '$m_2 m_1 m_1 m_3 m_1 m_4 m_2 m_2 m_4 \boldsymbol{m_2} m_1 m_1 m_3 m_1 m_1 m_2$'. However, a single error in reception with the second coding method can result in error propagation: '0111001100001**0**00001110011101'. This decodes to '$m_2 m_1 m_1 m_3 m_1 m_4 m_2 \boldsymbol{m_4 m_3 m_1 m_1} m_3 \boldsymbol{m_1 m_1} m_2$'. In other words, efficiency is gained at the expense of robustness. Such entropy coding methods are therefore used in systems where probability of error can be kept very small.

Frequency, Bandwidth and Power

As discussed in Chapter 3, an electromagnetic wave can be equally specified in a time or frequency domain. In the time domain a signal is specified as a continuous waveform, or a stream of discrete samples. Equally the signal is specified by the frequency components and their corresponding amplitudes. The amplitude of signal components in a time or frequency domain can be used to calculate the overall signal power. For example, signal power is calculated by averaging the square of signal amplitude over time: that is if A is the instantaneous amplitude, power is equal to the average of A^2 over time. For waveforms of interest in telecommunications, power in most cases is equal to the average of $A^2/2$. Signal power is measured in watts, in honour of James Watt (1736–1819) the inventor of the steam engine. It is an important measure in evaluating the quality of signal in the presence of noise: a high ratio of signal power to noise power (signal to noise ratio, SNR) indicates a clear, high quality signal, and a low SNR indicates a poor signal, which may not be intelligible. Both signal and noise power are measured similarly and with the same unit of watts.

Thermal Noise

Thermal noise is a natural process and is produced in any physical device that exists in above absolute-zero temperatures, that is 0 K or −273 °C. The power of thermal noise N in watts is calculated from the following formula:

$$N = k_B TB$$

where k_B is the Boltzmann constant and equal to 1.38×10^{-23} J/K, T is the absolute temperature in kelvin and B is the signal bandwidth in hertz. For example, a voice signal transmitted at room temperatures (27 °C or 300 K), will be received with an additive noise signal of a power $N = 1.38 \times 10^{-23} \times 300 \times 4000 = 1.66 \times 10^{-17}$ W or −137.8 dBm.

Signal and noise power are calculated at a telecommunications receiver. Noise power is generally produced locally and is a product of molecular friction in the receiving device. Received signal power is a function of the transmitted power, transmitting and receiving device characteristics, antennas, and the signal loss in the communications medium. Calculations are done using a technique known as Link Budget. These calculations are usually done using the decibel (dB) scale, a logarithmic operation which makes dealing with very large or very small numbers easier.

Decibel Scale

The decibel scale is a logarithmic scale which is used to represent and manipulate a wide range of numbers encountered initially in telecommunications (and nowadays a number of other fields). For example, sound level ratios may vary from 1 to 10^{12}. Clearly it is cumbersome to write a number with 12 digits or more. A logarithmic representation is more compact and easier to manipulate: multiplications and divisions can be replaced by additions and subtractions.

The conversion from a real to a decibel scale is simple:

'X' is the decibel representation of a real number 'x' if

$$X = 10 * \log_{10}(x)$$

It follows that

$$x = 10^{(X/10)}$$

For example, if $x=100$, then $X=20$ dB, or if $x=2521$, then $X=34.02$ dB. Also, if $X=30$ dB, then $x=1000$, or if $X=13.2$ dB, then $x=20.89$.

As noted above for two real numbers x and y we have:

$$x * y \equiv X + Y \text{ and } x/y \equiv X - Y$$

That is, multiplication in the real domain is equivalent to addition in the decibel domain, and division in the real domain is equivalent to subtraction in the decibel domain.

In telecommunications signal power levels may also be represented using the decibel scale. Usually a signal power level is written with respect to a power unit such as 1 watt (W) or 1 milliwatt (mW). Using the decibel scale, the representation takes the form dBW or dBm to specify which unit has been used. For example, a power level of 4 W or 4000 mW can be written using the decibel scale as:

$$10*\log_{10}(4\ W)=6\ dBW$$

or

$$10*\log_{10}(4000\ mW)=10*\log_{10}(1000)+10*\log_{10}(4\ mW)=36\ dBm$$

Link budget calculations are very important in system design. They may be used to determine the distance over which communications can be made, or calculate the amount of necessary power that yields a desired receiver SNR. As Case Study 4.1 on the Voyager probe demonstrates, these calculations are common practice in any telecommunications system design, as they determine whether a channel can be established and the amount of information that may be transferred.

Example 4.5

Consider a satellite that sends signals to an Earth station as shown. What is the received power to noise ratio at the Earth station receiver for the following parameters?

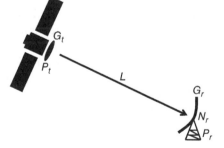

- Transmit Signal power$=P_t=40\ W=46\ dBm$
- Satellite transmitter antenna gain$=G_t=15\ dB$
- Earth station receiver antenna gain$=G_r=30\ dB$
- Path loss$=L=196\ dB$
- Noise level at the receiver $N_r=$-120 dBm

Answer

- The received signal power P_r is:

$$P_r=P_t+G_t+G_r-L=46+30+15-196=-105\ dBm$$

- The received SNR is:

$$\text{Received SNR}=P_r-N_r=-105-(-120)=15\,dB$$

Example 4.6

Calculate the power link budget for a device 1 km from a base station for the following parameters:

- Transmit signal power $= P_t = 33$ dBm
- Transmitter antenna gain $= G_t = 7$ dB
- Mobile device antenna gain $= G_r = 1$ dB
- Noise level at the receiver $N_r = -111$ dBm
- Required receiver $SNR_r = 7$ dB
- Path loss at 1 km $= L = 138$ dB

Answer

- Receiver SNR at 1 km:

$$\text{SNR at } 1\text{ km} = P_t + G_t + G_r - N - L = 33 + 7 + 1 - (-111) - 138 = 14 \text{ dB}$$

$$\text{Link budget} = \text{SNR at } 1\text{ km} - \text{required SNR}_r = 14 - 7 = 7 \text{ dB}$$

This means that an extra 7 dB signal loss may be tolerated before telecommunication performance is degraded.

Case Study 4.1: Voyager Mission

Two Voyager space probes were launched in 1977 by the National Aeronautics and Space Administration (NASA) agency of the United States government to explore Jupiter, Saturn, Uranus and Neptune and continue on to explore the interstellar space beyond the solar system. Their mission was to examine the atmospheric, magnetic and spectral properties of outer solar system planets, and communicate the collected information back to Earth. Clearly a telecommunications system needed to be designed that ensured such information could be correctly transmitted back to Earth.

The power systems on-board the crafts are designed to last many years and support telecommunications over the long distances to the Earth. Indeed, the probes sent valuable information on the structure of the outer planets, and have now left the solar system while continuing to send back important information. For example, as one probe crossed into interstellar space it observed that the Sun's magnetic field becomes warped and creates 'bubbles' at the edge of the solar system [6].

The design of the Voyager on-board telecommunications system needed to take into account the path loss experienced by a signal travelling over 15 billion km. Depending on the frequency of operation this path loss is in the order of 10^{30} times, or 300 dB. Strong antennas on-board the crafts as well as on the Earth ensure a highly directional line of sight is maintained between the transmitter and receiver, and signal power is collected as much

Table 4.2 Link budget calculation parameters for the downlink of Voyager 2's telemetry system

1. Transmission power	40.9 dBm
2. Probe antenna gain	48.1 dB
3. Path loss (7.3×10^9 km) in the 8 GHz band	308.2 dB
4. Receiver antenna gain	73.7 dB
5. Receiver noise power	−170.6 dBm
6. Signal suppression	−6.2 dB
7. Received SNR = (1+2−3+4−5+6)	18.9 dB

as possible. Nevertheless, the received signal is very weak, and its ratio to noise at the receiver is very small. Nevertheless this signal power is enough so that information and control command transfer still takes place between the probe and the Earth. Over the years stronger antennas have been deployed to improve the link budget of this telecommunications link [7]. The link budget calculation parameters of a probe at present are listed in Table 4.2.

Such a calculation gives insight on how communications with the probe may be carried out and the range at which the probe may be controlled. Furthermore, as will be described below, link budget analysis can be used to calculate the maximum possible transmission rate on the link, or channel capacity. This limit determines the amount of information that may be transmitted back to Earth and therefore the specification of sensors and other devices on-board the probe.

Case Study Questions

- What factors decided the type of sensors to be used on-board the Voyager probes?
- How did NASA engineers decide on the information telecommunications systems of the Voyager probes?
- What factors were needed for a link budget analysis in this system? Which parameter varies?
- Speculate whether the probes would be launched in the absence of understanding the information theory principles.
- Why is it that communications can be carried out even when the received signal power is significantly below that of noise?

Maximum transmission rate calculation uses the second of Shannon's contribution to the field of information theory, namely channel capacity.

Channel Capacity

One requirement for telecommunications system design is knowledge of the source information generation rate as discussed above, which can then be used to dimension the network to transport the information. Clearly the more information generated by a source, the wider the pipe

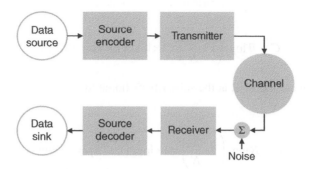

Figure 4.6   A simple telecommunications block diagram

needs to be to carry the stream. Moreover, the pipe needs to be clear and reasonably free of noise: it is a common experience to discontinue a conversation on the phone when the line is too noisy.

Figure 4.6 shows a simplified telecommunications block diagram. Shannon showed that the capacity of such a system is a function of two parameters: the bandwidth of the channel and the relative strength of the signal compared with noise. He formulated an equation, now known as Shannon's theorem that calculates the maximum possible transmission rate, or capacity(C) for a band-limited channel in the presence of noise:

$$C = B \log_2 \left(1 + \frac{S}{N} \right)$$

where B represents the channel bandwidth, and S/N is the ratio of the power of the desired signal to that of the background noise.

Shannon theorem teaches that capacity is directly related to bandwidth: in essence the wider a pipe, the more water can pass through it. It also teaches that as signal power relative to noise power increases, that is the cleaner the pipe becomes, the more information it can carry. In turn this means that a faster transmission of signals becomes possible. Conversely, the lower the signal power relative to noise power, the dirtier the channel becomes and therefore communication becomes less reliable. This equates to the experience that one has to speak more slowly to get a message across when the channel is not clean enough (or the receiver is hard of hearing – reduced received signal power).

Shannon theorem also implies that any channel where signal power exists (i.e. the power of the signal is larger than zero) can be used for telecommunications. Capacity is always positive and more than zero so long as bandwidth (B) and signal power (S) are not zero. (By definition bandwidth and power do not have negative values.)

Example 4.7

If a DSL system uses a signal bandwidth of 1 MHz in its downstream communications, and the signal power to noise power ratio at a subscriber's home is 7, what is the maximum information transmission rate this system can support?

Answer

$$C = B\log_2\left(1 + \frac{S}{N}\right) = 1 \times \log_2(1+7) = 3\,\text{Mbps}$$

What if we can increase the SNR at the subscriber's home to 15?

Answer

$$C = B\log_2\left(1 + \frac{S}{N}\right) = 1 \times \log_2(1+15) = 4\,\text{Mbps}$$

If a deep space probe uses a signal bandwidth of 100 MHz for its communications, and the SNR at the Earth's receiver antenna is 10^{-6}, what is the maximum information transmission rate the system can support?

Answer

$$C = B\log_2\left(1 + \frac{S}{N}\right) = 100 \times \log_2(1+10^{-6}) \approx 144\,\text{bps}$$

Shannon capacity is also known as Shannon limit since it calculates the maximum possible information transmission over a telecommunications channel. The limit in bits per second transmission with respect to a unit of bandwidth (1 Hz) is drawn vs S/N in Figure 4.7. Note for small values of S/N:

$$log_2(1 + S/N) \approx S/N$$

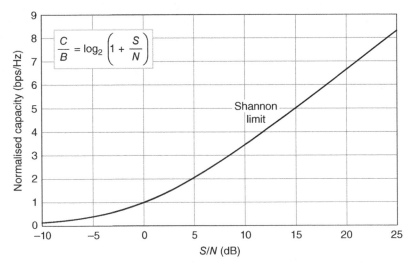

Figure 4.7   Normalised capacity vs S/N

Therefore the Shannon limit increases linearly as the signal power to noise power ratio increases. In contrast at large S/N:

$$log_2(1 + S/N) \approx log_2(S/N)$$

Therefore the Shannon limit increases logarithmically as the signal power to noise power ratio increases. The normalised capacity vs S/N is drawn in Figure 4.7.

Example 4.8

Considering the Voyager's channel bandwidth of 340 kHz to 1.35 MHz, the capacity of the telecommunications channel is limited. The Shannon theorem can be used to calculate the maximum possible transmission capability for the system over different distances. This limit is then used to design the telecommunications system so it can transmit data at rates as fast as possible. The system so designed has performed to expectation, indeed better because of ground system improvements, transmitting information back to the Earth at rates of 115.2 kbps from Jupiter, 44.8 kbps from Saturn, 29.9 kbps from Uranus and 21.6 kbps from Neptune. As the distances increase and as the craft's electrical power supplies diminish possible transmission rates decrease. Nevertheless, the systems are expected to communicate at rates of as low as 40 bps until 2057 (Table 4.3). This precise and efficient design of the Voyager telecommunications systems would not have been possible without the principles taught by Shannon's theorem.

The information theory as formulated primarily through the research work at the Bell Laboratories by a number of researchers, of whom Claude Shannon is the prominent example, led to the development of digital telecommunications and the information and communications technology (ICT) we are familiar with today. It enabled humankind to generate, process, analyse and disseminate information in unimaginable size and scope. The information the public and private organisations gather form one of their, if not the, most important assets. Information is a fundamental human need, and its mathematical formulation has revolutionised the way our society works. Indeed, the extent of information theory now extends to a variety of fields including social sciences, biology and economics [10].

Table 4.3 Telemetry link transmission rate for Voyager 2 probe [7]

Telemetry link transmission rate (bps)	Year
7200	1998
1400	2011
600	2017
160	2029
40	2057

Case Study 4.2: GSM vs CDMA

The early 1990s saw competition between two multiple access technologies for mobile communications systems. The first generation (1G) of mobile systems had been a success and the industry was looking to use digital technologies for the second generation (2G) of mobile phones. A number of standards were developed and rolled out, but soon the main competition was between Global System for Mobile Communications (GSM) and code division multiple access (CDMA). GSM was championed by a group of European manufacturers including Ericsson (where the author worked) and CDMA by mainly a group of US manufacturers led by Qualcomm Inc. While the TDMA-based GSM standards were well understood, CDMA was based on a relatively unknown technology. Regardless, the competition focused around which of these two technologies provided more 'capacity', that is which one was capable of supporting more simultaneous voice communications over the same bandwidth.

Capacity calculation and comparison is not trivial. Different technologies treat noise and multiple-user interference differently and therefore the comparison estimates varied significantly. The CDMA camp at one stage claimed a 21 times performance improvement compared with the analogue systems of the first generation. This was great news for capacity constrained mobile operators in the developed countries. The GSM camp disagreed with these claims, and calculated the improvements of the CDMA system more in the range of 3–4 times, and similar to that of GSM.

This capacity-focused disagreement continued for a number of years. Meanwhile both GSM and CDMA systems captured significant market share over the 1990s and mobile telephony became a great success because of the contributions of both camps. The marketing hype was put to test in practice, and it turned out that the two technologies had more or less the same capacity performance, as Shannon's theorem would have predicted [8, 9].

The success of both technologies owed as much to business and policy factors as technology advantages. Technologically, both GSM and CDMA had created a significant value proposition compared with the existing 1G analogue systems. From a business viewpoint, the marketing efforts of both camps worked to create an ecosystem of equipment manufacturers and operators which increased the overall value proposition for new operators. This was particularly important for the CDMA camp as they did not have the same extent of policy support as GSM, which had the backing of the European community. The CDMA camp is believed instead to have benefited from US government support and its influence on its allied countries. To close the loop, it is interesting that both GSM and CDMA benefited from national and regional government policies which funded or encouraged their technological developments.

This entire episode compares with the development of the 2G Japanese PDC standard, which similar to both GSM and CDMA provided multiple capacity improvement compared with the 1G analogue systems. Indeed PDC produced even higher capacity improvements (at the expense of voice quality) compared with both GSM and CDMA. The business ecosystem was similarly successful, although it was created mostly within Japan. International expansion of the technology was however unsuccessful as no other country adopted PDC. While technically a success as it continued to be used for some two decades, it has now been discontinued.

Case Study Questions

- Why did user capacity matter in the comparison between 2G and 1G standards?
- What was a good metric to compare different 2G standards? Is this still a relevant metric?
- What challenges did the CDMA camp face in marketing its technology?
- How are the information theory principles helpful in separating fact from marketing hype?
- How did the global extent of GSM/CDMA adoption help shape the present day mobile standards?

Review Questions

1. What is the entropy of a source which outputs, every second, one of 5 possible messages, $\alpha, \beta, \gamma, \delta$ and κ with probability of $p_\alpha = 1/2, p_\beta = 1/4, p_\gamma = 1/8, p_\delta = 1/16$ and $p_\kappa = 1/16$? Suggest a coding method for this system. What is the source information rate?
2. Find a 100% efficient encoding method for the source in question 1.
3. The following tweets were posted by Steve Martin[2] on 16 December 2010. What is the information content of each tweet [11]?

Steve Martin   Dec 16, 2010
New Years Eve falls on December 31st this year.

Steve Martin   Dec 16, 2010
So New Years Day will be on January 1st, pushing
Valentine Day to February 14th.

Steve Martin   Dec 16, 2010
The 4th of July will fall on the 3rd of March.

4. 'A picture is worth a thousand words' goes the saying:
    (a) How much information is contained in a picture if the picture size is 250×400 pixels, and each pixel may take one of 256 colours with equal probability?

    (b) Assume there are 60 000 words in a dictionary, and the 1000 words may be selected randomly from this dictionary. How much information do these 1000 words convey?

    (c) What is the flaw in the above comparison?

5. Assume the size of the photograph shown is 400×300 pixels. Each pixel may be one of 176 different shades of red, each with an equal probability of 1/256; or one of 10 shades of blue, each with a probability of 1/32. If the colours are spatially randomly distributed:

    (a) What is the entropy of this picture?

    (b) What is the total information represented by this picture?

6. A digital watch display is designed as follows:

    (a) What is the total information this watch provides?

    (b) How can this display be designed better if information theory is taken into consideration?

7. Which one of the following two statements has more information and why?

    (a) 'The last visit by Halley's Comet was in 1986'.

    (b) 'There was a full moon on 2 November 2009'.

8. Which one of the following two statements has more information and why?

    (a) '*The Hurt Locker*' won the best movie of 2009 at the Oscars.

    (b) '*Fearless*' won the best album of 2009 at the Grammys.

9. How are 'frequency' and 'bandwidth' different?

10. What is the maximum information transmission capacity of a 2 MHz channel with a SNR of 9?

11. The signal from a space probe arrives with a SNR of 0.001. If the utilised bandwidth is 50 MHz, what is the minimum time required for a picture of 16 Mbits to be transmitted to Earth?

12. The Square Kilometre Array (SKA) (pictured)[3] is a globally funded radio telescope system being rolled-out in Australia, New Zealand and South Africa. SKA will have a total collecting area of approximately 1 km². It consists of a large number of dish antennas spread over a vast geographic area. Several design option were proposed, with one being the following:

    3000 dish antennas, each with a gain of 60 dB in the 500 MHz to 10 GHz range. The array design (as in the photograph [12]) enables operators to focus on one section of the sky. The total antenna gain is equivalent to 3000 multiplied by the individual antenna gain.

    Assume an alien civilisation living on a planet 10 light years away from the Earth uses radars that transmit short duration high-power radio pulses of energy that are radiated into space by an antenna with a gain of 30 dB at a frequency of 10 GHz. Assume that the band-

[3] SKA Project Development Office and Swinburne Astronomy Productions [CC BY-SA 3.0 (https://commons.wikimedia.org/wiki/File%3ASKA_overview.jpg)].

width of the pulse is 10 MHz and transmission peak power is 67 dBm. Assume received noise power is −104 dBm. How long does it take for 1 bit of information to be carried over this channel? Speculate what that 1 bit of information may be.

References

[1] Tekniska Museet. https://www.flickr.com/photos/tekniskamuseet/6832884236, accessed 20 August 2015.
[2] Shannon, C.E. (1948) A mathematical theory of communication. *Bell System Technical Journal*.
[3] Huffman, D. (1952) A method for the construction of minimum-redundancy codes. Proceedings of the IRE, pp. 1098–1101.
[4] Ziv, J. and Lempel, A. (1977) A universal algorithm for sequential data compression. *IEEE Transactions on Information Theory*.
[5] Deutsch L.P. (1996) DEFLATE Compressed Data Format Specification. IETF RFC 1951.
[6] http://www.huffingtonpost.com/2011/06/10/nasa-voyager-bubbles-solar-system-heliosphere_n_874733.html?icid=maing-grid7%7Cmain5%7Cdl1%7Csec3_lnk3%7C69958, accessed 20 August 2015.
[7] Ludwig, R. and Taylor, J. (2002) Voyager Telecommunications. http://descanso.jpl.nasa.gov/DPSummary/Descanso4--Voyager_new.pdf, accessed 20 August 2015.
[8] Raith, K. and Uddenfeldt, J. (1991) Capacity of digital cellular TDMA systems. *IEEE Transactions on Vehicular Technology*.
[9] Gilhousen, K.S., Jacobs, I.M., Padovani, R., Viterbi, A.J., Weaver Jr, L.A. and Wheatley III, C.E. (1991) On the capacity of a cellular CDMA system. *IEEE Transactions on Vehicular Technology*.
[10] Hidalgo, C. (2015) *Why Information Grows: The Evolution of Order, from Atoms to Economies*, Basic Books Publishers.
[11] https://commons.wikimedia.org/wiki/File:Steve_Martin.jpg, accessed 20 August 2015.
[12] https://commons.wikimedia.org/wiki/File:SKA_overview.jpg, accessed 20 August 2015.

5

From Analogue to Digital

Preview Questions

- How are analogue signals converted to digital signals?
- What is sampling frequency and Nyquist rate?
- What is the difference between CD audio and MP3 audio?
- How is a still picture converted into discrete data? How is video source-coded?
- How are MP3 and MPEG standards related?

Learning Objectives

- The principles and techniques for analogue to digital signal conversion
- Source coding to ensure a minimal number of bits are used to represent signal information
- Principles behind voice source coding
- Principles behind audio source coding
- Principles behind image and video source coding

Broadband Telecommunications Technologies and Management, First Edition. Riaz Esmailzadeh.
© 2016 Riaz Esmailzadeh. Published 2016 by John Wiley & Sons, Ltd.
Companion Website: www.wiley.com/go/BTTM

Historical Note[1]

The origins of Sampling Theory can be traced to the invention of telegraphy and efforts to maximise a telegraph line's utilisation. Since there is a limit to how fast a telegraph operator may key-in messages, a question arose as to whether it is possible to use the same line by two or more telegraph operators. This was in effect multiplexing two or more telegraph calls on the same circuit and is known as time division multiplexing (TDM). Proposals for TDM of telegraphy are recorded as early as 1848, only 4 years after Morse's telegraph system formally went into operation. A rotating switch (commutator) connected a telegraph line to several operators in turn, and a matching rotating switch at the receiver distributed the incoming multiple stream to listening operators. The commutators 'sampled' the input of operators: and the speed of sampling had to be set to a level to ensure the telegraph data transmitted by each operator was fully captured. The speed setting of the commutators was based on practice, with little theory formulated.

The importance of TDM technology grew with the invention of telephony and the need for high capacity lines which connected telephone exchanges. In 1903, William Miner patented a rotating commutator as shown in Figure 5.1. The challenge again was to set the speed of the 'sampling' commutator in order that the reproduced voice was of an acceptable quality, as a slow rotating commutator resulted in distorted voice reproduction. Miner showed through trial and error that 'the best results [were obtained] with a rate of 4300 per second'. Miner concluded that sampling needs to be done at twice the voice's upper frequency component – his voice frequency was filtered at slightly above 2 kHz. This is the first time a report is made of an optimal sampling rate for voice signals.

The challenge for formulating a theoretical solution was to determine the necessary sampling rate. At such a rate a receiver would be able to interpolate the received samples and reconstruct the original signal. It was intuitive that the sampling rate must be proportional to the highest frequency of the signal. Furthermore, it had been shown by practitioners that the sampling rate must be at least twice the highest frequency. The theoretical basis for this sampling rate was found through the independent work of several scientists in the 1920s, including H. Nyquist

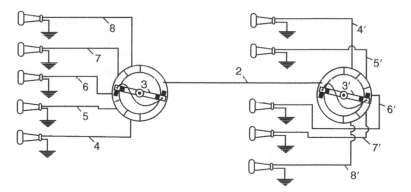

Figure 5.1   Miner's voice sampling commutator as drawn in his patent application. Source: Ref. [2]

---

[1] This Historical Note is based on a paper by Luke [1].

working for Bell Laboratories in the US, R. Hartley also working at Bell Laboratories, and V. Kotelnikov working at the Moscow Power Engineering Institute, Russia. These researchers proved what the practitioners already knew, that a sampling rate of twice the highest frequency was sufficient for exact reproduction of a signal.

With the formulation and publication of the sampling theorem, it was noted by mathematicians that J. Lagrange had formulated the principle of approximating a continuous signal with discrete samples as early as 1765.

H. D. Luke, in his paper on the historical development of the Sampling Theorem [1], recounts the contributions made by different individuals and fields of knowledge/art. He concludes:

> "The numerous different names to which the sampling theorem is attributed in the literature – Shannon, Nyquist, Kotelnikov, Whittaker, to Someya – gave rise to … discussion of its origins. However, this history also reveals a process which is often apparent in theoretical problems in technology or physics: first the practicians put forward a rule of thumb, then the theoreticians develop the general solution, and finally someone discovers that the mathematicians have long since solved the mathematical problem which it contains, but in 'splendid isolation' " [1].

Figure 5.2 shows a telecommunications block diagram. A source generates information which is carried to a receiver. In binary digital communications, the source information needs to be converted to a series of 1's and 0's which are then digitally modulated and send over a medium. This conversion process is known as source coding, and may apply to analogue information such as voice, audio and video, or may apply to discrete data streams such as text or numbers. The source coding process, which is the focus of this chapter, coverts an analogue signal to an equivalent digital signal. In many cases it also removes redundancies and produces a data stream with a size close to the source entropy as may be perceived by a human recipient.

A number of source coding techniques exist, classified by their efficiency vis-à-vis the information source entropy. Because telecommunications equipment needs to inter-operate, the choice of a source coding technique is usually determined by an international standard. The standards decisions are generally based on the technical performance of the source coding technique as well as business issues such as equipment complexity and backward compatibility. Intellectual property rights, patent restrictions and such are also important considerations. There may also exist policy considerations: development of international

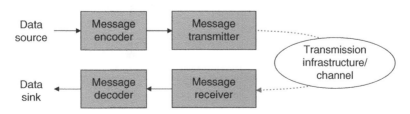

Figure 5.2   Telecommunications block diagram

standards requires collaboration from national standardisation bodies. Generally the market for most, if not all, telecommunications products is global, and therefore many national agencies need to take part and approve such standards.

This chapter discusses how continuous *analogue* information may be represented in a discrete *digital* form, and how such representation may be done with a minimal number of bits. We show that the mechanisms for representing such signals follow principles developed by Shannon as described in Chapter 4. Popular source coding techniques for data, voice, audio, image and video signals are described. We demonstrate how these techniques are used in popular information storage and communications applications such as zip programs, voice over internet protocol (VoIP), mobile phones, MP3 players, digital photography and video recording.

Analogue Signals

The amplitudes of natural phenomena and waveforms change as a function of time. Temperature, humidity, wind speed, soil moisture and electrocardiogram all rise and fall with time. Some may change more slowly and some may change faster than others, but all change as a function of time and in a continuous manner. Figure 5.3 is an example: it shows the temperature level in Adelaide on a typical summer day. The historical minimum and maximum temperatures, or temperature peak and trough, in a typical Adelaide summer are between a low of 5 °C and up to 48 °C. On this particular day, the temperature changed between a minimum of 27.2 °C and a maximum of 35.2 °C. It can be seen that the seismic activity changes more rapidly compared with temperature, as peaks and troughs occur more frequently over a measurement period. The measure of how often peaks and troughs occur in a signal is referred to as the frequency of the signal and measured in hertz (Hz).

Any signal may be recorded and represented in analogue as a continuous waveform. The signal can also be transmitted to a remote location using analogue telecommunications technologies. The signal can equally be represented, and communicated, as a stream of discrete numbers which represent the signal amplitude at specific measurement intervals. This series of numbers can then be transmitted to a remote location in discrete form using *digital*

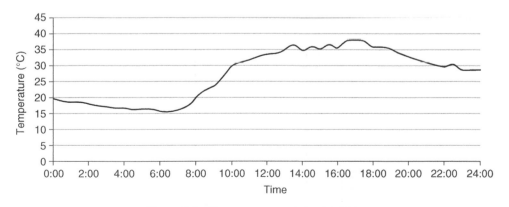

Figure 5.3   Temperature graph for Adelaide

Table 5.1 The temperature table – sampled every hour

Time	0:00	1:00	2:00	3:00	4:00	5:00	6:00	7:00	8:00	9:00	10:00	11:00	12:00	13:00	14:00	15:00	16:00	17:00	18:00	19:00	20:00	21:00	22:00	23:00	24:00
Temperature (°C)	19.9	18.9	18.4	17.3	16.9	16.8	16.1	16.5	20.2	23.9	30.3	32.1	33.9	35.1	34.9	35.5	35.8	38	36.2	35.3	32.9	31.3	29.9	28.8	28.9

Table 5.2 The temperature table – sampled every 2 h

Time	0:00	2:00	4:00	6:00	8:00	10:00	12:00	14:00	16:00	18:00	20:00	22:00	24:00
Temperature (°C)	19.9	18.4	16.9	16.1	20.2	30.3	33.9	34.9	35.8	36.2	32.9	29.9	28.9

telecommunications technologies. For example, the temperature graph in Figure 5.3 can be shown as a set of numbers corresponding to regular measurements every half an hour. This is shown in Table 5.1. The process of measurement at regular intervals is known as *sampling*, and the frequency of sampling, in this case once every hour, is known as the *sampling rate* or *sampling frequency*.

The same temperature graph may be represented as a set of discrete numbers corresponding to measurements every 2 h, and demonstrated in Table 5.2.

It can be argued that the two representations are quite equal in the information they convey. However, if both representations as shown in Table 5.1 and Table 5.2 are equivalent, then obviously Table 5.2 is more efficient as it needs fewer data points to convey the same temperature information. It is therefore important to determine the necessary sampling frequency for representing a particular analogue signal to ensure that *all* information is preserved efficiently.

Another example shown in Figure 5.4 is of a voice clip. The amplitude of human voice is an analogue, continuous waveform. Such a waveform may also be sampled at regular intervals and represented as a series of discrete numbers.

The representation of an analogue signal using a discrete set of numbers is known as digital representation; and the process as analogue to digital conversion (A to D in short). The reverse process is known as digital to analogue conversion (D to A).

As noted above, sampling frequency needs to be sufficiently fast to accurately represent the analogue signal. As demonstrated above in the temperature example (Table 5.1 and Table 5.2) sampling must also be efficient and use a minimum number of bits in its representation. Figure 5.5(a) shows a section of an analogue signal with a length of 1 s. The signal has three peaks and two troughs, and therefore can be estimated to have a frequency of ~3 Hz, although one can note that the rate of peaks and troughs is increasing over time. The signal is sampled at regular intervals at a sampling rate of 40 Hz in Figure 5.5(b), and the resulting discrete representation is shown in Figure 5.5(c). The representation appears to be a close replica of the analogue waveform. Another representation is shown in Figure 5.5(d) where the sampling frequency is 20 Hz or half as much compared with Figure 5.5(c).

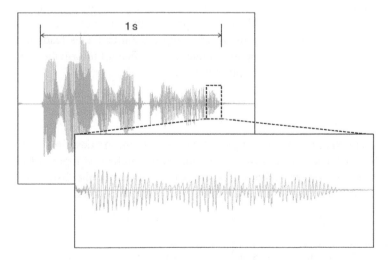

Figure 5.4   A one second voice clip as a time domain sound waveform

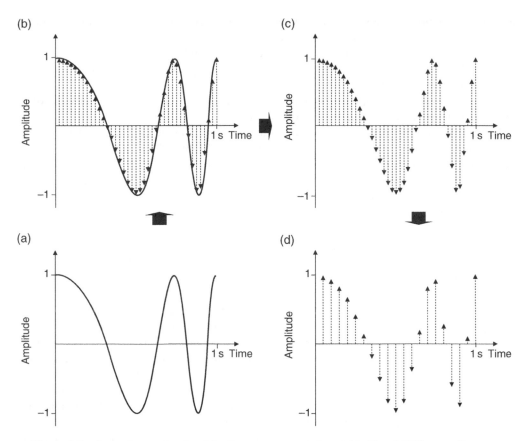

Figure 5.5   An analogue signal and its discrete representations with 40 and 20 Hz sampling rate

It is clear that Figure 5.5(d) is more efficient as it needs half as many samples to represent the analogue signal. However, it does not represent the analogue signal as closely as Figure 5.5(c). Is it therefore an equally accurate representation? This leads to the question whether there exists a minimum sampling frequency at which all information of an analogue signal can be preserved in its digital representation.

Nyquist Theorem[2]

The necessary sampling rate is given by the Nyquist theorem. As described in the Historical Note, Harry Nyquist, a scientist at Bell Laboratories, researched on topic of transmission of telegraph signals over telephone lines. His aim was to determine the fastest possible rate at which telegrams could be sent over band-limited telephone lines. In order to determine this he

---

[2] This theorem is also known as the Nyquist–Shannon theorem. Several other scientists are considered to have arrived at the same result independently.

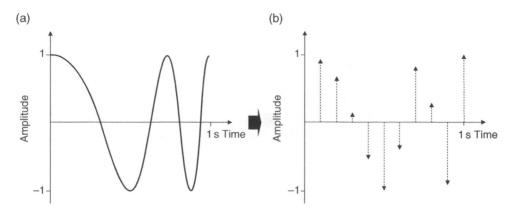

Figure 5.6 Analogue waveform to digital using a 10 Hz sampling rate

needed to find how analogue and discrete signals were related to each other in terms of the necessary transmission band [3].

Nyquist showed that for a digital representation of a band-limited signal, sampling must be done at a rate of more than twice the largest frequency component. Given the Nyquist rate, and exact measurement of the signal at the sampling point, the analogue signal can be perfectly represented in a digital form.

The band-limited condition is important as signals may have high frequency components. For example, human voice has frequency components are mostly in the range of 300–3400 Hz, as shown in Figure 5.4. However, some higher frequency components do exist, especially for people who have higher voices. However, since telecommunication channels are band limited, the normal practice is to filter the analogue signal to within that limited band by removing all higher frequency components. For voice this means limiting voice to 4000 Hz. While higher filtering can lead to higher fidelity reproduction, it comes at the expense of extra bit rate.

Nyquist theorem therefore mandates that a voice signal must be sampled at a rate of larger than $2 \times 4000 = 8000$ Hz. If we assume that the highest frequency component of the analogue signal in Figure 5.5(a) is 5 Hz, then the sampling rate can be as low as 10 Hz, which will result in a discrete representation as shown in Figure 5.6. Note that this representation is equally accurate as those of Figure 5.5(c) and (d). That is, sampling at any higher rate does not yield any advantage for the analogue to digital conversion process.

As discussed before, telecommunications is the science and technology of conveying information to a remote location. From this point of view continuous analogue waveform or discrete numbers are equivalent. That is, the sending of the analogue waveform in Figure 5.6(a) and the following set of numbers {0.913410317, 0.662541770, 0.137126512, −0.543347179, −0.980775131, −0.398801013, 0.823661318, 0.286615541, −0.917042669, 0.995046175} work equally.

The accuracy of discrete representation is however limited by how many digits are used. For example, sending {0.913410317, 0.662541770, 0.137126512, −0.543347179, −0.980775131, −0.398801013, 0.823661318, 0.286615541, −0.917042669, 0.995046175} is more accurate than sending the same information rounded to only three digits {0.913, 0.663, 0.137, −0.543, −0.981, −0.399, 0.824, 0.287, −0.917, 0.995} to represent the analogue signal of Figure 5.6(a).

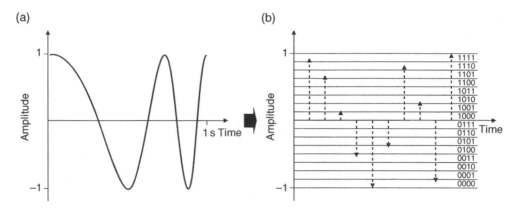

Figure 5.7   Quantisation into binary levels

The error thus introduced into discrete digital representation is known as quantisation noise. The required degree of accuracy is generally determined by the application which uses the received information, and/or by the applicable standard.

In practice, digital transmission is carried out using binary numbers (that is using only 0's and 1's) rather than decimal. The signal levels are converted to binary numbers using a set number of quantisation levels. This is demonstrated in Figure 5.7. Here, the maximum–minimum range of the input signal (−1 to 1) is divided into 16 levels, each level represented by binary numbers in the range of 0000 to 1111. Each discrete sample is then converted to a binary number based on the amplitude of the signal. The analogue signal can then be transmitted using binary bits as {1111, 1101, 1001, 0011, 0000, 0100, 1110, 1010, 0000, 1111}. The receiver can reproduce the analogue signal with the knowledge of the sampling rate, quantisation levels and the minimum–maximum range.

The accuracy of digital representation clearly depends on the number of quantisation levels. The example in Figure 5.7 uses 16 levels, which yields a certain degree of accuracy while some information is lost. Here the first and the last samples are both represented by 1111, whereas their amplitude is clearly different. This can be remedied through increasing quantisation levels, which increases the degree of accuracy and reduces quantisation noise. This increase is however at the expense of a larger data size which may take a longer time to transmit.

Therefore, two factors determine the number of quantisation levels. One is the necessary accuracy for a particular application. The other is telecommunications systems capacity and its associated device structure and the protocols. In practice the size is also determined by equipment standards. Computer systems usually use memory and processing structures based on units of one or more bytes, where each byte is 8 bits. Quantisation levels are therefore designed to produce binary digits of multiples of 8 bits. Voice for example uses 256 quantisation levels, and produces 8 bits per sample ($2^8 = 256$). Music (as in compact discs, CDs) uses 65 536 quantisation levels, and produces 16 bits per sample ($2^{16} = 65\ 536$).

Furthermore, quantisation level boundaries may be set to produce equal probability of signal occurrence within each band. This usually produces finer quantisation levels at lower amplitudes and coarser levels at higher magnitudes. An example of this is given in relation to digitisation of voice signals below.

Why Digital?

While capable of perfectly reproducing an analogue signal, a digital representation is still less than perfect in practice for two reasons. One is because of the filtering required to limit the signal bandwidth, which discards some information. The other is imperfect representation of the signal sample due to limited quantisation levels, which may lead to reproduction quality impairment. This partly explains why aficionados of vinyl records (analogue) have long bemoaned the lower quality of music CDs (digital). So why is digital so popular?

There are a number of technical and business reasons, but the most important technical reason is the ability of digital communications to remove the impact of thermal noise.

Noise impact is cumulative: it is added to a signal at each stage of reception as it progresses through a telecommunications system. For example, a voice call may go through several exchanges between the caller and the called parties as illustrated in Figure 5.8. The electronic devices at each stage add noise to the signal, and the cumulative result can be quite large. A long-distance telephone call quality worsens the more exchanges it passes through.

The impact of noise can be demonstrated by considering the analogue waveform of Figure 5.9. The amount of noise added at each stage (from top to bottom) is noise with a power equal to one-tenth of the signal. It can be seen that as noise accumulates, the analogue signal is more and more distorted.

The added noise is a random signal. While filtering and other processing may reduce the noise power, they cannot completely remove it. As a result, noise accumulates at each stage, and degrades signal quality. Analogue long-distance phone calls were famous for being noisy. The contrast in voice quality between long-distance and local calls was so great that people used to say that a call was so clear 'as if one was talking just across the town'.

Digital communications however enables noise-free *signal reproduction*. Consider the binary digital and noise signals in Figure 5.10. While the noise cannot be calculated and removed, the received signal can be detected as whether a 1 or a -1 (denoting a 0) was transmitted. When that decision has been made, the signal can be reproduced without any noise. In this way noise accumulation is stopped.

Digital telecommunications are not immune to noise however. If the level of additive noise is high, signal detection may become erroneous. Figure 5.11 and Figure 5.12 demonstrate this problem. When the power of signal compared with the power noise (SNR) is low, some bits may be erroneously detected. This gives rise to an important parameter in digital telecommunications, which is how many errors are received as noise level increases. This parameter is known as the bit error rate (BER).

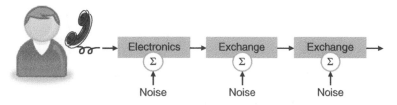

Figure 5.8 Noise added at each stage of a telephone call

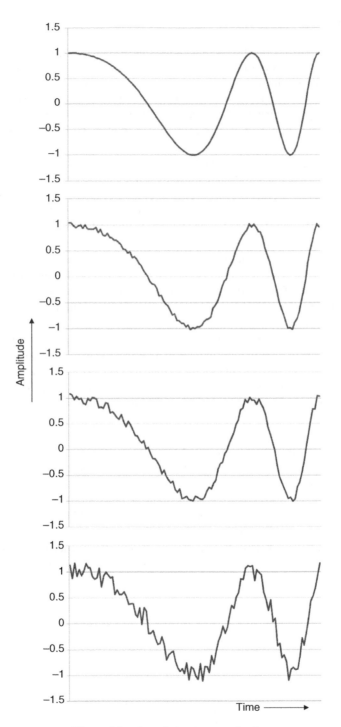

Figure 5.9   Cumulative impact of noise

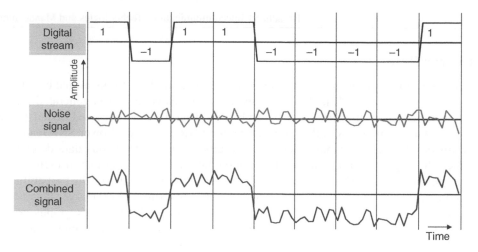

Figure 5.10 Digital data and additive noise (signal to noise ratio, SNR = 10 dB)

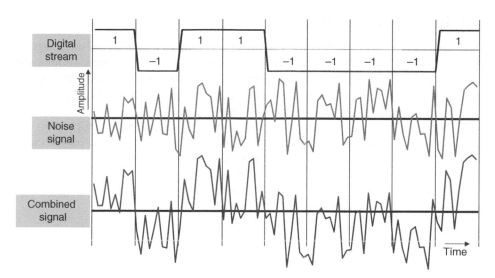

Figure 5.11 Digital data and further additive noise (SNR = −3 dB)

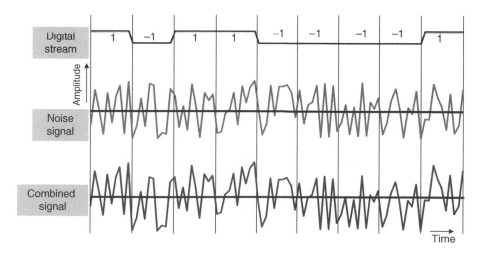

Figure 5.12 Digital data and even more additive noise (SNR = −10 dB)

Bit Error Rate

BER is the ratio of bits detected erroneously to the total number of transmitted bits. In the case of thermal noise, BER is drawn against the SNR. Noise power is generated locally and is a function of temperature and bandwidth. Therefore, as received signal power decreases due to path loss or other attenuation, the SNR decreases. Lower SNR leads to a more distorted received signal, which then results in higher erroneous data detections. A typical SNR graph for a common digital modulation scheme is drawn in Figure 5.13. The graph shows that if signal power to noise power ratio is one (0 dB), the probability of erroneous reception is ~7%, but if the signal power is five times (7 dB) larger than noise, then the probability of erroneous reception is reduced to ~0.08%. BER graphs are an important measure of quality and are widely used in the evaluation of telecommunications technologies.

A to D Business Advantages

Digital telecommunications popularity is not merely a result of technological advantage over analogue, which is in fact debatable as perfect representation of analogue signals is impossible because of quantisation loss. However, robustness against noise provides an important advantage to digital representation as discussed above. Equally important are business/economic advantages of digital telecommunications.

Moore's law [4] predicted that microprocessing speeds increase by a factor of two every 18 months. This prediction has held for the past 30 or so years. Over the same period, the cost of such processing has also decreased by the same ratio or more. Since microprocessors work well with digital signals, costs of digital processing and storage devices have decreased significantly vis-à-vis analogue devices. Digital technologies also facilitate easier information systems integration between business and trade partners.

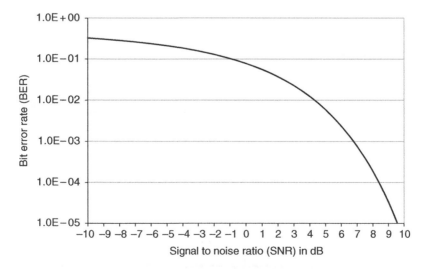

Figure 5.13   A typical BER curve

There are also disadvantages to digital communications. Two of these are noted above, namely the loss of quality due to filtering and quantisation noise. One important complexity associated with digital communications is the need to follow well defined protocols. Transmission of information in bursts of 1's and 0's must be standardised so machines can understand each other. These include the size of signal representation in binary form, quantisation levels and their boundaries, error detection and handling, data frame sizes, and so on. Such detailed standardisation is in contrast with analogue communications where standards were mainly needed for signalling and call set-up purposes.

Case Study 5.1: Integrated Services Digital Networks

Voice communications in a digital form is common in many forms today. Mobile telephony is digital, and so are voice over IP applications such as Skype. However, this has come about only recently and most of these systems date to the mid- to late1990s. The first major digital voice communications standard, Integrated Services for Digital Network (ISDN) was only developed in the 1980s. ISDN was designed to carry digital voice over twisted pair copper telephone wires and was expected to eventually replace the analogue telephone standard.

ISDN used a waveform digitisation of voice signals. It filtered voice at 4 kHz, and sampled it at the Nyquist rate of 8 kHz. Each sample was quantised using 8 bits, corresponding to $2^8 = 256$ quantisation levels, resulting in a bit rate of 64 kbps (kilobits per second). Such digitisation was shown to deliver high quality voice communications with minimal signal processing requirements. Twisted copper wire pairs were shown to be capable of supporting 144 kbps transmission rate over long distances, thereby providing two telephone lines of 64 kbps (called Bearer or B channel) and a data and signalling channel of 16 kbps (called Data or D channel) to a subscriber. The 16 kbps data connectivity from a home was deemed sufficient for most subscriber applications, which at the time included network computing and fax machines. This basic connection was called a Basic Rate Interface (BRI) of 2B+D [5].

The first set of ISDN standards was published through the auspices of CCITT (Comité Consultatif International Téléphonique et Télégraphique), later named the ITU-T, an agency of the United Nations. The standards envisaged higher transmission rates of 30B (2.048 Mbps or E1 carrier) and 23B+D (1.544 Mbps or T1 carrier) for business data communications applications. Most national operators anticipated ISDN services to become main-stream, and started the planning process. (The author's first job after graduation from university in January 1988 was in a group in Telstra tasked with planning the roll-out of ISDN networks in Australia.)

ISDN standards were designed to replace analogue phone services to subscriber premises. The roll out of this new system required new digital equipment at customer premises and local exchanges. A number of operators around the world embarked on building ISDN networks and by the early to mid-1990s many were operational. The Japanese operator NTT installed ISDN public payphones with data connectivity slots.

ISDN's value proposition was higher quality digital voice as well as data connectivity. By the late 1990s, 64 kbps ISDN links were actively marketed for data connectivity as the internet became more popular. However, the growth of internet services and their need for much higher transmission rate on the one hand, and the development of Digital Subscriber Links (DSL) standards on the other made ISDN as a data communications service obsolete. Meanwhile for telephony services the inter-exchange links have become digital, while the link between customers premises and a local exchange has largely remained analogue. ISDN systems failed to find a suitable application and are being phased out.

Case Study Questions

- Why was ISDN standard developed?
- What were the motivations of operators in rolling out ISDN systems?
- What were the technology–business–policy elements necessary for ISDN success?
- What do you see as the reasons for ISDN failure? Could these have been foreseen?
- In hindsight, what would you have done differently?

Data Source Coding

The analogue to digital conversion process does not take into consideration the information content of an analogue signal as determined by information theory. As a result, redundancies in the discrete data stream remain and the transmitted number of bits representing a signal can be more than necessary. A number of source coding techniques have been developed that reduce the number of required bits and increase transmission efficiency.

Source Coding

Discrete data source coding generally requires a lossless algorithm, which can be done with either fixed length or variable length techniques. However, analogue information sources can only be represented digitally as an approximate value, since exact discrete representation requires infinite data size. The degree of information loss depends on whether the code follows the waveform of the analogue signal or analyses the analogue waveform into parts which are then encoded and synthesised at the receiver.

Voice Source Coding

Voice coding techniques can follow the analogue waveform or use signal synthesis. Waveform coding starts by an analogue to digital conversion and may be followed by further compression using techniques such as entropy coding. Synthesis coding starts from human physiology and voice generation, and how its component parts may be represented by a series of discrete numbers. Major waveform coding techniques include pulse code modulation (PCM) used in ISDN and adaptive differential pulse code modulation (ADPCM) used in a number of standards including the Japanese Personal Handy-phone System (PHS). A number of synthesis coding techniques exist which are widely used in VoIP and mobile phone applications.

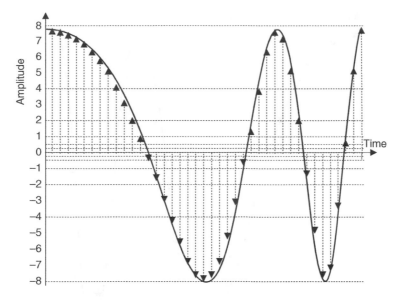

Figure 5.14 Logarithmic voice coding quantisation

We will discuss Adaptive Multi-Rate (AMR) coding standards which are used in wireless telecommunication 3G and 4G standards.

Pulse Code Modulation

ISDN standards used a special modulation technique called Pulse Code Modulation (PCM). This technique is a pure A to D conversion waveform source coding. Voice is sampled at 8 kHz and quantised at 8 bits per sample, resulting in a 64 kbps source coding rate. Quantisation levels are mostly logarithmically distributed with finer (more) levels at smaller amplitudes and coarser (fewer) levels at higher amplitudes as illustrated in Figure 5.14. This quantisation level distribution increases the dynamic range of voice, giving more accuracy to smaller signals and less to larger signals. Such a quantisation technique reduces the signal to distortion ratio as noise impacts smaller signal samples.

Adaptive Differential Pulse Code Modulation

The PCM described above is an inefficient form of an information theory perspective since it does not take the correlation between neighbouring samples into account. In practice consecutive samples of an analogue signal are highly correlated and therefore a technique that decorrelates this interdependency can reduce the number of bits required to represent the signal.

One such method is to only encode the difference between two consecutive samples. This is illustrated in Figure 5.15. The difference between consecutive samples can be encoded with fewer bits since the range of the signal is much smaller than the overall signal range (since it does not vary largely between two consecutive sampling intervals).

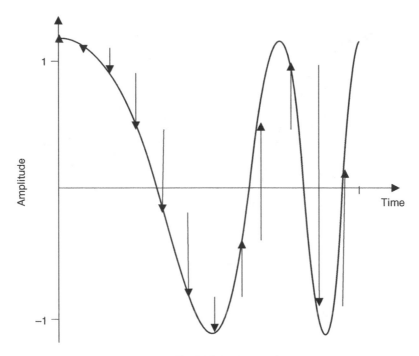

Figure 5.15   Differential encoding

The difference between two consecutive samples of voice can be entropy coded as the range of possible values is smaller and well understood. The ADPCM standard samples are represented by 4 bits, and therefore the total coding rate is equal to $8000 \times 4 = 32$ kbps. Both PCM an ADPCM coding use A to D conversion and are lossless.

Synthesis Coding

An alternative class of voice coding is through voice analysis into its components. These methods are lossy in that some information is lost in the source coding process. Synthesis coding uses the fact that human voice generation occurs because of the sound created by our vocal cords which is then formed into speech as it travels through our vocal tract (throat and mouth). Our speech consists of a 'buzzer' sound generated by the vibrating vocal cords, and 'hissing' and 'popping' sounds as air passes through stationary vocal cords and is shaped by our tongue, lips and throat. The voice decoder finds the intensity and pitch of the buzzer sound and the residue of the signal when the buzzer sound is filtered out [6].

The calculation of filter values is done through a synthesis process as illustrated in Figure 5.16. Here a codebook representing a range of possible pitches, intensity and residue is used to synthesise a sound similar to the input voice every 20 ms. A feedback loop adjusts the filter parameter until the synthesised and actual sounds differ minimally. The numbers representing the pitch, intensity and filter parameters are transmitted to a receiver which then synthesises the received voice.

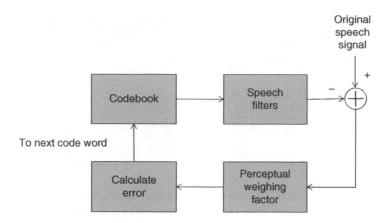

Figure 5.16   Voice analysis coding

Table 5.3   A number of commonly used voice codecs and their applications

Voice codec	Application	Bit rate (kbps)
G.729	VoIP	8
PCM	ISDN	64
ADPCM	VoIP	32
AMR	Wireless 3G and 4G standards	4.75–12.2
AMR Wideband	Wireless 3G and 4G standards	6.6–23.85

The synthesis process reproduces the components of the voice and passes them through the filter with parameters from the received information. This voice coding technique can yield very high compression rates, and is used in many applications. An example is AMR, the voice coder technology used in GSM, 3G and 4G standards. AMR encodes voice at variable rates from 12.2 kbps down to 4.75 kbps, showing significant compression compared with a full digital rate of 64 kbps. This comes however at the expense of sound quality, that is, the lower the bit rate is the higher the loss of information. Loss of information leads to lowering of the quality of the voice. Moreover, the quality loss is accentuated when reception errors occur. A 64 kbps lossless voice coding is very robust in the presence of large reception errors in that the quality is not significantly degraded. However, a 4.75 kbps coded stream can become unintelligible very quickly as transmission errors increase.

A number of voice coders have been developed by a number of organisations and standardised by ITU. Table 5.3 lists a number of commonly used codecs and their applications. Most codecs convert a voice stream filtered at 4 kHz, except for AMR Wideband (AMR-WB) which encodes a voice stream filtered at 7 kHz [7].

Comparison of Voice Coding Techniques

As discussed above, different voice coding techniques result in different bit rates, and therefore transmission requirements. Clearly a lower rate is beneficial from a business point of view as transmission resources can be used more efficiently. However, this reduced

Table 5.4 Perceptual quality of a number of voice codecs [10]

Narrowband AMR		Wideband AMR	
Rate (kbps)	MOS	Rate (kbps)	MOS
4.75	2.8	6.60	3.2
6.70	3.1	8.85	3.5
7.40	3.2	12.65	4.0
10.20	3.3	15.85	4.1
12.20	3.4	18.25	4.2

transmission rate comes at the expense of perceptual voice quality. A number of voice quality measures informing this trade-off have been standardised. One such measure is the perceptual evaluation of speech quality (PESQ) scale which compares different voice coding schemes with a benchmark of a clean analogue reference [8]. Another method is the mean opinion score (MOS) which averages the perception of listener on the scale of 1 (bad) to 5 (excellent). Table 5.4 shows a typical MOS comparison for several voice coders. As expected the perceptual quality degrades as lower bit rates are used to represent voice. The new wideband AMR codecs also can have a much better quality of voice reproduction and are used in new wireless standards such as 4G voice over LTE (VoLTE) [9].

Lower call quality generally results in unsatisfactory communications and may result in early call termination, therefore reducing potential operator revenue. Indeed it has been shown that telephone calls tend to last longer if the voice quality is high, and shorter if the quality is low [11]. Adaptive rate coding allows a network operator to use the best quality source coding schemes during periods of low traffic to deliver the highest possible perceptual quality. When the network is congested, the coding rate can be reduced to accommodate more calls.

Case Study 5.2: Skype

Skype is primarily a VoIP application, launched in 2003 to enable peer-to-peer telephony. Initially designed for personal computer (PC)-to-PC communications, enhancements have been made to enable instant messaging, phone-PC telephony, mobile application, teleconferencing and video telephony. Furthermore, Skype has entered the WiFi hotspot market, and is estimated to provide Skype subscribers access to 2 million hotspots globally. As of August 2013, the number of subscribers was more than 300 million [12].

PC-to-PC Skype telephony, or the technology of VoIP, has remained free and has been the main reason behind its subscriber growth. VoIP services existed long before Skype. The first VoIP services were offered in the mid-1990s, and ITU VoIP standard H.323 was published in 1996 [13]. By the early 2000s several VoIP services existed, specially targeted at the corporate market. A service ecosystem emerged to manage high quality VoIP services: for example Genista Corporation where the author worked in the early 2000s developed a tool to measure the perceptual quality of VoIP communications. However, the VoIP market size remained quite small until Skype was introduced.

The Skype offering was revolutionary in several ways. One way was a simple user interface which made it easy for anyone to use the service. A second way was a peer-to-peer protocol which made it possible to place and receive calls from behind corporate firewalls, and which facilitated a decentralised database of on-line subscribers, a protocol already used by Skype founders Friis and Zenstrom for their music file-sharing site KaZaa. However, the most important factor was the voice codec used to transfer analogue voice to digital stream. The quality of sound was very good compared with the international analogue telephony service. Added to this was the fact that one could make free international calls with ease. Skype was a great success: it was soon acquired by the on-line auction company eBay in 2006 for $2.6 billion to enhance their peer-to-peer sales platform. The company has changed and Skype is now owned by Microsoft after an $8.5 billion deal in 2012 again aimed at enhancing Microsoft applications such as internet messaging and email.

Skype is reported to use a number of voice codecs including G.711 (64 kbps), G.729 (8 kbps), proprietary SVOPC (16 kbps) and AMR-WB (16 kbps). High quality voice source coding has been one of the most important factors in maintaining Skype's customer base and continued success.

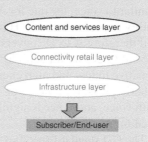

Skype also offers phone services to and from landline and mobile telephones through its Skype-out/Skype-in services. Skype and similar companies such as Viber, WhatsApp and WeChat are known as over-the-top (OTT) services. These companies are active in the content and services layer of the broadband telecommunications ecosystem. They utilise the broadband infrastructure as retailed by an ISP and replace services traditionally provided by telephone and mobile operators. Revenue lost to OTT service providers can be very significant; it is reported that more than one-third of international voice calls is now carried over the Skype platform (Figure 5.17) [14].

Case Study Questions

- Why was VoIP market uptake slow?
- How was Skype successful in establishing its VoIP service and attracting customers?
- How did Skype-out and Skype-in services help the strategic goals?
- Why did Skype enter the WiFi hotspot market?
- Why do you think Microsoft acquired Skype?

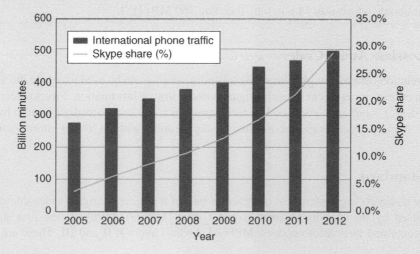

Figure 5.17   Skype share of international telephony (based on data from TeleGeography [14])

Audio Coding

Similar to voice, analogue audio is encoded with both lossless and lossy source coders. Several lossless coding techniques exist, with the most widely used being the CD standard. MP3 coders are the most well-known of the lossy audio coding techniques.

CD Standard

Similar to voice, the rate for A to D conversion of audio signals is also determined by human ability to perceive sound. Human audible sound is in the frequency range of 20–20 000 Hz, and therefore A to D conversion must sample sound at greater than $2 \times 20\,000$ samples per second to ensure all audible frequencies are captured. Furthermore, the human ear is very sensitive to low and high energy level sound and the audible range is from 10^{-9} to 10^4 atmospheric pressure, a range of 10^{13}, or 130 dB. Since the range is so much wider compared with voice, audio needs significantly more quantisation levels compared with the 256 levels used for voice.

The CD standard specifies sampling rate at 44 100 samples per second, slightly above the audible frequency range, in order to accommodate digital sound for the analogue video recorders then in use [15]. To accommodate the audible range, the standard uses $2^{16} = 65\,536$ quantisation levels. The bit rate of a CD recording is therefore 44 100 samples/s * 16 bits/sample * 2 (for stereo recording) to be equal to 1.411 Mbps. For 60 min of music the total number of bits required is calculated as follows:

$$= 1.411 \text{ Mbps} * 3600 \text{ s} = 5.08 \text{ gigabits} \approx 620 \text{ megabytes} \left(\text{MB}\right)\left(1 \text{ kilobyte} = 1024 \text{ bits}\right).$$

CD Size

The audio CDs were initially designed to have a diameter of 11.5 cm, equal to the diagonal length of a compact audio tape. This size provided an audio capacity of approximately 60 min of music. The size was increased to the present 12 cm diameter because an executive at Sony decided that CDs should be able to contain his favourite recording of Beethoven's 9th symphony, which was 74 min long (needing 760 MB) [15].

Free Lossless Audio Codec

The correlation between neighbouring digital audio samples can be used to reduce the number of bits necessary to represent an audio signal without loss of information. Free lossless audio codec (FLAC) is one such audio coding technique which can compress digital audio by up to 50–60%. A FLAC decoder can reconstitute the digital audio without loss of information [16].

MP3 Standard

A source audio coding system was designed as part of a project on digital video coding, and standardised by the Moving Pictures Expert Group (MPEG). MPEG-1, the first standard release contained three audio codecs, MPEG-1 Audio Layers I, II and III. These are better known as MP1, MP2 and MP3, respectively. All three standards are lossy coding of audio, that is, the encoded audio is not an A to D conversion of the input analogue audio signal.

MP1 encodes audio by sampling at 32, 44.1 and 48 kHz and produces output at variable rates from 32 to 448 kbps. It was used in digital compact cassettes, a now discontinued product. MP2 also encodes analogue audio by sampling at 32, 44.1 and 48 kHz, and produces a compressed digitised audio stream with variable rates between 32 kbps and 384 kbps. MP2 is a part of digital audio broadcasting (DAB) and digital video broadcasting (DVB) standards and is commonly used for professional recording by radio stations. The MP2 standard is also used in the video CD (VCD) system.

Of the three MPEG-1 audio standards MP3 is the most popularly known and used. It has applications in audio coding for recording, storing and sharing music and is used in most portable music players. It is also commonly used for streaming music. Similar to voice compression standards such as AMR, MP3 also analyses audio signals by dividing the input signal into sub-bands of many different frequencies. The encoder then filters out information that is perceptually insignificant: any sound that cannot be perceived by the human auditory system is discarded. For example if a particular frequency of a signal has a large amplitude, it can mask its neighbouring frequencies which have lower amplitudes. This phenomenon is known as auditory masking. Encoding such masked information is not necessary as our ears cannot hear it. Using this principle, an MP3 source coder can reduce the number of bits used by only providing information that is perceptually significant to the human ear [17]. The sub-band information is further Huffman coded to minimise the number of bits used. The MP3 standard specifies sampling rates of 16, 22.05, 24, 32, 44.1 and 48 kHz, and produces bit rates in a 32–320 kbps range. These bit rates represent significant reductions compared with an audio CD coding rate of 1.411 Mbps.

It should be noted that audio encoders such as MP3 work in a different way to voice encoders such as AMR. Voice coders analyse the generation of voice signals by the human vocal system in order to efficiently encode it. In contrast audio coders analyse how our auditory system perceives audio in order to efficiently encode it. MPEG standards of which MP3 is part are globally ratified through the International Organisation for Standardisation (ISO). In general the standards specify the format of the data fields output from an MP3 encoder, and the operation of the decoders. Different vendors are therefore free to enhance the quality of audio encoding, so long as a standard decoder can replay the encoded audio. This ensures that any MP3 player can play back an audio recording by another vendor's MP3 encoder.

Table 5.5 shows a list of commonly used lossless audio compression standards and Table 5.6 gives a list of lossy compression standards. A comprehensive list of all audio codecs can be found in the literature [18, 19, 20, 21, 22].

Table 5.5 Lossless compression standards

Lossless audio codec	Application	Bit rate	Standard
CD	Storage/playback	1411 kbps	IEC 60908
ALAC	Storage/playback	50–60% of CD	Apple Open Source
FLAC	Storage/playback	50–60% of CD	XIPH.ORG Open Format

Table 5.6 Lossy compression standards

Lossy audio codec	Application	Bit rate (kbps)	Standard
MP3 (MPEG-1 and MPEG-2)	Storage/playback/ streaming	32–320	ISO/IEC 11172-3 ISO/ IEC 13818-3
AAC (MPEG-2 and MPEG-4)	Storage/playback/ streaming	32–320	ISO/IEC 13818-7 ISO/ IEC 14496-3
Vorbis	Storage/playback/ streaming	45–500	XIPH.ORG Open Format
Windows media audio (WMA)	Storage/playback/ streaming	24–768	Microsoft Proprietary

Case Study 5.3: iTunes

The MP3 technology was developed at the German company Fraunhofer-Gesellshaft, and their first German patent was issued in 1989. The technology was integrated into the MPEG-1 standard in 1992 and published a year later. MP3 system's high compression capability and relatively high quality of the reproduced audio presented a credible audio recording and playing alternative to the CD standard. Software products were soon developed and the first MP3 music and portable MP3 players appeared in the market in 1999 [23]. However, major music producers did not sell music in the MP3 format because of file copying and difficulty of digital rights management (DRM). While MP3 files sharing on platforms such as Napster and later on KaZaa gained popularity around the turn of the century, a mainstream market for these products was not established.

In January 2001, Apple Inc. introduced iTunes to record music in, as well as convert digital audio (CDs) into, the MP3 format. This was soon followed in October 2001 by the introduction of the iPod product to store and play MP3 files. This became the first popular MP3 music player, supported by a well-known brand and a platform where music could be produced and purchased. In addition, Apple's strategy in disaggregating music albums into individual songs and selling them for as low as 99 ¢ through the iTunes platform complemented their iPod product. The iTunes platform and iPods became very popular in a very short time.

Steve Jobs famously promoted iPod as a portable music player that could carry 1000 songs. With a reasonably high audio quality, iPod's competitive advantage over portable CD players was great. The market for CDs as well as CD players has greatly shrunk since their peak in 2000.

Apple has built on the success of the iTunes music platform by adding further features such as music file metadata (such as information on music and CD covers), and video such as films and TV programmes. Furthermore, radio programmes are now commonly 'podcast' in MP3 format. With the release of the iPhone in 2007, the iTunes platform has been used for music and video streaming, and application management.

Apple is now a major player in the broadband tele-communications ecosystem as a content provider due to its strategic use of MP3 technology. The main advantage of MP3 over CD was its much smaller file size at the expense of some quality degradation. Efficient source coding offered by MP3 was exploited and packaged in a way acceptable to music producers and easy to use by consumers. The ensuing success of iPhone and iPad devices is in part due to iPod and MP3.

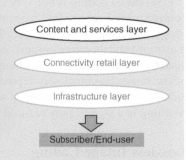

Case study Questions

- Why was the MP3 standard developed?
- What are the main reasons for the slow popular uptake of MP3 devices?
- Why could Apple popularise MP3? How was the network-mediated platform concept applicable?
- How has the iTunes platform expanded?
- How is Apple a player in the content and services layer of the broadband ecosystem?

Image Coding

A digital image consists of pixels of different colours (Figure 5.18). As discussed in Chapter 4, the number of possible colours a pixel may take determines how many bits are required to represent a pixel. For example if 256 possible colours exist then 8 bits ($2^8=256$) are required to represent one pixel. If such an image has dimensions of 40×40 pixels, then the total number of bits required to represent this image is 40 * 40 * 8 = 12 800 bits. Such a coding technique is used for example in uncompressed bitmap (resulting in files with .bmp extension).

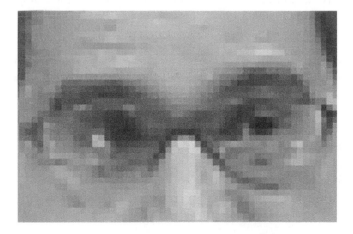

Figure 5.18   Pixels making up an image

Pixel-based coding is not very efficient as it does not take the correlation between neighbouring pixels' colours into account. Furthermore, it does not take the perceptual content of the image into account. For example, the human visual system is insensitive to rapid pixel colour transitions and therefore insignificant perceptual information may be discarded with little or no visual impact. While such a technique leads to loss of some information, the perceptual clarity of the image can be closely reproduced. Based on this principle, a number of image coding techniques have been developed to facilitate both lossless and lossy compression, thereby resulting in a significantly reduced number of bits required to represent an image.

Lossless Image Coding

Lossless image coding is possible if correlation between adjacent pixels is taken into consideration. Information theory teaches that the total information content of an image can be calculated by the probability that an individual pixel takes a particular colour, summed over the total number of pixels. An entropy coding method can then produce an efficient coded message. If one can calculate the probability that a pixel takes a particular colour, then an efficient coded message can be produced.

As can be observed from Figure 5.18, the probability that a pixel takes a particular colour is highly correlated with its neighbouring pixel's colour. For example, a grey pixel is highly likely to be next to another similar grey coloured pixel.

One way to decorrelate the image is by calculating the difference between neighbouring pixels. We can start from the top-left pixel, note the colour and assign a specific code. The next pixel has a slightly different colour, the difference can be measured and a code assigned. One can move across horizontal and vertical axes and measure the difference in pixel colours, assign a code to the difference and compile them into a file which represents the image information. Such an encoding method can be significantly more efficient than when pixel correlation is not considered. The encoded image size can be further reduced if an entropy coding technique, such as Huffman coding, is used to encode the inter-pixel difference values. This is the principle behind the image coding Portable Network Graphic (.png file format) method for lossless image compression. PNG compression is useful for images where pixel-to-pixel variation is small or zero, such as text or graphic images, for example for computer screen dumps as shown in Figure 5.19.

horizontal and vertical axes and measure the difference in pixel colours, assign a code to the difference and compile them into a file which represents the image information. Such an encoding method can be significantly more efficient than when pixel correlation is not considered. The encoded image size can be further reduced if an entropy coding technique, such as Huffman coding is used to encode the inter-pixel difference values. This is the principle behind image coding Portable Network Graphic (.png file format) method for lossless image compression. PNG compression is useful for images where pixel-to-pixel variation is small or zero, such as text or graphic images, for example for computer screen dumps as shown in Figure 5.19.

Figure 5.19   A screen dump of the author's screen as text is being typed

Lossy Image Coding

Images such as Figure 5.19 can be compressed with a large degree of efficiency. However, many images such as photographs of nature usually have large pixel transitions and therefore lossless differential coding does not yield significant compression. Further compression is achieved if some loss can be tolerated.

Similar to audio and voice, lossy compression techniques take human perceptual ability into account. Our eyes can distinguish small colour transmissions across pixels but high frequency transitions are not well perceived. Image compression is principally carried out by removing these perceptually insignificant data.

The process of identifying and removing high frequency components of an image is carried out in the frequency domain. First an image is transformed from the spatial domain into the frequency domain, then higher frequency transitions are identified and removed according to a desired compression level.

Frequency domain representation of a spatial image is done through the discrete cosine transformation (DCT) technique. DCT in effect translates an image into a sum of many possible pixel transitions. Figure 5.20 shows an 8x8 pixel grey-scale image. This image may be represented as sum of possible 8 pixel grey-scale transitions as shown in Figure 5.21.

The DCT technique converts the spatial domain image of Figure 5.20 into a weighted value of all frequency components as shown in Figure 5.21. The image file can be represented in two ways. One is in the spatial domain where the grey-scale of each pixel is found. The other way is in the frequency domain where the weight of each frequency transition is found. Both equally represent the same image [24].

Figure 5.22 shows two images. Figure 5.22(a) has few black to white transitions: it has relatively less information and can be source coded with relatively fewer bits. Figure 5.22(b) has many more transitions and is encoded using a larger file. A lossy encoder can remove some of these high frequency components without significantly reducing the perceptual quality of image reproduction. The images to the right are the DCT produced images in the frequency domain. As expected, Figure 5.22(c) has less high frequency information, and the white dots are concentrated in the top left part of the image. Figure 5.22(d) shows more high frequency throughout the DCT result [25].

The frequency domain representation facilitates a lossy compression of an image. For example the image may have small significant high frequency transitions, which are those shown in the lower-right corner of Figure 5.21. These may be discarded with little loss of perceptual quality. Progressively more and more of the high frequency components may be deleted, resulting in smaller size files but with lower quality. An example of an image with three levels of compression is shown in Figure 5.23. As it can be seen, the sharpness of edges is lost as more and more high frequency transitions are removed.

8×8

A

Figure 5.20 An 8×8 pixel grey-scale image. Source: https://commons.wikimedia.org/wiki/File: Letter-a-8×8.png

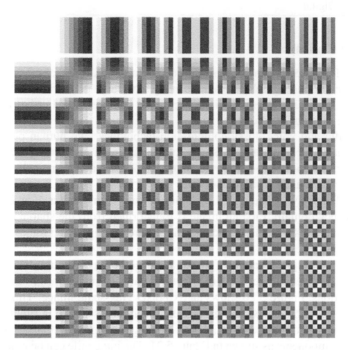

Figure 5.21 The set of possible 8 pixel grey-scale transitions. Source: https://commons.wikimedia. org/wiki/File:DCT-8x8.png

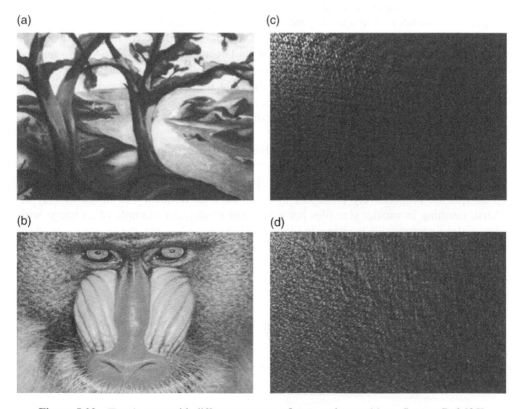

Figure 5.22 Two images with different amounts of grey-scale transitions. Source: Ref. [25]

Figure 5.23 Three compression levels, with file sizes (from left to right) of 83.2, 9.5 and 4.8 kilobytes. Source: Ref. [26]

Video Coding

As video signals are simply a sequence of images, video coding uses image coding techniques such as described above. Video coding also takes advantage of temporal similarities between consecutive image frames to reduce the number of bits required to encode moving images.

A to D conversion of video while possible is impractical. Broadcast analogue TV uses 6–8 MHz of bandwidth. Using PCM digital encoding with a 2^{24} quantisation range will lead to: 8*2*24 = 384 Mbps. A DVD can store 4.5 gigabytes of data, and at this rate can provide storage for 94 s of digital video. Even transmission of individual image frames is impractical. If an image is encoded with 500 kilobytes, a 30 frames/s video requires 15 Mbps transmission rate.

Since consecutive image frames are highly correlated, video encoding needs to decorrelate these images to reduce the total number of required bits. The principle is therefore similar to differential coding described in relation to voice, audio and image coding: once an image has been coded, subsequent images can be coded by calculating their difference from the initial image.

All these standards encode video by producing three types of frames through prediction and differential encoding based on the DCT technology and temporal (inter-frame) differential coding. Three kinds of frames are specified: intra-coded frames (I-frames), predictive-coded frames (P-frames), and bi-directionally predictive-coded frames (B-frames). I-frames are in effect JPEG files using the DCT technology. I-frame data are independent of previous or following frames. P-frames are temporally several frames after an I-frame and are calculated through a predictive differential process: the data for the P-frame is calculated by extrapolating the I-frame data. Since the extrapolation is not perfectly accurate, the difference between the actual value and the extrapolated value is calculated, encoded and transmitted as the value of the P-frame. As the difference is generally of a smaller size it provides more compression as P-frames take advantage of I-frame information. In practice several P-frames are extrapolated and differentially calculated.

B-frames represent the information for the frames between I- and P-frames. Their data is calculated based on prediction and differential coding. An interpolation between I- and P-frames is made to estimate the B-frame data. Again since the actual and estimated values are not exactly equal, the difference between the two is encoded and transmitted as B-frame. The calculation relationship between of I-, P- and B-frames is illustrated in Figure 5.24.

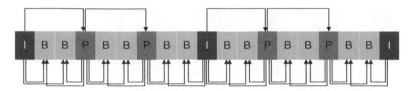

Figure 5.24   Relationship between I-, P- and B-frames

Table 5.7   Common MPEG video compression standards and applications

Codec	Standard	Applications
MPEG-1	H.261	Video CD DAB
MPEG-2	H.262	DVD, Blu-Ray DVD DVB
		Blu-Ray DVD
		Video streaming (YouTube, Vimeo, etc.)
MPEG-4	H.264	Video telephony (e.g. FaceTime)
		HDTV broadcasting
		XAVC (4K)
MPEG-H	H.265	High efficiency video coding 8K ultra high definition TV

A number of video coding standards have been developed since 1988 within the MPEG group under the auspices of ITU [27]. These standards have been instrumental in facilitating video conferencing, digital TV, video telephony and video streaming application. A list of the main standards and their applications are shown in Table 5.7.

Although natural signals are continuous and analogue, they may be represented in a discrete, practical manner to be perceived 'natural' by a human observer. Discrete or digital representation of signals allows for better preservation and processing of information and enables the realisation of theoretical telecommunications capacity limits. Furthermore, considering our limited perceptual capability, digital representation and processing enables an efficient but acceptable representation of analogue signals. These are some of the principles behind the recent digital transformation of our society.

Review Questions

1. Why do you think sampling rate is known as the Nyquist rate despite contributions from many scientists prior to his formulation?
2. What is sampling rate and how is it calculated?
3. What is quantisation level and what factors are considered in its determination?
4. How does digital transmission compare with analogue? Why has digital become popular?
5. Why do errors occur in digital communications?
6. Calculate the sampling rate for audio music, considering that our ears can detect sounds of up to 20 kHz. What do you do if music has frequency components up to 25 kHz?

7. How is information theory applied for services intended for human subjects?
8. What is the difference between waveform coding and synthesis coding?
9. What are the business benefits of AMR codes? Why are there 8 different code rates?
10. What are AMR-WB codes and why were they developed? What are their applications?
11. What were the business advantages of using Skype for long-distance telephone calls? How did source coding technologies enable these?
12. What are the parameters in deciding which voice coder to use for an application?
13. How was the size and capacity of CDs decided?
14. How is audio synthesis coded? How is it different from voice synthesis coding?
15. Give examples of audio lossless and lossy source coders? What are the applications for each?
16. How was MP3 developed and what business difference did it make?
17. What are the impacts of international policy decision making in video coding standards? Why do national governments cede decision making to these bodies?
18. Why is neighbouring pixel colour and lumina correlation important in image coding?
19. How do lossless image compression techniques work?
20. How do lossy image compression techniques work?
21. How is information theory useful in image and video coding?

References

[1] Luke, H.D. (1999) The Origins of the Sampling Theorem. IEEE Communications Magazine.
[2] Miner, W.M. (1903) Multiplex Telephony. US Patent 745 734 (26 February).
[3] Antoniou, A. (2007) On the Roots of Digital Signal Processing—Part II. IEEE Circuits and Systems Magazine (Fourth Quarter).
[4] Moore, G. E. (1965) Cramming More Components onto Integrated Circuits. Electronics Magazine.
[5] Decina, M. and Scace, E. (1986) CCITT Recommendations on the ISDN: A Review. *IEEE Journal on Selected Areas in Communications*.
[6] http://www.data-compression.com/speech.html, accessed 21 August 2015.
[7] ITU G series for AMR.
[8] ITU. ITU-T recommendation P.862. http://www.itu.int/rec/T-REC-P.862, accessed 21 August 2015.
[9] Tanaka, I. and Koshimizu, K. (2012) *Overview of GSMA VoLTE Profile*. *NTT Docomo Technical Journal*.
[10] Holma, H. and Toskala, A. (2005) WCDMA for UMTS: HSPA Evolution and LTE, 3rd edition, John Wiley & Sons, Ltd.
[11] ITU. R-Model in ITU Recommendation G-107. http://www.itu.int/ITU-T/recommendations/rec.aspx?id=9730, accessed 21 August 2015.
[12] http://blogs.skype.com/2013/08/28/skype-celebrates-a-decade-of-meaningful-conversations/, accessed 21 August 2015.
[13] ITU. ITU-T Recommendation H.323. http://www.itu.int/rec/T-REC-H.323/en/, accessed 21 August 2015.
[14] http://www.telegeography.com/press/press-releases/2013/02/13/the-bell-tolls-for-telcos/index.html, accessed 21 August 2015.
[15] Peek, H.B. (2010) The Emergence of the Compact Disc. IEEE Communications Magazine (January).
[16] https://xiph.org/flac/comparison.html, accessed 21 August 2015.
[17] Watanabe, K. (2008–2009) Objective Perceptual Audio Quality Measurement Methods. Broadcast Technology no. 35.
[18] http://en.wikipedia.org/wiki/Lossy_compression#Audio, accessed 21 August 2015.
[19] http://msdn.microsoft.com/en-us/library/windows/desktop/gg153556(v=vs.85).aspx, accessed 21 August 2015.
[20] Xiph.Org Foundation. (2015) Vorbis I Specification. http://www.xiph.org/vorbis/doc/Vorbis_I_spec.html, accessed 21 August 2015.

[21] http://mpeg.chiariglione.org/, accessed 21 August 2015.
[22] MacOSForge.org. Apple Lossless Audio Codec. http://alac.macosforge.org/, accessed 21 August 2015.
[23] http://inventors.about.com/od/mstartinventions/a/MPThree.htm, accessed 21 August 2015.
[24] http://dvd-hq.info/data_compression.php, accessed 21 August 2015.
[25] Khayam, S.A. (2003) The Discrete Cosine Transform (DCT): Theory and Application. http://www.lokminglui.com/DCT_TR802.pdf, accessed 21 August 2015.
[26] http://en.wikipedia.org/wiki/JPEG#mediaviewer/File:JPEG_example_JPG_RIP_100.jpg, accessed 21 August 2015.
[27] ITU. ITU-T H Series: Audiovisual and Multimedia Systems. http://www.itu.int/net/itu-t/sigdb/spevideo/Hseries-s.htm, accessed 21 August 2015.

6

Error Control Coding

Preview Questions

- Is it possible to know whether bits have been detected erroneously?
- Is it possible to design an error-free communications system? Is there a limit?
- How do cyclic redundancy check (CRC) codes work?
- What are turbo codes?
- What are LDPC codes?

Learning Objectives

- Importance error control coding in digital telecommunications
- Error detection coding techniques, in particular cyclic redundancy check (CRC) codes
- Different forward error correction coding systems, including block and convolutional error codes
- The Shannon limit and the error-free communications threshold
- Basic processes of data transmission and reception

Broadband Telecommunications Technologies and Management, First Edition. Riaz Esmailzadeh.
© 2016 Riaz Esmailzadeh. Published 2016 by John Wiley & Sons, Ltd.
Companion Website: www.wiley.com/go/BTTM

Historical Note

Error-free transfer and storage of information has been an important aspect of *telecommunications* for millennia. An example is the process of copying holy books (pictured)[1] such as the Bible, where scribes had developed elaborate techniques to ensure the holy text was exactly replicated [1]. These techniques included counting the number of words and letters in each paragraph and comparing them with the original text. Further accuracy was obtained through counting the number of occurrences of an individual letter of alphabet on each page and checking it against the original. Clearly, any discrepancy indicated errors, which could then be searched for and corrected. The accuracy of replicated holy text was clearly of great importance for which such an effort and expense were made. This transfer of information to remote places and time needed to remain true to the original and therefore justified spending time and resources.

Similar techniques can be applied to discrete telecommunications of data. For example the number of characters sent by a telegraph can be appended to a message to facilitate an error detection process at the receiver. Again this extra transmission incurs cost, which can be justified as it ensures the received message is reliably equal to the transmitted message.

Clearly such error detection techniques are not perfect. Words may be misspelled, or alphabets may be switched resulting in undetected errors. The counting process is also imperfect and therefore an error may remain undetected. All these undetected errors are discovered later on. Miscounting may also lead to false positives: a correctly transcribed message is detected as erroneous. Errors commonly occur in printed books, resulting in inclusion of an erratum after post-publication reviews.

With the growth of digital telecommunications associated with computer systems, a number of techniques were developed in the 1950s to reduce errors in discrete information transmission and storage. These are generally based on binary arithmetic and the field is known as *error control coding*. These techniques depend on the concept of code distance first defined by Richard Hamming, an American mathematician and a professor at the University of Louisville, who also worked on the Manhattan Project before joining Bell Laboratories in 1947 where he worked with Claude Shannon [2].

Shannon's theorem in the mid-1940s had already demonstrated that practical error-free communication is possible if the information transmission rate is less than the channel capacity. The main application of this theory was for noisy channels, including those of satellite and space communications where transmission power is limited, and long distances mean received power level is very small. The theorem had also shown that error correcting codes could be designed to reduce the probability of error and therefore the necessary transmission power for achieving a desired quality of service. Hamming designed a set of error

---

[1] Philippe Kurlapski at fr.wikisource (transferred from fr.wikisource): https://commons.wikimedia.org/wiki/File%3AAccueil_scribe.png.

correction and detection codes known as Hamming codes in 1950. He was followed by other mathematicians who developed more powerful codes during the 1950s, codes which were soon used in digital telecommunication of information over noisy channels.

A growing need for error-free communications especially with the emergence of the internet and electronic commerce, as well as the transmission power efficiencies error correction codes provide, has led to the development of stronger and more robust codes. These codes are now essential parts of all standards, from space to mobile to personal telecommunications systems. Furthermore, advances in computer processing power have made it possible to implement very complex codes to provide near-optimal performance as predicted by Shannon's theorem. Error control coding techniques have found applications in error-free storage as well as telecommunications, ensuring that holy text replicates, as well as other information, remain identical to the sources from which they are copied.

In Chapter 5 we showed that digital telecommunications facilitate the removal of noise through regeneration of data. We further discussed that this is subject to correct detection of a received bit in the presence of noise and interference distortion; and that the rate of erroneous detection, or bit error rate (BER) increases as the ratio of signal power to noise and interference power decreases. Given that there always exists a non-zero probability of erroneous detection, several questions may be asked. Is it possible to know if any detection errors have occurred in a received data block? What BERs may be tolerated (analogous to how much noise may be tolerated e.g. in a voice call)? Is it possible to reduce detection errors by correction?

In general, communications errors occur frequently. In everyday conversation we may be misheard and asked to repeat what was said. At times we realise we must have heard things incorrectly: 'a sentence just does not make sense'. We may say things twice, or say things in different ways, to make sure the message is accurately conveyed. Sometimes we are capable of tolerating a certain level of error by automatically correcting what was misheard, or filling the gaps where things were not heard. Many of these logical concepts are applied to communication of binary data so that erroneous data reception may be minimised. These result in a class of mathematical processes which facilitate detection or correction of reception errors, known as error control coding. The field is divided into two main sub-fields of error detection coding and error correction coding.

Error detection coding takes at its input a block of data bits and through a mathematical process calculates an extra sequence of bits (called parity bits) which are then appended to the original block. These bits do not carry any *information*, and are sometime called redundant bits. The resulting bit sequence, the information and parity bits, is referred to as a 'coded word' or a 'codeword'. The receiver applies the same mathematical process to the received data block and calculates its own set of parity bits. If no error has occurred, then the same set of parity bits is obtained. However if there are errors, it is very unlikely that the same set of parity bits are obtained and therefore the receiver can detect error occurrence. Note that the error detection decoding process does not yield information on where and how many errors have occurred. The decoding process is in effect the logical equivalent of conversational 'what I just heard does not make sense – *something* is wrong'.

Example 6.1

An example of error detection coding is single parity bit coding. In this method a parity bit is added to the transmitted bit sequence in order to make the number of bits odd. For example a data block '0011011' is encoded to become '00110111' to have an odd number of 1's. Similarly '0011001' is encoded to become '00110010'. Since the receiver expects a bit sequence with an odd number of 1's it can flag that an error has occurred should a sequence with an even number of 1's be detected. This method is also known as odd parity bit coding. Similarly an even parity bit coding may be designed, where the number of 1's in a bit sequence is set to be even. Parity bit coding techniques follow a set protocol which specifies the number of input bits to the encoder. A typical input size is a 7-bit sequence, with an output codeword of 8-bit length.

The error correction coding process similarly takes at its input a block of data bits. The size of the input data block may be fixed or variable depending on the class of the error correction code used. The encoder calculates an output sequence through a mathematical process and produces an extra set of parity bit sequence. This encoding process creates a mathematical relationship between the data, in that if one or more bits are received in error, then an error event may be detected. The mathematical relationship existing between the data block and the parity bits further means that a *most likely* transmitted sequence may be found corresponding to the received sequence, yielding a correct detection with high probability. Note that this process yields a most likely sequence if errors occur. Error correction decoding is the logical equivalent of the 'what I just heard does not make sense – it *must have* meant such and such'.

Example 6.2

An example of an error correction coding technique is a repetition code. In this method a transmitted data bit is repeated n times. At the receiver side, any one of the transmissions may arrive in error. However, an overall error detection only occurs if the number of errors is larger than $n/2$. For example, if n is equal to 3, a 1 is transmitted as a '111', and a 0 as a '000' codeword. That is two redundant, non-information bits are added. If a receiver detects a '010' then it can conclude that it is more likely that a 0 than a 1 was transmitted. This is because it is more likely that one bit error '000' to '010' than two bit errors '111' to '010' have happened. Through this logical process the receiving decoder in effect *corrects* the '010' to '000'. Note that the receiver does not know with full certainty that indeed the middle bit was erroneously received. It only decides between the more likely of two possible error scenarios.

Both above coding methods are constructed based on creating a mathematical relationship within the codeword. In Example 6.1 the relation is construction of a codeword with an odd number of 1's, and in the relation is the construction of a multi-bit codeword. Both methods are capable of detecting/correcting a one bit error. However this capability comes at the expense of inefficiency: only 7 out of 8 bits in the error detection case carry data, an efficiency of 7/8. Similarly only 1 out of 3 bits carry information in the error correction case, an efficiency of 1/3. The ratio of information bits to the total bits transmitted is called coding rate and is a measure of a code's efficiency.

Since parity bits do not convey any data, they represent an extra 'cost' as energy and time is used for their transmission. The benefit gained however is significant: error detection coding is beneficial as it ensures digital communications reliability, and error correction coding is beneficial as it reduces the amount of overall energy needed to achieve a certain level of error rate performance. This will be discussed further below.

The error control coding field comprises a number of classes of codes. Historically each of these codes was designed for a specific application, from deep space to terrestrial mobile, to optical fibre telecommunications. Accordingly, the complexity and performance of each code differs for different applications. In the following sections a summary of the underlying error control coding principles is given in order to better describe the coding methods used in broadband telecommunications systems.

Codeword Sets

Assuming a set of possible data sequences, the operation of an encoder is to take at its input a sequence of bits and create a set of codewords using a well-defined function. For example Figure 6.1 shows a set of all possible 3-bit sequences. If this sequence is transmitted without any parity bits, all received bit sequences can be valid, and therefore no mechanism exists to detect errors. Figure 6.2 however shows all possible 4-bit words with valid odd-parity codewords in bold. As we can see there are eight valid 4-bit codewords as a subset of all 16 possible 4-bit sequences. The receiver can therefore detect whether errors have occurred by observing if the received sequence is a valid codeword. For example all valid codewords are shown in bold in Figure 6.2: if a codeword, such as '0011' is received, then a detection error has occurred.

Similarly the repetition encoder of the example above limits the numbers of possible codes to only 2 out of 8 possible 3-bit sequences. Figure 6.3 shows all possible single bits, and Figure 6.4 all possible 3-bit codewords. The function of the encoder is to map the input bit sequence to a subset of all 3-bit sequences. The receiver is then designed to check the received

{000,001,011,010,110,111,110,100}

Figure 6.1   All possible 3-bit words

{0000,**0001**,0011,**0010**,0110,**0111**,0101,**0100**,1100,**1101**,1111,**1110**,1010,**1011**,1001,**1000**}

Figure 6.2   All possible 4-bit words, with valid odd-parity codewords shown in bold

{0,1}

Figure 6.3   All possible 1-bit codewords

{**000**,001,010,011,100,101,110,**111**}

Figure 6.4   All possible 3-bit words, with the two *valid* codewords shown in bold

bit sequence against valid codewords and thereby decide whether an error has occurred and if so try to correct the error using a maximum likelihood mechanism.

As discussed, the above coding techniques are capable of detecting or correcting one error. Two or more errors will go either undetected or erroneously detected. More sophisticated coding techniques have been developed that are capable of detecting/correcting many errors. All these techniques use mathematical processes based on modulo-2 binary arithmetic.

Modulo-2 Operation

The mathematical processes used in most coding schemes of interest to broadband telecommunications are based on modulo-2 binary arithmetic, the rules of which are shown in Figure 6.5. Only addition and multiplication are shown here as subtraction is equal to addition and division can easily be derived from multiplication and addition.

Example 6.3

What is the modulo-2 addition of '011' and '111'? What is the modulo-2 multiplication of '011' and '111'?

Answer

```
    011           011
  + 111         × 111
  -----         -----
    100           011
                 0110
                01100
                -----
                01001
```

Codeword Distance

Modulo-2 operation facilitates the measurement of an important metric in error control coding, the codeword distance, defined as the number of places two codewords differ. For example the following two odd parity bit codewords '11010101' and '01010111' differ in two places. Two codewords '000' and '111' differ in three places and therefore the codeword distance is 3. Codeword distance can be calculated using modulo-2 addition and counting the number of 1's of the sum. For example the modulo-2 addition of '11010101'

```
Modulo-2 Addition:          Modulo-2 Multiplication:

   0 ⊕ 0 = 0                    0 ⊗ 0 = 0
   0 ⊕ 1 = 1                    0 ⊗ 1 = 0
   1 ⊕ 0 = 1                    1 ⊗ 0 = 0
   1 ⊕ 1 = 0                    1 ⊗ 1 = 1
```

Figure 6.5   Modulo-2 arithmetic rules

and '01010111' equals '10000010', and since the sum has two 1's the codeword distance is 2. Codeword distance is also known as the Hamming distance, named in honour of Richard Hamming.

Codeword distance is an important measure of an error detection coding method as it shows the number of error events that can lead to false error detection. For example '11010101' may be transmitted and because of two error events be detected as '**0**1010111'. However since the resulting codeword is valid, that is has an odd number of 1's, the error event goes undetected by the decoder. It follows that increasing the codeword distance decreases the probability that a decoding error may occur. However, increasing the code distance can mostly be achieved by increasing the number of parity bits, resulting in reduced code efficiency and increased cost. Appropriate code complexity and performance level are usually determined for different applications and specified within relevant standards.

It follows that a coding technique's effectiveness in detecting or correcting errors is found from the minimum distance between its codewords. The minimum codeword distance is found by calculating the distance between any two possible codewords generated by an encoder. It is denoted d_{min} and is used to find the maximum number of errors that the coding technique can detect or correct. For an error detection coding technique the largest number of errors that may occur without a possible false error detection is $d_{min} - 1$. For an error correction coding technique the largest number of errors that may be corrected is less than or equal to $0.5 \times (d_{min} - 1)$.

Common Error Control Codes

As discussed in previous chapters, broadband telecommunications standards focus on delivery of data blocks from a content provider through a retailer network and different types of infrastructure such as wireless, optical fibre and satellite. The characteristics of different infrastructure mean that processes leading to error occurrence are different. As a result the error control coding techniques used in different standards differ. The following sections describe the coding techniques widely used in these systems.

Cyclic Redundancy Check Codes

A popular error detection coding technique is the cyclic redundancy check (CRC) code. CRC codes are used in many standards including those of third and fourth generation of mobile communications systems (3G and 4G), fixed broadband standards based on digital subscriber line (DSL) and broadband satellite communications systems. CRC coding can work with variable input data sizes, which is in contrast to parity bit coding techniques.

The principle behind CRC coding can be explained by the following division process. Assume data block '124354' is to be transmitted. Let us calculate the integer division $124354 \div 241$, where 241 is a prime number:

$$124354 = 515 * 241 + 239$$

The data '124354' is then appended by the remainder of the above division, namely '239', and the resulting data block '124354 239' is transmitted. The receiver examines the received data '124354 239', performs the division by the same prime number '241' and checks to see if the remainder is the same as that contained in the received block. Should the remainder of the division be different from '239' then an error must have occurred in transmission.

As discussed above, CRC codes use binary modulo-2 arithmetic and can operate on a binary data block of indefinite length. This sequence of binary data is divided by a modulo-2 binary prime number,[2] to produce a remainder between zero and the prime number, which is then appended to the binary sequence to produce a CRC codeword. This operation results in the added sequence becoming modulo-2 divisible by the prime number. To ensure that the original data are transmitted, the data block is appended by a number of zeros before division. The error detection decoder performs the modulo-2 division by the same prime number and if the remainder is zero it checks that no error has occurred. If the remainder is non-zero it follows that an error has occurred.

Example 6.4

The binary sequence '11010101' is to be transmitted. Use '1011', a module-2 prime number, as a divisor and calculate the remainder. Form the CRC codeword sequence and verify that it is divisible by the prime number.

Answer

'11010101' is appended by '000' to produce '11010101000'
11010101000:1011=11110110 and remainder '010'
CRC codeword is then formed as 11010101010 which consists of the original data sequence '11010101', and parity bits '010'. You can check that '11010101010' is indeed divisible by '1011'.

The CRC divisor is usually written as a modulo-2 polynomial and characterised by the largest degree polynomial. For example four CRC polynomials are used in the 3G standards as shown in Figure 6.6. The polynomial lengths (prime number size) are 24, 16, 12 and 8. The length of a CRC polynomial is generally selected to yield a desired level of undetected error probability. This is the probability that a codeword may be corrupted during transmission into another valid codeword (in other words, an error occurs and is not detected.) This error probability is a function of the length of the input bit sequence and therefore is variable.

$$g_{CRC24}(x) = x^{24} + x^{23} + x^6 + x^5 + x + 1$$
$$g_{CRC16}(x) = x^{16} + x^{22} + x^5 + 1$$
$$g_{CRC12}(x) = x^{12} + x^{11} + x^3 + x^2 + x + 1$$
$$g_{CRC8}(x) = x^8 + x^7 + x^4 + x^3 + x + 1$$

Figure 6.6   3G standards CRC polynomials of lengths 24, 16, 12 and 8 [2, 3]

---

[2] Modulo-2 binary prime numbers are similar to decimal prime number in that their divisors are only themselves and 1.

Also, generally the larger the size of the CRC polynomial, the greater is its ability in detecting whether errors have occurred. Regardless, CRC techniques provide very good performance and their undetected error probability is very small. The reason that several CRC polynomials have been standardised is because different applications have different degrees of sensitivity to error events and therefore use an appropriate polynomial size.

Case Study 6.1: Error Detection in Ethernet

An experimental system of decentralised communication between computers was built and demonstrated by engineers at Xerox Palo Alto Research Center led by Robert Metcalf. They named it Ethernet, as they perceived its communication facility to be ether and shared among many computers. Ethernet was designed to allow computers to communicate with each other autonomously and without a centralised control. The medium connecting the computers was a coaxial cable and the equipment could operate at 3 megabits per second (Mbps) [4].

Each Ethernet packet was designed as shown in Figure 6.7. It began with a synchronisation sequence, followed by the addresses of the sender and the receiver, the data payload and ended with a checksum.

An Ethernet packet is transmitted over a medium where transmission errors may occur, so error control coding is necessary. The checksum is a CRC parity sequence of length 16 bits, which can detect possible reception errors with very high probability. The exact error detection probability is difficult to find as it depends on the distribution of errors. However one rule of thumb indicates that a CRC-16 code can detect nearly 99.999% errors.

In computer communications Ethernet packets are dropped when an error is detected. This saves resources (time and power) as a receiving computer does not need to process these corrupt packets. Since CRC is not perfect, other error detection mechanisms are necessary to detect those errors which go undetected at this stage. It may appear that one in 100 000 undetected errors is sufficiently low. However when one considers that a typical Ethernet link can carry more than 200 million packets per hour, then this undetected error performance is quite inadequate.

Present Ethernet standards employ a CRC-32 code, which according to estimates in one paper only lets between one in 16 million to one in 10 billion errors go undetected [5]. Again any packet arriving with error is discarded by the Ethernet receiver and is not passed to the higher layer applications. While the CRC performance is not perfect, it helps improve the system efficiency significantly.

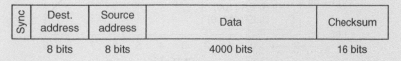

Figure 6.7   The first proposed Ethernet packet structure. Source: Ref. [4]

Bit Error Rate and Block Error Rate

As demonstrated in Case Study 6.1, in digital communications data are usually sent in blocks of a predetermined size. As shown, a block of data must be correctly received before it can be passed on to higher layers to complete the telecommunications process. Block Error Rate (BLER) – also known as Frame Error Rate (FER) – is a measure of the rate of erroneously received blocks (e.g. percentage of blocks received in error). Clearly BLER is related to BER discussed in Chapter 5. It is also a function of the size of the data block.

Generally a data block is received erroneously if there is at least one un-corrected bit error in the block. If errors happen randomly, then BLER (P_B) is a function of BER (P_e) and the number of bits in a packet (N):

$$P_B = 1 - (1 - P_e)^N$$

An approximation can be done for very low BER levels to arrive at BLER as follows:

$$P_B \approx N P_e$$

Clearly the larger the block size is, the higher the BLER becomes as BER rises. This is because there are more bits in a block and therefore the higher the probability that at least one of them may arrive in error. Figure 6.8 shows the BLER for blocks of 10, 100 and 1000 bits

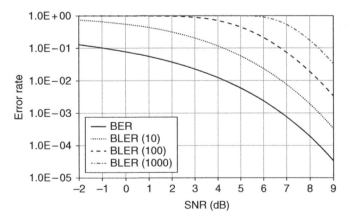

Figure 6.8   A typical BER and BLER

plotted against the signal to noise ratio (SNR). For a SNR of 8 dB nearly all blocks are in error and have to be discarded. In addition the block size can be in the order 10–20 000 bits in some standards. Clearly one needs to reduce BER as much as possible and reduce the randomness of the process of error occurrence. These can be accomplished through forward error correction codes.

Error Correction Coding

Most error correction coding techniques, also known as forward error correction (FEC) codes of interest to broadband telecommunications are based on modulo-2 binary arithmetic and generator polynomial. Two classes of error correction codes exist, block codes and convolutional codes, and both classes can perform near capacity limits as derived from the Shannon theorem. Block codes work with fixed input block sizes, and are therefore suitable for fixed data block sizes. They can be generated through binary polynomial multiplication or matrix operation. Convolutional codes do not impose any limits on the input data block size and can also be generated using binary polynomial multiplication. They are more commonly generated using shift registers and modulo-2 adders. Both classes of codes can be designed with a large degree of variation of complexity and efficiency and are widely used in telecommunications standards.

Hamming Codes

Hamming codes are an early class of block codes with well-defined complexity and error correction capability. They were invented by Richard Hamming and are a good example on how error correcting codes work. One Hamming code is the (7,4) code which takes at its input a block of 4 bits and produces a 7-bit length codeword [2]. This may be implemented using a generator matrix as shown in Figure 6.9. The possible 4-bit messages, arranged as a 16×4 matrix is multiplied by a 4×7 generator matrix to generate 16 possible codewords shown as a 7×16 matrix.

Essentially the Hamming encoder takes a 4-bit word at its input and maps it to a 7-bit codeword, where only 16 codewords are valid from a set of 128 possible words. These are shown

$$
\begin{pmatrix}
0\ 0\ 0\ 0 \\
0\ 0\ 0\ 1 \\
0\ 0\ 1\ 0 \\
0\ 0\ 1\ 1 \\
0\ 1\ 0\ 0 \\
0\ 1\ 0\ 1 \\
0\ 1\ 1\ 0 \\
0\ 1\ 1\ 1 \\
1\ 0\ 0\ 0 \\
1\ 0\ 0\ 1 \\
1\ 0\ 1\ 0 \\
1\ 0\ 1\ 1 \\
1\ 1\ 0\ 0 \\
1\ 1\ 0\ 1 \\
1\ 1\ 1\ 0 \\
1\ 1\ 1\ 1
\end{pmatrix}
\times
\begin{pmatrix}
1\ 0\ 0\ 0\ 1\ 1\ 0 \\
0\ 1\ 0\ 0\ 1\ 0\ 1 \\
0\ 0\ 1\ 0\ 0\ 1\ 1 \\
0\ 0\ 0\ 1\ 1\ 1\ 1
\end{pmatrix}
=
\begin{pmatrix}
0\ 0\ 0\ 0\ 0\ 0\ 0 \\
0\ 0\ 0\ 1\ 1\ 1\ 1 \\
0\ 0\ 1\ 0\ 0\ 1\ 1 \\
0\ 0\ 1\ 1\ 1\ 0\ 1 \\
0\ 1\ 0\ 0\ 1\ 0\ 1 \\
0\ 1\ 0\ 1\ 0\ 1\ 0 \\
0\ 1\ 1\ 0\ 1\ 1\ 0 \\
0\ 1\ 1\ 1\ 0\ 0\ 1 \\
1\ 0\ 0\ 0\ 1\ 1\ 0 \\
1\ 0\ 0\ 1\ 0\ 0\ 1 \\
1\ 0\ 1\ 0\ 1\ 0\ 1 \\
1\ 0\ 1\ 1\ 0\ 1\ 0 \\
1\ 1\ 0\ 0\ 0\ 1\ 1 \\
1\ 1\ 0\ 1\ 1\ 0\ 0 \\
1\ 1\ 1\ 0\ 0\ 0\ 0 \\
1\ 1\ 1\ 1\ 1\ 1\ 1
\end{pmatrix}
$$

Figure 6.9 Hamming code (7,4) generated using modulo-2 matrix multiplication

0000000 0000001 0000010 0000011 0000100 0000101 0000110 0000111
0001000 0001001 0001010 0001011 0001100 0001101 0001110 **0001111**
0010000 0010001 0010010 **0010011** 0010100 0010101 0010110 0010111
0011000 0011001 0011010 0011011 0011100 **0011101** 0011110 0011111
0100000 0100001 0100010 0100011 0100100 **0100101** 0100110 0100111
0101000 0101001 **0101010** 0101011 0101100 0101101 0101110 0101111
0110000 0110001 0110010 0110011 0110100 0110101 **0110110** 0110111
0111000 0111001 0111010 0111011 0111100 0111101 0111110 0111111
1000000 1000001 1000010 1000011 1000100 1000101 **1000110** 1000111
1001000 **1001001** 1001010 1001011 1001100 1001101 1001110 1001111
1010000 1010001 1010010 1010011 1010100 **1010101** 1010110 1010111
1011000 1011001 **1011010** 1011011 1011100 1011101 1011110 1011111
1100000 1100001 1100010 **1100011** 1100100 1100101 1100110 1100111
1101000 1101001 1101010 1101011 **1101100** 1101101 1101110 1101111
1110000 1110001 1110010 1110011 1110100 1110101 1110110 1110111
1111000 1111001 1111010 1111011 1111100 1111101 1111110 **1111111**

Figure 6.10 All possible 7-bit words, with valid Hamming code (7,4) codewords shown in bold

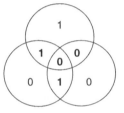

Figure 6.11   Representation of a Hamming decoder

in bold in Figure 6.10. The minimum code distance can be verified to be $d_{min} = 3$: this code is capable of detecting two and correcting a onebit error.

Another way to explain the (7,4) Hamming code's mathematical process is shown in Figure 6.11. Here three intersecting circles are drawn and the 7-bit received codeword of the Hamming encoder is written into the seven sections of the circles. For example the codeword '1010001' is shown here. The inner numbers '1010' (shown in bold) are the data bits and the outer numbers '001' (shown in black)are the parity bits. Each individual circle contains 4 bits. The receiver rule is that each circle must include an *even* number of 1's. This means that if the receiver detects that one or more circles have an odd number of 1's, it will change one bit (and one bit only) to restore the rule. If only one circle has an odd number of ones, it changes the outer bit (which is not shared with other circles.) If two circles have an odd number of 1's then it changes the bit that is shared by the two circles. Similarly if all three circles contain an odd number of 1's, then the bit in the centre is changed to restore the rule. This example further illustrates the point that error control coding techniques create mathematical relationships between data bits and parity bits to enable the receiver to detect/correct reception errors.

The performance of the Hamming (7,4) coding technique may be studied by comparing the BER with and without coding. As it can be seen from Figure 6.12 the coded system performs better by some 0.5 dB at BER of 10^{-4}. This demonstrates the value of a mathematical inter-relationship between the transmitted bits, and signal processing at the receiver to reduce the required energy for a desired telecommunications quality.

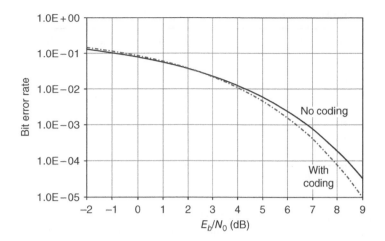

Figure 6.12 Hamming (7,4) BER performance

Error correcting code BERs are generally drawn against a normalised SNR measure, E_b/N_0. This is to account for the fact that extra non-information bits are used to facilitate error correction. The extra power consumed to transmit these bits needs to be taken into consideration. E_b is the energy per information bit and N_0 is noise power per unit of bandwidth. The relation between SNR and E_b/N_0 can be written as:

$$\frac{S}{N} = \frac{f_b}{B} \cdot \frac{E_b}{N_0}$$

where B is the signal bandwidth and f_b is the source information rate. The improvement in BER resulting from error correction is called 'coding gain' and is the main metric for evaluating an FEC code's efficacy. The coding efficiency also depends on the number of extra bits added: this is indicated by the coding rate or the ratio of information bits to the total transmitted. For example the Hamming (7,4) code has a coding rate of 4/7.

The effect of FEC coding in reducing BLER is twofold. One is the reduction of underlying BER which reduces the overall probability of erroneous reception. The other is skewing the error events to burst errors. As Hamming (7,4) code can correct all single-bit errors, block error events are multiple-error events, and therefore the overall number of erroneously received blocks decreases.

Shannon Limit

As shown in Chapter 3, Shannon's capacity formula calculates the amount of information that may be carried over a band-limited channel. The theorem's application to error correction coding shows that if transmission rate is less than channel capacity, then one can design FEC codes to provide arbitrary small error rates at a certain E_b/N_0 level. This E_b/N_0 level is known as the Shannon limit, and is a function of coding rate.

Table 6.1 Shannon limit for common coding rates

Coding rate	Shannon limit: E_b/N_0 (dB)
1/4	−0.8
1/3	−0.5
1/2	0.2
2/3	1.0
3/4	1.6
6/7	2.6
9/10	3.2

Source: Ref. [6]

Historically the Shannon limit has been very difficult to achieve but in recent years two codes have been designed with near Shannon limit performances. The Shannon limits for common FEC coding rates are shown in Table 6.1. Note that as coding rate increases, the number of extra bits used to generate a mathematical relation between information bits, and thereby the ability to correct errors, decreases. This leads to a higher required power (E_b/N_0) to achieve (virtually) error free communications.

Convolutional Coding

Convolutional codes are the other class of FEC codes and similar to block code were first introduced in the 1950s. Decoding of convolutional codes is complex but these codes became popular with the invention of the trellis decoder in the mid-1960s. Convolutional codes use memory elements, or shift registers, and modulo-2 adders to generate parity bits as a function of a number of previously input bits as shown in Figure 6.13. The shift registers keep their input for one clock period (one time delay, D) before passing it on to the next memory element at the next clock cycle and as a result, each parity output bit is a function of several input bits (here three input data bits – which is technically referred to as a code with constraint length of 3). For most systems of interest to this book, the shift registers are set to 0 at the beginning of the encoding process. The memory elements create a mathematical relationship between the parity bits and the data bits which can then be used at the receiver to correct errors which may occur during the transmission process.

The convolutional code shown in Figure 6.13 is a rate 1/2 code, and by convention is denoted by its number of output bits, number of input bits and the constraint length as (2,1,3). Convolutional codes may be designed with a wide range of coding rates and constraint lengths. For example 3G standards use convolutional codes with coding rates of 1/2 and 1/3 with constraint lengths of 9.

Example 6.5

For the (2,1,3) convolutional encoder of Figure 6.14, and for input bit sequence of '110010' what are the output bit sequences?

Answer

Data bits (top branch): '**000**10011'
Parity bits (bottom branch): '**01**111001'

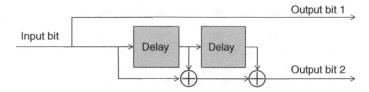

Figure 6.13   A constraint length 3 convolutional encoder

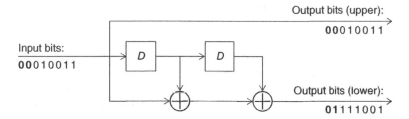

Figure 6.14   Example of coding with the constraint length 3 convolutional encoder

In contrast to block codes which take at input a fixed number of data bits, the input size of convolutional codes is indefinite and generally depends on applications and standard. However, when the data block does end, the shift registers must be reset to 0. This is carried out by several 0's to the end of the data sequence. These are referred to as tail bits, and are necessary for the proper operation of the decoder at the receiver side. For example placing two zeros at the end of the input bit sequence of '110010', to make it '11001000' ensures that both shift registers return to the 0 state. The outputs of the convolution encoder therefore become:

Data bits (top branch): '**00**010011'
Parity bits (bottom branch): '**01**111001'
The two parallel output sequences are converted to a serial sequence and transmitted as:
  0001011101001011.

Viterbi Decoder

Convolutionally encoded codewords can be decoded using a trellis decoder or Viterbi decoder (named after Andrew Viterbi) [7]. A trellis decoder traces the received codes and calculates its Hamming distance from all possible sequences that an encoder may produce. For the encoder of Figure 6.14 the trellis diagram is shown in Figure 6.15. The circles indicate the state of the shift registers and branch out to another state based on the input data bit. For example at the start the shift register state is '00'. This state may remain '00' if a 0 is input or may change into state '10' if a 1 is input. Similarly the trellis may move from state '10' to '11' or '01' depending on the input data. The output of the decoder after each state transition is indicated on the link connecting the two states. For example if the encoder changes from state '00' to '10', then the

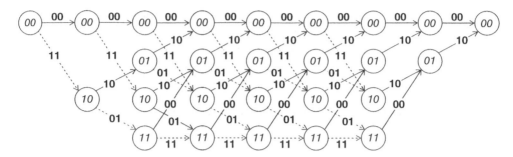

Figure 6.15 A trellis decoder

output of the encoder is '11'. At the end of the data block, two 0 tail bits ensure the state of the decoder returns to the '00' state.

At each state the received bits and the branch bits are compared and their Hamming distance is calculated. When two branches merge into a state the total Hamming distances of each branch (called branch metric) are compared and the branch with the lowest metric is selected as the most likely transmitted codeword, and duly decoded [6].

Example 6.6

Assume the sequence '111000' has been received. Draw and calculate the surviving branch.

Answer

As shown in Figure 6.16, the trellis decoder starts at state '00'. Only two transitions to states '00' and '10' are possible corresponding to data bits 0 and 1, and encoder outputs of '00' and '11'. As the first received bits are '11' the Hamming distance is 2 for the top branch and 0 for the lower branch. In the second data bit interval, similar transitions may be shown and the branch metric calculated. At the end of the third bit interval the two branches converge in state '01' and the accumulated branch metrics are 0 for the lower branch and 4 for the higher branch. It is likely that the lower branch sequence was transmitted corresponding to a '110' input bit sequence.

A typical BER vs E_b/N_0 performance of a Viterbi decoder for a convolutional (2,1,3) code is shown in Figure 6.17. A coding gain of 3.5 dB can be observed at BER of 10^{-4}. This is equivalent to a transmission power reduction in the order of 2.2 times. As the ratio of the number of parities to data bit increases and more shift registers are used to increase the inter-relationship between more input data bits and output parity bits the performance of the convolutional codes improves, resulting in coding gains in the order of 6 dB (power savings of 4 times).

A number of convolutional codes are used in various telecommunications systems and various standards. For example, the 3G standards use two convolution (2,1,9) and (3,1,9) codes, shown in Figure 6.18. The 4G standards of mobile communications also specify a slightly different class of convolutional codes.

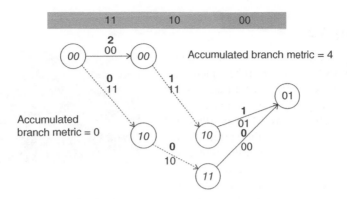

Figure 6.16   Branch metric calculation

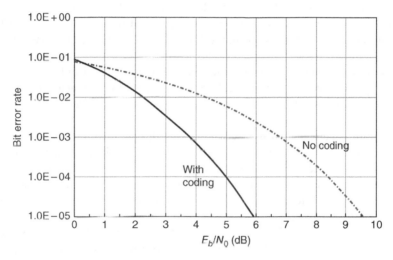

Figure 6.17   BER for a convolutional $(2,1,3)$ code

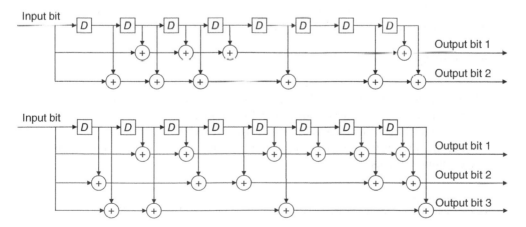

Figure 6.18   Convolutional $(2,1,9)$ and $(3,1,9)$ codes used in the 3G standards

Case Study 6.2

The first application for error control codes was in the NASA's Mariner probes (pictured)[3] in the early 1960s specifically for transmission of data collected by their sensors. In particular, Mariner 4 which flew by Mars in 1964 had cameras to take photos of the surface of the planet. Due to the large size of the photos and the low SNR at the Earth's receiver, the most accurate (and perhaps the only practical) way

of transmitting these photos back was to convert them into a digital stream and transmit them digitally. The channel capacity however was small, and therefore photos were converted to digital and stored locally on a digital tape which had a capacity of 20 photos. These photos were then transmitted to Earth as a digital file with a certain block size. Error control coding techniques, including CRC and FEC were used to ensure these pictures were received error-free. For example Mariner 9 used a (32,6) block code. Such transmission would have been impossible considering the large path loss and the small transmission power on-board the probe. Furthermore, the FEC coding technique ensured that BLER remained small, resulting in fewer retransmissions to recover erroneous data blocks.

Space communications have since relied on digital communications to ensure information collected by space probes can be transmitted back to Earth. The advantages are many; including the fact that Shannon's theorem can inform the telecommunications rate even when the received SNR is very small. Error control coding techniques are also used extensively for different space telecommunications applications including telemetry and control signalling.

For example the Voyager 2 space probe takes 800×800 pixel digital photos, and uses 8 bits to quantise each pixel. However since most of the picture is of black space, the difference between adjacent pixels is coded resulting in up to 60% image compression. The error control codes used in the Voyager probe is a concatenation of convolutional (2,1,7) and block codes – Reed–Solomon code (255,223) for the state of the art in error correction at the time and within 2 dB of the Shannon limit [8, 9].

Case Study Questions

- Why is it difficult to transmit analogue photos from space probes?
- What are the advantages of digital technology for space telecommunications?
- How does FEC coding assist with system link budgeting?
- How did the FEC techniques help with the overall BLER performance?
- What are the business benefits of source coding and channel coding in space telecommunications?

---

[3] NASA (public domain): https://commons.wikimedia.org/wiki/File%3AMariner_3_and_4.jpg.

Turbo Codes

Another class of codes used in the 3G wireless standards are turbo codes. Turbo codes are an iterative class of codes based on convolutional codes, invented by Berrou, Glavieux and Thitimajshima and first presented in 1993 [10]. The code's near-Shannon limit performance created a sensation and soon iterative decoding became the focus of FEC coding research.

The name 'turbo' was used since the operation of the decoder is somewhat similar to a turbo engine, which takes exhaust energy in and feeds it back into the engine. In the same way two parallel turbo decoders take turns to improve the reliability of a received bit, and pass it between each other to further improve on the reliability.

The turbo encoder proposed by Berrou *et al.* [10] is shown in Figure 6.19. The input data are sent with two sets of parity bits each calculated separately using identical convolutional encoders. One encoder outputs parity bits based on the received order of the input bits, whereas the other encoder is fed a rearranged sequence of the input bits and therefore outputs a different sequence of parity bits.

The receiver decodes the received bits once using one set of parity bits. The order of the decoded bits is then rearranged and decoding is done once again using the other parity bits. This process is iterated several times, recursively improving error correction as bits with higher reliability are fed back into the decoder. Each iteration improves the performance of turbo codes until the theoretical limits of Shannon's theorem are reached.

The performance of the Turbo coding as presented by Berrou *et al.* [10] is shown in Figure 6.20. Note that the Shannon limit, or the best possible performance, is 0.19 dB for a rate 1/2 code. This limit is approached to within 0.5 dB by a turbo decoder after 18 iterations at BER of 10⁻⁴.

Turbo codes were not included in the initial releases of the 3G standards as they were perceived to have unacceptable processing complexity. Voice communications need to be carried

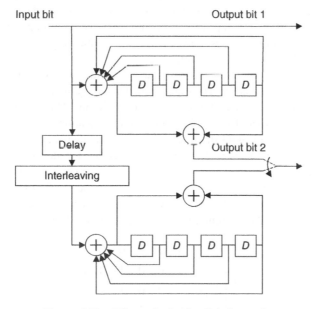

Figure 6.19   3G standard turbo (3,1,4) encoder

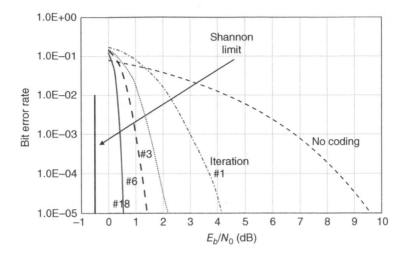

Figure 6.20   BER results for the 3G standard turbo code [10]

out in real-time and the processing delay needed for turbo decoding due to the necessary iterations was considered impractical. Turbo codes were incorporated in later releases particularly for data communications applications. The fourth generation standards use a rate 1/3 turbo code for all user data channels [11].

Low Density Parity Check Codes

Another class of codes used in broadband telecommunications systems are low density parity check (LDPC) codes. These codes were developed by Robert Gallagher as part of his PhD thesis in 1960 [12], but due to high processing complexity were not implemented in practice. They were rediscovered in the 1990s, and have since become very popular because of their performance approaching the Shannon limit.

LDPC codes are a class of block codes similar to Hamming codes, and are encoded and checked using matrix manipulation. They are characterised by their size of the parity check matrix, usually in the order of several thousand rows and columns. This means that the input data block needs to be of same order of size which limits their application to very fast communication links such as gigabit and 10 gigabit Ethernet. LDPC parity check matrices are further characterised by a low ratio of 1's compared with 0's, which begets the 'low density parity check' name.

Similar to all other error control codes, the LDPC encoder generates a mathematical relation between the data and parity bits which can be used by the receiver to correct errors. The mathematical relation of the LDPC code may be explained using the so-called 'bipartite' diagram shown in Figure 6.21. Assume the parity check matrix H is as follows:

$$H = \begin{bmatrix} 0 & 1 & 1 & 1 & 1 & 0 & 0 & 1 \\ 1 & 1 & 1 & 0 & 0 & 1 & 0 & 0 \\ 0 & 0 & 1 & 0 & 0 & 1 & 1 & 1 \\ 1 & 0 & 0 & 1 & 1 & 0 & 1 & 0 \end{bmatrix}$$

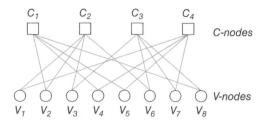

Figure 6.21   The bipartite diagram for an LDPC decoder [13]

Two types of nodes exist in the diagram: constant nodes (C-nodes), corresponding to the columns of H, and variable nodes (V-nodes) corresponding to the rows of H. The links connecting C-nodes and V-nodes in the bipartite diagram exist if the parity check matrix element is 1. For example the element in the intersection of row 3, column 6 is 1 and therefore C-node c_3 is connected to V-node v_6 by a link. In contrast the element in the intersection of row 2, column 5 is 0 and therefore C-node c_2 and V-node v_5 are not connected.

The decoder places a received codeword on the variable nodes. The receiver rule is that the links ending in a constant node must modulo-2 add to zero. For example [1 0 0 1 0 1 0 1] is a valid codeword. We place this codeword on v_1 to v_8 nodes, that is v_1=1, v_2=0, and so on. The modulo-2 sum at c_1 is 0+1+0+1=0; at c_2, 1+0+0+1=0; at c_3, 0+1+0+1=0; and at c_4, 1+1+0+0=0. The bipartite diagram confirms that [1 0 0 1 0 1 0 1] is indeed a valid codeword. Now let us assume that one bit has arrived in error: [1 0 0 1 1 1 0 1]. The constant node summation at c_1 is now 0+1+0+1=**1**; at c_2, 1+0+0+1=0; at c_3, 0+1+0+1=0; and at c_4, 1+1+0+0=**1**. The error must therefore come from a variable node connected to both c_1 and c_4 and which yields the error position of v_5. In practice the LDPC decoder iteratively goes through the bipartite diagram and attaches probabilities to the variable node data on whether it is a 0 or a 1 to satisfy the modulo-2 sum rule. Similar to iterative turbo-decoding each iteration increases confidence on 0 or 1. This process continues until a certain degree of confidence in accuracy is reached. LDPC codes have been shown to virtually perform to within 0.2 dB of Shannon's theorem for LDPC codes of 10 000–100 000 codeword lengths. The BER for an LDPC code (4599,4227) is shown in Figure 6.22. Note that because of very low coding rate of 0.92, the Shannon limit is quite high and equal to 3.5 dB. Of greater interest is the BLER shown here. BLER is approximately 50 times the BER value. This is a 30 times BLER performance improvement over the BLER ≈ 4599 × BER value that a random error process would yield. The error correction process delivers most packets error-free: leaving erroneous packets with large numbers of errors.

Rate Matching (Repetition and Puncturing)

Both block and convolutional encoders generate a finite number of output bits as a function of the number of input bits and coding rate. In addition most convolutional codes require a number of tail bits to reset the shift register to the initial all 0 state. For example if 20 data bits are input to the 3G convolutional encoder (3,1,9) of Figure 6.18 which uses 8 shift registers, the output will be (20+8)×3=84 bits.

Many standards are designed with fixed transmission time interval (TTI) specifications. For example 3G mobile communications systems primarily use a 10 ms TTI. All user data and

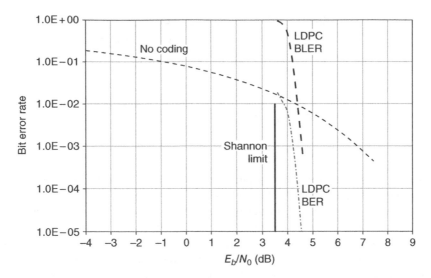

Figure 6.22   BER and BLER for an LDPC (4599,4227) code. Source: Ref. [6]

control information must be organised so it can be transmitted in an integer number of TTIs. At times the FEC encoder output can be designed to perfectly fit one or more TTIs. At other times the number of bits may be slightly larger or smaller than the payload size of a TTI. Since the TTI needs to be a specified number of bits, a process of 'rate matching' needs to be carried out. This can be done by adding some extra bits to an encoder output through repeating some of the bits. This process is known as 'bit repetition'. Alternatively some bits are deleted from an encoder output in order that the TTI may be perfectly filled. This process is known as 'puncturing'. Repetition of bits adds some extra energy to the data block and improves its correct detection probability. Puncturing on the other hand removes some energy and therefore degrades the performance. Note that because of the mathematical relation between the codewords correct detection is still possible even if some output bits are missing.

Example 6.7

A control channel in a 3G standard has a transmission time interval of 10 ms and can carry 90 bits. The control information block is 27 bits long. Either of the standard specified encoders shown in Figure 6.18 may be used. What degree of 'repetition' or 'puncturing' should be used?

Answer

If the rate 1/2 encoder is used, the number of output bits will be: $(27+8)\times 2 = 70$ as 8 tail bits are needed to return the encoder to its initial all 0 state. Therefore 20 output bits need to be repeated to fill the transmission time interval.

If the rate 1/3 encoder is used, the number of output bits will be: $(27+8)\times 3 = 105$. Therefore 15 output bits need to be punctured to fit the output within the transmission time interval.

It can be shown that the BER performances of the two techniques are similar.

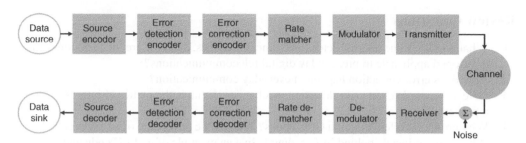

Figure 6.23   A digital telecommunications bit sequence processing order

It follows that the exact position of the repeated or punctured bit must be known to both encoder and decoder. A decoder generally uses the repeated bits to increase the reliability of the received symbol (more energy is received as the output is received twice). Conversely, a marker is placed by the receiver where data has been punctured. The decoder treats this bit as unknown (neither a 0 nor a 1) and proceeds to calculate the codeword Hamming distance without considering the punctured position.

Data Flow Order

The order of information transmission can therefore be described as in the following. First, a source encoder converts continuous or discrete source information into a structured and efficient bit sequence. This data sequence then undergoes a CRC process to generate a parity check sequence. The combined sequences are then FEC encoded, followed by a rate matching process to ensure the total bit sequence fits within the transmission time interval of the system. These binary bits are then transmitted using a process of digital modulation which is the subject of the next chapter.

On the receiver side, the received bit sequence is rate dematched and then FEC decoded. This process ensures that a highly reliable data sequence is obtained with ideally no errors. A CRC decoder then checks to determine whether any errors have occurred and if not the resulting decoded sequence is passed to a data sink. This is the process chain implemented in almost all digital telecommunications system in use today (Figure 6.23).

Error-free communication of information has been of great value throughout the ages: from preservation of holy texts, to maintenance of transactional integrity. Information theory predicts the transport capacity of a telecommunications channel. Channel coding realises this by creating mathematical relationships between data bits and redundancy bits. The structure thus developed allows a receiver to detect anomalies, and possibly correct errors. The extent of such error detection and correction is a function of the complexity of the code and ultimately the Shannon limit.

Review Questions

1. What was the value of error detection to ancient scribes, and how are the techniques they developed applicable to present day digital telecommunications?
2. How does error correction happen in everyday communication?
3. Give an example of an error detection mechanism. What is the metric of interest in such a process?
4. Why are there different length CRC codes? What is the advantage of a longer code?
5. What is the principle behind FEC coding? Give an example of an FEC code and explain how this principle is used.
6. What is the Shannon limit and what is its importance?
7. What is BLER and does it compare with BER? Which of the two is of greater importance to digital telecommunications?
8. What is rate matching and why is it used?
9. List two applications where turbo codes are used.
10. List two applications where LDPC codes are used.
11. Why does error correction decoding occur before error detection decoding?
12. What do business value error detecting codes bring to telecommunications?
13. What do business value error correcting codes bring to telecommunications?

References

[1] Johnson, P. (1993) *A History of the Jews*, Phoenix, cited in http://en.wikipedia.org/wiki/Scribe, accessed 21 August 2015.
[2] Hamming, R.W. (1950) Error detecting and error correcting codes. *Bell System Technical Journal*.
[3] 3GPP Standards 25.212 Series. http://www.3gpp.org/DynaReport/25212.htm, accessed 21 August 2015.
[4] Metcalfe, R.M. and Boggs, D.R. (1976) Ethernet: Distributed Packet Switching for Local Computer Networks. Communications of the ACM.
[5] Stone, J., Greenwald, M., Partridge, C. and Hughes, J. (1998) Performance of checksums and CRCs over real data. *IEEE/ACM Transactions on Networking*.
[6] Lin, S. and Castello, D. J. (2004) *Error Control Coding*, 2nd edition, Pearson Prentice Hall Publishers.
[7] http://www.ieeeghn.org/wiki/index.php/Oral-History:Andrew_Viterbi#Linkabit_and_M.2FA-COM.3B_development_and_consumers, accessed 21 August 2015.
[8] Costello, D.J. *et al.* (1998) Applications of error-control coding. *IEEE Transactions on Information Theory*.
[9] Ludwig, R. and Taylor, J. (2002) Voyager Telecommunications. descanso.jpl.nasa.gov/DPSummary/Descanso4--Voyager_new.pdf, accessed 21 August 2015.
[10] Berrou, C., Glavieux, A. and Thitimajshima, P. (1993) Near Shannon limit error-correcting coding and decoding: Turbo-codes. Proceedings of the ICC, May, Geneva, Switzerland.
[11] LTE 4G Standards. http://www.3gpp.org/technologies/keywords-acronyms/98-lte, accessed 21 August 2015.
[12] Hardesty, L. (2010) Explained: Gallager Codes. MIT News.
[13] Leiner, B. (2005) LDPC Codes – a Brief Tutorial. http://www.bernh.net/media/download/papers/ldpc.pdf, accessed 21 August 2015.

7

Digital Modulation

Preview Questions

- How are digital signals modulated?
- How can you speak twice as fast? Four times as fast? Six times as fast?
- Which digital modulation techniques are commonly used in wireless communications?
- How is the efficiency of a digital modulation technique measured?
- How are different digital modulation techniques and bit error rate related?

Learning Objectives

- Digital modulation
- Amplitude and phase modulation and their combination
- Multi-level modulation
- Limitations on power requirements for higher level modulation
- Digital modulation and Shannon capacity relationship

Broadband Telecommunications Technologies and Management, First Edition. Riaz Esmailzadeh.
© 2016 Riaz Esmailzadeh. Published 2016 by John Wiley & Sons, Ltd.
Companion Website: www.wiley.com/go/BTTM

Historical Note

Discrete data telecommunications require technologies similar to analogue signals as data need to be transmitted over a physical medium using electrical currents or electromagnetic waves. Modulation of discrete data onto a carrier signal dates back to electrical telegraphy and on–off keying: a telegraph *key* (pictured)[1] connects and disconnects the flow  of electricity and through this process information is communicated [1]. The word 'key' defines to this day the digital modulation techniques used to transmit discrete signals.

Modulation of discrete data onto an electromagnetic wave carrier was shown by Guglielmo Marconi (pictured)[2] in 1895. His experiments led to the development of the wireless telegraph, and he is famously credited as the inventor of wireless telecommunications. The technique Marconi used was on–off keying through which discrete data modulated a carrier signal. These techniques can be considered a form of analogue *amplitude modulation* where a specific amplitude is used to represent a period of on, and a zero amplitude to represent the period of off. In essence the carrier signal's amplitude is varied to transmit information. Amplitude modulation of discrete signals is known as amplitude shift keying (ASK).

In addition to amplitude modulation, and in a similar fashion as analogue signal transmission, discrete information can also be embedded in the phase or the frequency of the carrier electromagnetic wave. Frequency modulation for transmission of discrete data was reported in 1947 in devices developed by Cable and Wireless [2]. The system used a frequency shift keying (FSK) technique for telegraph transmission using two distinct frequencies to communicate periods of mark (on) and space (off). Frequency modulation for data transmission was then reported by Doelz and colleagues at Collins Radio in 1956 following the work of Gilbert on information theory at the Bell Laboratories [3, 4].

Phase modulation for data transmission followed and was first reported by Cahn in 1959 [5]. A researcher at Ramo-Wooldridge Corporation, Cahn proposed and evaluated a phase shift keying (PSK) modulation technique where bits of information could be mapped to multiple phases of a carrier signal. The multiple PSK or m-PSK enabled a transmitter to send multiple bits of data in one modulation symbol interval. A theoretical comparison of the three digital modulation techniques, ASK, PSK and FSK was presented in a paper by Arthurs and Dym [6].

A combined modulation technique using both ASK and PSK was proposed by Smith in 1972. In this method a data bit is identified by both its phase and amplitude, thereby enabling further flexibility in representing multiple data bits in one modulation symbol [7]. These techniques have come to be known as multiple quadrature amplitude modulation (m-QAM). Multiple amplitudes, multiple phases or a combination of both are now widely used in both fixed and wireless telecommunications systems.

Digital Modulation

As discussed in Chapter 3, a process of analogue to digital conversion is used to represent a continuous signal as a series of discrete binary numbers. 'Digital' numbers cannot be transmitted over physical media in their pure numerical form since a carrier electromagnetic wave is continuous and analogue. Therefore binary symbols –1's and 0's – must be modulated and carried over telecommunications channels in a similar way to analogue signals. There exist several digital modulation techniques which are designed for communication of discrete information represented by 1's and 0's. While modulation principles are the same, digital modulation techniques differ from analogue in a number of significant ways as will be discussed below.

Analogue modulation was discussed in Chapter 3 in relation to voice telecommunications. Voice telecommunications as well as audio and video broadcasting were until recently carried out using analogue technologies such as amplitude or frequency modulation, and the bandwidth utilised for the modulated signal was equal to (in the case of amplitude modulation, AM) or greater than (in the case of frequency modulation, FM) the original baseband signal. In contrast, digital modulation techniques take at their input discrete symbols. These symbols may be 1's and 0's, or a set of real numbers each representing one or more bits of data. For example the symbol set may be $\{-2, -1, 1, 2\}$ where each symbol represents two bits of data: $\{00, 01, 10, 11\}$. Such configuration allows more data bits to be transmitted over a unit of bandwidth, and therefore higher transmission rates given a fixed frequency resource. As will be seen below, transmission rates can be increased by a factor of two or more using different digital modulation techniques if channel conditions are favourable. This allows for transmitting at channel capacity rates as predicted by the Shannon theorem: more bits/s data rates per unit of bandwidth may be carried by the channel as the received signal quality improves (see Figure 4.7). In contrast, analogue modulation techniques cannot utilise the bandwidth in a similar manner: a 4 kHz bandwidth voice signal needs a minimum 4 kHz channel.

A carrier frequency can be represented as a cosine waveform, $A \cos (2\pi f t + \theta)$, specified by its amplitude (A), frequency (f) and starting phase (θ). The modulation process is about varying the amplitude, frequency or phase of the carrier signal according to the input data (e.g. 0 or 1). These techniques are, respectively, known as ASK, FSK and PSK. Broadband telecommunications systems commonly use ASK and PSK, or a combination of the two: that is, transmission data may be embedded within the amplitude A, phase θ, or both. FSK is not used and therefore we will not describe it further.

In ASK, the carrier amplitude A may take one of $A_1, A_2, ..., A_n$ values to represent one or more bits of data. Similarly, in PSK the carrier phase θ may take one of $\theta_1, \theta_2, ..., \theta_n$ values to represent one or more bits of data. As noted above, a combination of the two may also be used: specific values for both A and θ may be used to represent two or more bits of data. This technique is known as quadrature amplitude modulation (QAM). ASK is commonly used in wired broadband systems such as Ethernet, and PSK and QAM techniques are used both in wired (xDSL) and wireless (mobile and satellite) telecommunications systems and applications.

The design choice generally depends on the application and the complexity of the modulation technique, as well as the required bit error rate (BER) performance of the method with respect to the signal to noise ratio (SNR) at the receiver.

Amplitude Shift Keying

ASK techniques use different amplitude levels to represent 0's and 1's. For example a two-level ASK system is shown in Figure 7.1. As before the transmitted signal can be written with its carrier frequency as a cosine waveform, with a time-varying amplitude $A(t)$. The information signal is embedded in the carrier's amplitude with $A(t) = a$ representing a 1 and $A(t) = b$ representing a 0. As shown in Figure 7.1, the carrier frequency is modulated by these two different amplitudes, and one bit of information is transmitted at every transmission time interval. The receiver side extracts the information by measuring the amplitude of the received carrier signal and comparing it to a reference level to detect whether a 1 or a 0 has been transmitted. In some systems one of the levels is set to zero, and the resulting ASK modulation technique is referred to as on–off keying (OOK). This method is used widely in optical fibre telecommunications.

Figure 7.2 shows the binary ASK signal constellation, where the x-axis shows the amplitude of the signal, and a and b are the signal levels corresponding to 0 and 1. The receiver measures the amplitude of the received signal and compares it with the mid-point value $(a+b)/2$. If the received signal amplitude is smaller than $(a+b)/2$ it determines that a 0 was transmitted, and if larger, a 1. This mid-point value is also known as the detection threshold.

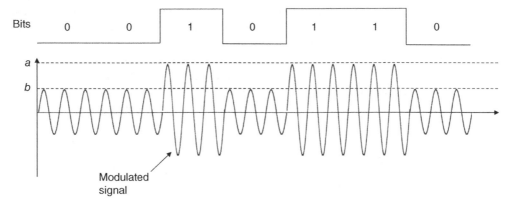

Figure 7.1 Binary amplitude modulation

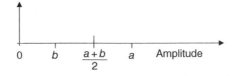

Figure 7.2 Binary amplitude shift keying modulation signal detection threshold

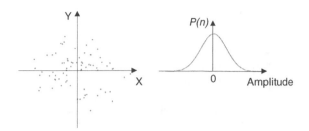

Figure 7.3   Noise signal and its distribution

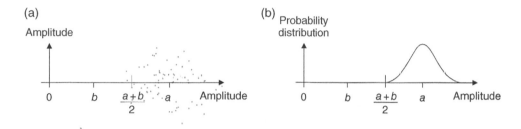

Figure 7.4   Binary ASK signal with noise (a) and its distribution (b)

As discussed in Chapter 3, digital telecommunication systems are robust against noise, however detection errors occur if noise and interference levels are large compared with received signal. That is, a transmitted 0 may be detected as 1 because of added noise and interference. Consider the ASK signal constellation in Figure 7.2. Additive thermal noise as discussed in Chapter 4 is a random process with a normal distribution. Typical noise signal points (a total of some 70 dots, each of which represent the noise signal at one sampling time) and their x-axis normal distribution are shown in Figure 7.3. These noise signals are added to the received signal amplitude (a or b) and as a result the total value of the received signal is different from the ASK signal constellation point.

Errors occur if the $(a+b)/2$ boundary is crossed. This is illustrated in Figure 7.4. The received signal is the sum of the transmitted signal (a) and the noise values as shown in Figure 7.4(a). It can be observed that some signal points cross the $(a+b)/2$ boundary and as a result are detected erroneously. Figure 7.4(b) shows the distribution of the received signal, a normally distributed variable. The area under the distribution curve between $-\infty$ and $(a+b)/2$ is the probability that an error occurs. This is the probability that a signal transmitted as an a is detected as a b, and therefore in error because of additive noise. The error is shown in more detail in Figure 7.5.

The larger the noise level compared with the signal, the more spread is the probability curve and the larger the area under the normal distribution curve. This is illustrated in Figure 7.6. As can be expected errors occur more frequently and therefore the BER is higher.

Since thermal noise is only a function of the operating temperature and bandwidth, it is the signal amplitudes a and b which determine the SNR. Bit error rate curves are usually drawn against the normalised SNR as energy per information bit (E_b) to normalised noise (N_0) or E_b/N_0. The BER for ASK is calculated by considering the cumulative normal distribution of noise and the detection threshold and is shown in Figure 7.7.

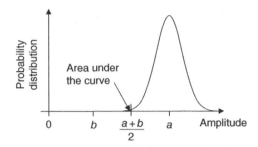

Figure 7.5   The principle behind the calculation of error probability

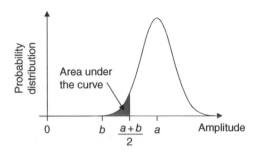

Figure 7.6   Probability distribution vs amplitude. More errors occur with a smaller SNR

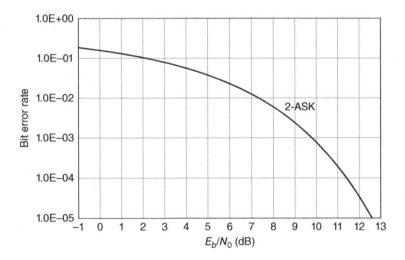

Figure 7.7   Bit error rate for ASK modulation

Multi-level ASK Modulation

Multiple levels may be used to represent several bits. For example, a four-level ASK can be designed with $A(t) = a$ representing 00, $A(t) = b$ representing 01, $A(t) = c$ representing 11, and $A(t) = d$ representing 10. This is shown in Figure 7.8.

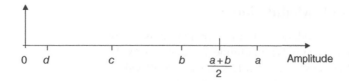

Figure 7.8 Four-level ASK

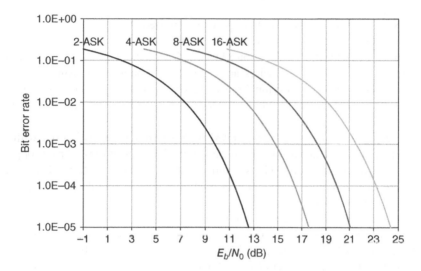

Figure 7.9 BER performance of multi-level ASK techniques in an optical system. Source: Ref. [8]

Note that again an error is detected only if any of the detection boundaries are crossed. Furthermore, it is possible that a transmitted c may be received as a if the added noise is large, and the combined signal and noise amplitudes cross the $(a+b)/2$ threshold. This will result in two bit errors as a transmitted '11' is detected as a '00'. This probability is however negligible compared with the more common single-bit error events. Also note that to achieve the same error rate with 4-ASK, the amount of transmission power needs to be larger than that needed with 2-ASK. The reason is that to maintain the same detection boundary, the signal constellation points need to be larger in amplitude, resulting in larger necessary transmission powers. Similarly 8-ASK, 16-ASK and so on may be designed, each of which requires significantly larger transmission power. The BER performances of different ASK techniques are compared in Figure 7.9. Note that to obtain a similar BER of 10^{-4}, 4-ASK requires nearly 5 dB (3 times) more power compared with 2-ASK. This is clearly inefficient, but nevertheless presents the telecommunications engineer with a mechanism to transmit more information in an available bandwidth. Therefore if transmission power is not a constraint, then 4-ASK allows the system to transmit at twice the rate of 2-ASK.

ASK modulation techniques are simple to implement and relatively simple to detect. However, they are less efficient in their BER performance compared with PSK techniques and are only used in applications that are not sensitive to power limitations, such as transmissions over wired lines used for telephony or local area networks.

Pulse Amplitude Modulation

ASK techniques are widely used in Ethernet systems where
they are better known as pulse-amplitude modulation (PAM).
For example Ethernet standard IEEE 802.3ab, 1000BASE-T
uses 5-level PAM where five different amplitude levels, $-2, -1$,
0, 1 and 2 V are used to represent 2 bits of data as well
as one extra level for error control purposes as shown in
Figure 7.10 [9].

Ethernet systems commonly use four pairs of twisted
wires known as category-5 as standardised by the
Telecommunications Industry Association (TIA) and accredited by the American National
Standards Institute (ANSI). These wires are capable of supporting transmission rates of 1
gigabit per second (Gbps) using a symbol rate of 125 M-symbol/s and four sets of twisted pair
copper wires [2 bits/symbol×125 M-symbol/s×4 wire pairs = 1000 megabits per second
(Mbps)].

A higher order modulation of 16-level PAM with 16 voltage levels is used with 833 M-
symbol/s transmission rate and a category-6 cable with four sets of twisted pair copper wires;
a maximum transmission rate of 10 Gbps is supported. A novel modulation scheme facilitates
each symbol to carry 3.5 bits, which yields the overall transmission rate of 10 Gbps (3.5 bits/
symbol×833 symbol/s×4 sets of wires) [9].

Clearly as more amplitude levels are used to represent different signals, the probability of
errors increases as more transitions may occur between different levels. A comparison between
2-level PAM, 4-level PAM and 8-level PAM is shown in Figure 7.11. The extra power used to
achieve the same BER is in the order of 4 and 4.3 dB, respectively. This extra power is justified
because of higher data transmission rates facilitated by the higher order modulations.

Usage of higher order modulation realises the higher capacity as shown by Shannon's
theorem. As SNR increases, the channel has the capacity to carry more information. A higher
order modulation accordingly facilitates multiple number of bits to be transmitted when
channel conditions are sufficiently suitable.

ASK is commonly used in optical fibre communications. An optical fibre is a transparent
medium made of glass or plastic which transports light pulses generated by a laser or light
emitting diode. One common telecommunications technique used by optical fibre systems
is on–off keying where reception of a pulse of light – 'on' - over a symbol period indicates a
1 and no-light – 'off' – indicates a 0. Some optical fibre systems use a technique where 1 is
represented by a strong power pulse of light, and a 0 by a weak power pulse of light. This is
a classic ASK system as shown in Figure 7.2 where two different amplitudes are used to
communicate a bit of data.

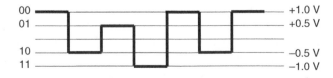

Figure 7.10   PAM-5 signal levels as used in 1000 BASE-T Ethernet systems

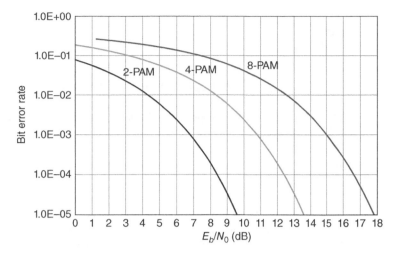

Figure 7.11 BER performance of 2-, 4- and 8-PAM modulation techniques. Source: Ref. [10]

Case Study 7.1: Ethernet

Ethernet was developed by Robert Metcalfe and his colleagues at the Xerox Palo Alto research laboratories in 1973 as a communications protocol to enable computers to connect with each other. The initial network used coaxial cables and modems for transmission rates of 100 kbps to 10 Mbps over distances of 100 m to more than 10 km [11].

With the growth of the internet, faster transmission rates became necessary and gradually faster Ethernet protocols were developed. These systems have been developed through the IEEE 802.3 working group and standards and specify a data transmission on a number of different media, including copper wires, optical fibres, and twisted pair copper wires categories 5, 6 and 7. A list of the standards, date, transmission rates and medium are shown in Table 7.1.

Table 7.1 Ethernet standards

Standard	Date	Tx Rate	Medium
Experimental Ethernet	1973	0.1–10 Mbps	Co-axial cable
802.3 10Base5	1983	10 Mbps	Co-axial cable
802.3u 100Base-Tx	1995	100 Mbps	Co-axial cable
802.3y 100Base-T2	1998	100 Mbps	Twisted pair
802.3ab 1000Base-T	1999	1 Gbps	Twisted pair
802.3ae 10GBase-SR	2002	10 Gbps	Fiber
802.3an 10GBase-T	2006	10 Gbps	Twisted pair
802.3ba 100GBase	2010	40–100 Gbps	Fiber
802.3bq 40Gbase-T	2016	40 Gbps	Twisted pair

Source: Ref. [9]

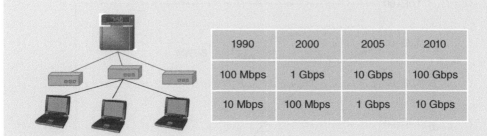

1990	2000	2005	2010
100 Mbps	1 Gbps	10 Gbps	100 Gbps
10 Mbps	100 Mbps	1 Gbps	10 Gbps

Figure 7.12 Ethernet transmission rate requirements over a period of time (10 Mbps to 10 Gbps) for a server to switch to a local node

A number of these standards were developed to operate over category 5 and later on category 6 twisted pair copper wires. Initially the local area networks of many offices used coaxial cables. These were however difficult to use and expensive. The category 5 wires are cheaper, the sockets and connecter are smaller and since the mid-1990s have been the common medium for office networking. Most computers and network equipment are manufactured with sockets associated with these wires.

However, as Figure 7.12 demonstrates, transmission rate requirements have been increasing over the years for the server to switch to a local node.

Higher order modulation techniques help the existing wiring infrastructure to support these faster transmission rate requirements. The bandwidth of category 5 and category 6 wires can be extended and higher modulation rates can be employed to satisfy faster required transmission rates. Office infrastructure can be re-used and only network cards need to be replaced. However, this is not necessary as the network equipment is regularly upgraded.

Case Study Questions

- Why are there different Ethernet standards?
- What has been the driver for higher transmission rate Ethernet standards?
- Why is the 'twisted pair copper wire' a suitable medium for Ethernet? What are the advantages and disadvantages vis-à-vis optical fibres?
- How does digital modulation technology support the local area network business?
- What do you see as the limit of the present infrastructure?

Phase Shift Keying

Referring back to the carrier signal $A\cos(2\pi ft+\theta)$, modulating information can also be embedded in the carrier phase θ. In phase modulation, the transmitted signal is written as $A\cos[2\pi ft+\varphi(t)]$ where $\varphi(t)$ denotes the offset phase of the signal at the beginning of transmission time interval, or time t. A binary phase shift keying (BPSK) signal constellation is illustrated in Figure 7.13 where the offset phase $\varphi(t)=0°$ denotes transmission of a 0 and $\varphi(t)=180°$ denotes transmission of a 1. The receiver again compares the received signal's phase against a boundary level, in this case amplitude$=0$, and detects whether a 0 or a 1 was transmitted.

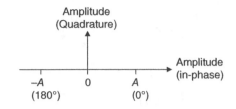

Figure 7.13   BPSK signal constellation

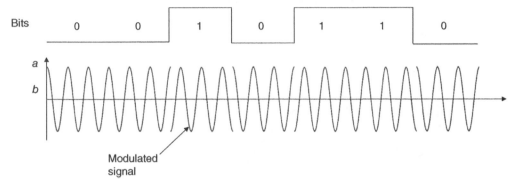

Figure 7.14   BPSK modulated signal

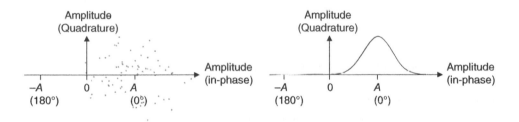

Figure 7.15   BPSK signal with noise, and its distribution

The transmitted signal for the same bit series as in the 2-ASK example is shown in Figure 7.14. Note that during each transmission time interval the signal phase indicates whether a 0 or a 1 is transmitted.

In a manner similar to ASK, additive noise and interference result in erroneous detection of transmitted data. The addition of noise and the distribution of the received signal are shown in Figure 7.15. Again, some received signal points cross the detection boundary and therefore are detected erroneously. The probability of error is similarly calculated to that of ASK, taking the normal probability distribution and the area under the curve from $-\infty$ to 0.

Similar to ASK, the PSK signal constellation can be designed to accommodate a number of phase levels to represent multiple bits in each transmission time interval. Figure 7.16 shows BPSK, 4-PSK (generally known as quadrature PSK or QPSK), 8-PSK and 16-PSK signal constellations. The axes are denoted as in-phase and quadrature (I and Q) to show the phase of the modulated signal. The in-phase axis represents the phase of a local oscillator which is

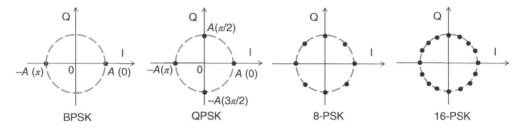

Figure 7.16   BPSK, QPSK, 8-PSK, and 16-PSK signal constellations

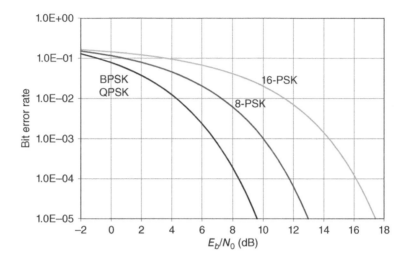

Figure 7.17   BPSK, QPSK, 8-PSK and 16-PSK BER performance

locked to (or is 'in-phase' with) the transmitter's carrier reference phase. The quadrature axis is a reference axis which is perpendicular to the in-phase axis.

QPSK, 8-PSK, and 16-PSK multi-phase modulation schemes facilitate transmission of multiple number of bits at any transmission time interval (QPSK: 2 bits; 8-PSK: 3 bits; and 16-PSK: 4 bits). In effect they enable the transmitter to speak two, three or four times as fast.

Multi-PSK techniques are efficient in supporting transmission of more data bits. However, similar to multi-level ASK they too suffer from relatively poor error performance. Again, to maintain a similar detection boundary the distance between two signal constellation points needs to be increased. This results in a higher level of required SNR for the same BER performance. The BER performance of BPSK, QPSK, 8-PSK and 16-PSK are compared in Figure 7.17. Note that 8-PSK systems require 3 dB higher transmission power (twice as much) compared with BPSK/QPSK for the same BER of 10^{-3}. The BPSK and QPSK performances are identical. QPSK can be considered as two BPSK systems superimposed on top of each other, with an orthogonal (90°) phase shift. Since each system behaves as a BPSK and does not interfere with the other, the overall performance is the same as a BPSK system.

BPSK and QPSK modulation schemes are the most efficient in terms of utilised power. Figure 7.18 compares BPSK and 2-ASK systems. As can be seen a 2-ASK system requires 3dB more power for a similar BER performance of 10^{-3}. QPSK and 8-PSK are commonly used

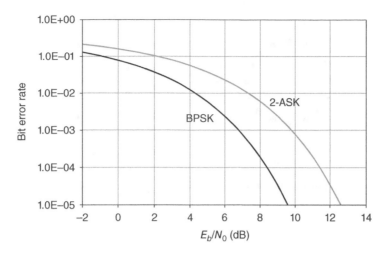

Figure 7.18   BPSK and 2-ASK BER performance

in wireless telecommunications standards, in particular QPSK is used in the 3G WCDMA (Wideband Code Division Multiple Access) standard and 8-PSK in the 2G EDGE [Enhanced Data Rates for Global System for Mobile Communications (GSM) Evolution] standard where power utilisation efficiency is of great importance [12].

Quadrature Amplitude Modulation

It can be noted from Figure 7.16 that when the number of points in a PSK signal constellations increase, such as in 16-PSK, the distance between constellation points becomes very small. A combination of both phase and amplitude levels can be used to increase inter-constellation point distance, thereby reducing the probability that additive noise may cause the detection boundary to be crossed and result in a detection error. Following from the transmit signal representation, $A(t)\cos[\omega\ t+\varphi(t)]$, both $A(t)$ and $\varphi(t)$ may vary as a function of time and be used to carry transmitted data information. This technique is known as QAM.

One commonly used QAM technique is 16-QAM. This constellation uses 3 different amplitude levels and 12 different phase levels for a total of 16 constellation points. Each point in the 16-QAM signal constellation represents 4 bits, enabling the transmission of 4 bits during each time interval. Figure 7.19 compares 16-QAM with 16-PSK signal constellations. It can be observed that for a similar signal power, inter-constellation symbol distance is larger for 16-QAM, and therefore the probability that a symbol is erroneously detected is lower. This is verified by the BER performance graph for the two systems shown in Figure 7.20, as approximately 4 dB separates the performances at BER of 10^{-4}. As might be expected, 16-QAM is used much more commonly than 16-PSK.

16-QAM is used in wireless standards such as 3G WCDMA and the digital subscriber line (DSL) family of standards for fixed telephone line systems. Higher order modulations such as 64-QAM and 256-QAM have also been standardised and are used in wireless and fixed applications. The 64-QAM constellation facilitates transmission of 6 bits per symbol and

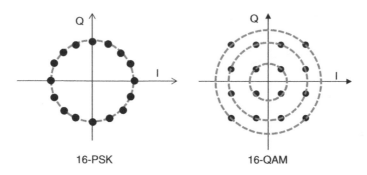

Figure 7.19   16-PSK and 16-QAM signal constellations

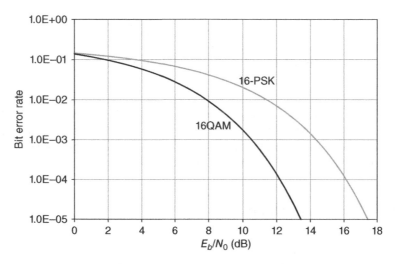

Figure 7.20   16-PSK and 16-QAM BER performance

is used in 3G WCDMA standard evolutions as well as the 4G Long Term Evolution (LTE) standard [13, 14]. Similarly, the 256-QAM constellation facilitates transmission of 8 bits per symbol and is used in DSL standards [15].

Case Study 7.2: WCDMA and HSDPA

The second generation of mobile tele-phony systems was standardised in the early 1990s. The GSM was one of a number of standards which facilitated digital voice communications. It also specified a nascent data communica-tions capability in the form of short message service (SMS) [12].

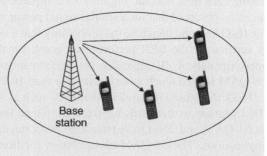

Voice communications require a fixed bandwidth; analogue voice communications are carried out using a fixed bandwidth, typically 4 kHz, and digital voice communications require a fixed data rate. For GSM, this data rate was approximately 30 kbps. Moreover all users need the same transmission rate irrespective of where they are located with respect to a cellular base station.

The third generation partnership project (3GPP) standards were also initially developed to deliver mobile telephony services. The first and second releases of the standards in 1999 and 2001 (Release 99 and Release 4) were mainly focused on the delivery of circuit-switched voice communications, although variable data transmission rates of 384 kbps to 2 Mbps were also specified. Again the standard specifications ensured that all users could equally access voice and data channels irrespective of their position in the wireless network [13].

Data communications however became a significant aspect of Release 5 of the 3GPP standards. These standards took into consideration where a receiver was located with respect to the base station. This was important because if a receiver was close, it could

receive the signal from the base station at a high power level and therefore its SNR would be large enough to support high channel capacity. A device located far away from the base station would experience lower power reception and therefore its channel capacity would be low. Higher order modulation techniques could therefore be used to deliver higher transmission rates to

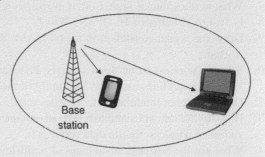

users closer to the base stations than users farther away. For example a user near the base station could use modulation schemes such as 16-QAM with an efficiency of 4 bits/s/Hz of capacity, whereas users near the cell edge would use QPSK modulation and experience a capacity of 2 bits/s/Hz. This adaptive modulation technique increased the overall network efficiency.

Case study Questions

- Why were the 2G and early 3G standards' focus on voice services?
- Why does voice service require equal bandwidth regardless of the location of the mobile device?
- Why does the received power get smaller as a mobile device gets further away from a base station?
- Why can data communications tolerate different transmission rates (in contrast to voice communications)?
- How do different modulation rates enable a higher resource utilisation efficiency?

The Shannon channel capacity formula teaches that under favourable channel conditions, faster information transfer is possible. That is, a system's transmission rate in bits per second per unit of bandwidth (Hz) increases as the quality of the received signal improves. It is impractical to take advantage of this capacity increase in analogue communications. In contrast, multi-level and multi-phase digital modulation techniques facilitate simultaneous transmission of multiple data bits, thereby realising channel transmission capacity as predicted by Shannon. Hence, digital modulation provides a mechanism to realise Shannon capacity limits. This is one of the most important advantages of digital over analogue.

Review Questions

1. Which parameters of a carrier wave may be used for transmission of discrete data?
2. How is ASK different from PSK? Which one is more efficient?
3. Where would you use ASK and why?
4. What is PAM-5 and where is it used?
5. How is the BER metric useful?
6. Why is QPSK superior to 8-PSK?
7. What is the value of higher order modulation techniques?
8. Is 16-QAM better or 16-QPSK? Why?
9. Draw the signal constellation for 64-QAM. Speculate where this modulation technique is used.
10. Why do wireless technologies use the PSK family of modulation?
11. Why do fixed lines use the ASK family of modulation?
12. How do higher modulation techniques realise the Shannon theorem capacity limits?
13. How do 3G and 4G standards take advantage of multi-level modulation schemes in provision of wireless mobile data services?
14. What are the business advantages of multi-level modulation schemes?

References

[1] https://commons.wikimedia.org/wiki/File:Swiss_Army_Telegraph_Key.jpeg, accessed 20 August 2015.
[2] Smale, J. A. (1947) Some developments in commercial point-to-point radiotelegraphy. *Journal of the Institution of Electrical Engineers - Part IIIA: Radiocommunication.*
[3] Doelz, M.L., Heald, E.T. and Martin, D.L. (1957) Binary Data Transmission Techniques for Linear Systems. Proceedings of the IRE.
[4] Gilbert, E.N. (1952) A comparison of signalling alphabets. *Bell Systems Technology Journal.*
[5] Cahn, C.R. (1959) Performance of digital phase-modulation communication systems. *IRE Transactions on Communications Systems.*
[6] Arthurs, E. and Dym, H. (1962) On the optimum detection of digital signals in the presence of white Gaussian noise. *IRE Transactions on Communication Systems.*
[7] Smith, J.G. (1972) On the Feasibility of Efficient Multi-amplitude Communication. National Telecommunications Conference.
[8] http://www.opticsinfobase.org/oe/fulltext.cfm?uri=oe-19-5-4280&id=210159, accessed 23 September 2014.
[9] http://www.ieee802.org/3/, accessed 23 September 2014.
[10] http://ece485web.groups.et.byu.net/ee485.fall.03/lectures/PAM_notes.pdf
[11] Metcalfe, R.M. and Boggs, D.R. (1976) Ethernet: Distributed Packet Switching for Local Computer Networks. Communications of ACM.
[12] http://www.3gpp.org/DynaReport/45-series.htm, accessed 23 September 2014.
[13] http://www.3gpp.org/DynaReport/25-series.htm, accessed 23 September 2014.
[14] http://www.3gpp.org/DynaReport/36-series.htm, accessed 23 September 2014.
[15] ITU. G.99x Series for DSL Standards. http://www.itu.int/rec/T-REC-G/en, accessed 23 September 2014.

8

Packetised Data Communications

Preview Questions

- What is packet switching?
- What information is required to deliver a data packet?
- What are protocols of data communications? How do different layers interact?
- How did the packet-switched communications standard come about?
- What are the performance measures for evaluating a data communications system?

Learning Objectives

- Packet-switched data communications
- History of the Internet
- Internet telecommunications technologies
- Queuing performance metrics
- Internet protocols

Historical Note

Internet development traces its roots back to the 1950s and the emergence of main-frame computers. Early computers (pictured) were stand-alone machines and did not have the means to communicate with other machines: commands, programming and data storage and transfer

Broadband Telecommunications Technologies and Management, First Edition. Riaz Esmailzadeh.
© 2016 Riaz Esmailzadeh. Published 2016 by John Wiley & Sons, Ltd.
Companion Website: www.wiley.com/go/BTTM

were carried out manually using magnetic tapes (pictured) [1] or punch cards (pictured).[1] As the usage of computers in business and government applications grew, interactions between computers, and between computers and network terminals became necessary. Clearly there was a need for a theoretical framework and network equipment to support computer communications, which needed to be developed.

On the theoretical side, while voice traffic characteristics were well researched, no such information existed for computer communications traffic. Voice traffic models show the average and variance of a telephone call: a voice call could be expected to last a certain length of time with a certain statistical behaviour. Moreover, the bandwidth necessary to support a voice call is fixed – 4 kHz, and given a number of subscribers, inter-arrival of calls is also well modelled. In contrast, the required time for transmitting a data file is variable and depends on the size of the file, as well as the capability of the telecommunications equipment. The traffic character is also different to that of voice: computer data are generated in bursts and have a short active length followed by long inactive periods. As a result of these, using the same circuit-switched technologies of voice communications which assigns a fixed bandwidth was inefficient because (a) the bandwidth could be excessive and (b) it could only be used for a very short period of activity and left idle for long periods. Furthermore, as computer communications are relatively delay insensitive, a store and forward system works well with data as real-time communications is not necessary. A queuing system which considered waiting time as well as different transmission speeds could work well for data transmission. Such a theoretical framework for computer data traffic and associated performance metric formulas was published by Leonard Kleinrock (pictured),[2] whilst a PhD student at MIT, in 1962 [2].

Computer communications also benefited from a US military project. Around the same time as Kleinrock's research was published, a project was funded by the United

[1] Computer room: NASA Ames Research Center (NASA-ARC) (NIX A-28284) https://commons.wikimedia.org/wiki/File%3AIBM_7090_computer.jpg. Punch card: Blue-punch-card-front.png: Gwern derivative work: agr (Blue-punch-card-front.png) https://commons.wikimedia.org/wiki/File%3ABlue-punch-card-front-horiz.png. Magnetic tape: Hannes Grobe 23:27, 16 December 2006 (UTC) (own work). Used under CC BY-SA 2.5 (https://commons.wikimedia.org/wiki/File%3AMagnetic-tape_hg.jpg).

States military to develop a data communication system that could withstand a nuclear attack by the then Soviet Union. It was feared that such an attack would destroy telecommunications hubs and cut off command and control from the battle field. The desired system was to have a large degree of redundancy, and be capable of working in networks with different bandwidths. The project was initiated and funded through the Advanced Research Projects Agency (ARPA) of the United States.

The merging of the two streams, research into packet-switched communications and funding for building such an experimental network gave birth to the Arpanet, a data communications network connecting several universities and research institutions in the late 1960s. Kleinrock who was by then a professor at UCLA was one of the architects of the Arpanet and helped build the first equipment realising packet-switched communications.

As minicomputers entered the workplace in the 1980s and intra-office and inter-office data communications became widespread, internetworking grew – and the entire interconnected data communications network came to be known as the Internet. Services such as email, World Wide Web, automated supply chain, social networks and so on are built over this platform [3, 4].

Data Communications

Electrical telegraphy and electrical telephony can be considered major technological breakthroughs. While they were based on prior discoveries and inventions and the inventors stood on the 'shoulders of giants', they still represented major milestones and advances in the history of telecommunications. Packet-switched mode of telecommunications can be considered alongside these advances as a major milestone.

Traditional telegraphy and telephony use the 'circuit-switched' modes of telecommunications. This means that an end-to-end circuit is first established between two parties over which a call is made. When the call ends the circuit is disconnected (switched off) so it can be used to carry another call between two different parties. The circuit may physically be a channel in a specific frequency band of media such as wire or air.

Telecommunications between computers or other machines on what has come to be known especially as the internet uses a technique known as 'packet switching'. In packet switching information is broken into blocks of data known as *packets*, which are then sent to an intended receiver independent of each other. Each packet is designed to be self-contained and may travel a different route to reach its destination. The receiver puts the packets back together and retrieves the transmitted information.

Packet-switched data communication applies equally to analogue, such as music, as well as digital, such as an email, information. In data communications, information is generated by or through a machine which sends it to another machine. As a result, it is characterised by well-defined protocols which indicate the beginning and end of calls, ensure errors are detected and dealt with, maintain a correct order of information transfer and so on.

---

[2] Photograph by Dan Anderson, Imagining the Internet Center, School of Communications, Elon University, Elon, North Carolina, USA. Copyright Elon University. CC BY-ND.

Data communications can therefore be defined as systems and protocols which facilitate transmission of mainly discrete but also analogue source information from one point to another. In effect data communications is the superset of digital communications, in that digital communications applies to an analogue information source, whereas data communications applies to all information transfer regardless of the source.

Data communications can also generally apply to mostly delay insensitive information transfer. For example, in contrast to telephony which must be carried out in real-time, transfer of emails for example can tolerate some delay. This facilitates transmission over links of diverse speeds and diverse routes: part of a message may be carried on a satellite link and another part on a submarine cable. One part of a message may be received later and another part earlier. A receiver's capability to re-order received data and re-construct the transmitted information in the correct order creates significant flexibility and robustness in a telecommunications system design. Special protocols have been designed to cater for delay sensitive service such as voice communications. These allow for prioritisation of real-time services and minimise delay and delay variations.

This chapter gives an overview of packet switching technologies and protocols.

Circuit Switching and Packet Switching

As discussed in the Historical Note at the beginning of this chapter, data communications technologies and protocols trace their origins back to the 1950s. With the rise of computers it became necessary to find ways to facilitate machine-to-machine transfer of data. The nature of such communications was significantly different to that of voice telecommunications and audio and video broadcasting. A voice call has a well-defined statistical probability and continues for an average length of 2–3 min, and therefore lends itself well to a circuit-switched mode of communications where a dedicated end-to-end circuit could be set up for the call. The process of such a call set-up is shown in Figure 8.1. The process involves connection between exchanges, and the called party, ensuring there are circuits available for an end-to-end connection, checking to see whether the called party is not busy, and waiting for the called party to answer the telephone. This process can take a considerable amount of time and resource. At the end of a call, control signalling is necessary to disconnect the call and allow the circuit to be used for another call. Other processes such a billing and verification are also active during a call. These control processes are all overheads on the actual 'information' communication. However, since an average call continues for several minutes, the overall overhead is comparatively small and does not reduce system efficiency significantly. Moreover, a voice call is bidirectionally balanced: the two parties are expected to talk and listen in equal proportions. This allows for two equal capacity links to be set up by the network which simplifies the overall design.

In contrast, a data communications 'call' is generally characterised by a short transfer of data followed by a long period of inactivity. A transmission may take a fraction of a minute, followed by an inactive period of several minutes before the next portion of data is generated by the source and transmitted. Under these conditions a circuit-switched approach can be very inefficient as the call set-up process may take much longer than the time required for the short burst of data transfer. Furthermore, the process for disconnecting the circuit is difficult to

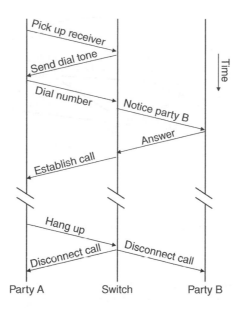

Figure 8.1   A circuit-switched voice call set-up

manage since an early termination may increase inefficiency if more short-burst communications appear. A late disconnection on the other hand may occupy a transmission line needlessly. Moreover, bidirectional data transfer is likely to be unequal due to the client–server traffic nature of most data communications. Therefore, while circuit switching is an effective technology for continuous connectivity, it is not effective for communications for intermittent information transfer. Switching a communication channel for a short duration as needed for a message is more efficient but a mechanism needs to be designed that removes the need for end-to-end channel establishment.

Packet-switched communication facilitates independent transmission of the data packets towards the destination. As a result it is not necessary to establish an end-to-end circuit: a transmitter only needs to know where a packet is to be sent next, as each packet can carry information that enables an intermediate node to determine where to forward the packet. In practice a packet goes through many intermediate nodes, each of which ideally brings the packet closer to its final destination.

The intermediate points are known as routers. Routers switch incoming packets individually based on the destination address of the packet and one or more forwarding rules. These rules are local: the packet forwarding may not take into consideration what the optimal global forwarding route might be. It only follows a set of rules set by the local administrator on which node a packet may be forwarded to, based on the final address. In general, the rules are summarised in a routing table, which lists all addresses and the corresponding set of possible next nodes. A simple packet-switched communication between two computers is illustrated in Figure 8.2. Packets are forwarded at each router and may take different paths before they arrive at the final destination.

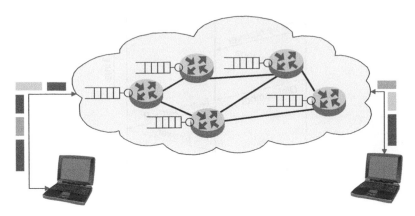

Figure 8.2   Packet-switched data transfer

The packet-switched system can be likened to sending letters through a postal network where a letter is processed at each distribution office.[3] No single post office determines the end-to-end route which a letter should take. Instead each distribution office determines based on a set of *local rules* to which centre the letter should be forwarded in order that the letter is finally delivered.

Packet switching can flexibly assign different forwarding routes based on the traffic situation of the local and regional network. Should a forwarding path be costly, overloaded or temporarily unavailable alternative routes may be selected, again based on a set of local rules. In the example of a post office, a letter may travel part of its way using a track and part of its way using an aeroplane with a much faster speed. This architecture provides for significant redundancy and robustness in the presence of accidental/planned link disruptions.

Packet switching provides significant advantages over circuit switching for intermittent information transmissions. One advantage is a more efficient usage of transmission medium as short and long messages from multiple transmitters are time-multiplexed together on one link. Another is the possibility of using links with different transmission speeds for the transfer of data as diverse links are utilised. Yet another advantage is the possibility of using diverse routes to transfer a message: two packets containing parts of a message may go through different paths as routers on the way to determine the next relay point. As discussed above, this redundancy was a major requirement of the US military in funding the ARPA project in the 1960s that led to the development of the internet: a communications system that could withstand the destruction of major switching hubs in the event of a nuclear war.

---

[3] Post van: OSX (own work) https://commons.wikimedia.org/wiki/File%3A2000-2004_Ford_Transit_(VH)_low_roof_SWB_van_(Australia_Post)_01.jpg.

Case Study 8.1: DHL and FedEx Packet-switched Networks

An early application for packet-switched networks was document transmission using facsimile (fax) machines. The invention of modern fax machines is credited to Xerox Company and its Long Distance Xerography (LDX) in 1964 and improved Magnafax in 1966. These were analogue image transmission systems and on average took 6 min to transmit a document page. In the mid-1960s Hochman and Weber, two scientists from Lockheed who had worked on digital image compression techniques for space probe telecommunications, founded Dacom to commercialise digital fax machines. Their digital fax machines connected to telephone lines and used in-band frequency tones to send digital signals. Digital fax improved on Xerox technology and could transmit a document page in less than 1 min. By the late 1970 many companies were making fax machines, prominently three Japanese companies: Canon, Ricoh and Konica. These early fax machines however were quite heavy and expensive and beyond what most companies needed or could afford [5, 6, 7, 8].

Courier and overnight document delivery companies felt threatened by this new mode of document delivery and started their own fax services. Two courier companies, DHL and FedEx, used the nascent packet-switched networks to provide fax services.

FedEx built a packet-switched network based on the X.25 standard to offer a US-wide fax service named Zapmail in 1984. Fax machines were installed at local FedEx offices where customer could fax a document for $3–4 per page, as well as in large companies. Overall it is reported that FedEx spent $350 million to roll-out this nation-wide service. However by 1986 the prices of fax machines had dropped significantly and were affordable for many businesses. The service was discontinued after only 2 years with FedEx incurring a $320 million write-off. One reason for the failure was the fact that Zapmail machines could not communicate with the standard 'group 3' fax machines and operated within a closed, limited network.

DHL also entered the fax machine business. However they focused on designing X.75 standard switches for packet-switched fax traffic through their NetExpress subsidiary. NetExpress then used a global packet-switched network for fax telecommunications, as well as marketing the switches to global operators to build an international network. Their focus of business was fax operators rather than retail customers. NetExpress operated successfully for many years but was finally discontinued. It is an example of an early successful packet-switched telecommunications business.

Case Study Questions

- Why do you think fax services were delivered over a packet-switched network?
- Why were Comité Consultatif International Téléphonique et Télégraphique (CCITT) standards X.25 and X.75 used?
- What do you see as the reason for FedEx rolling out a hardware network?
- What do you see as the, main reasons for the failure of the Zapmail service?
- How do you compare the strategies of FedEx and DHL?

Data Packet

An individual data packet needs to carry certain
information in order that a router can successfully
forward it to its final destination. Using the post office
analogy, if a letter contains the user data, the envelope
carries the information which ensures the data reaches
the recipient. The envelope must therefore carry the
address of the recipient as well as the sender. It must

further inform the post office how the letter is sent (express, standard, etc.), the date it is posted
and should it be returned after a certain waiting period and to whom, and so on. The envelope
may even indicate which post office it may not travel through. It may also carry information
on the language used, and whether any processing of the data (including encryption) has been
carried out.

A data packet is constructed similarly. Each data packet needs to carry information that
ensures its correct delivery. This information is sometimes known as control data. The structure of a data packet, including control and user data is shown in Figure 8.3.

A packet 'header' field acts as the 'envelope' and contains all control information to ensure
desired delivery of the data payload. This control information may include one or more of the
following fields:

- Source address
- Destination address
- Total packet length
- Number in the series
- Time to live
- Protocol (language) used

The user data (information) is the payload of the packet and is usually of a limited size and
depends on the protocol being used. A 'tail' field may be included to inform the receiver of certain
extra control information (such as to indicate that a data payload has ended, include a check sum
for error detection, and/or to act as a buffer between two successive packet transmissions).

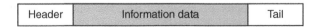

Figure 8.3   Structure of a packet-switched data transfer

Case Study 8.2: Internet Protocol and OSI

Packet-switched data communications in the late 1970s held the promise of a merger
between two fast growing industries: telecommunications and computers. A number of
companies, including network operators and computer manufacturers and service providers, had a major interest in the emergence of a global standard, and ensuring that the
new standard enhanced their business interests.

Two approaches emerged for the system architecture of the new packet-switched systems. Both methods broke down the processes involved in communicating messages. The former methods resulted in the development of Open Systems Interconnection (OSI) and the latter Transmission Control Protocol and Internet Protocol (TCP/IP). Both of these approaches support a number of 'protocols' and provide for a comprehensive control of communications to ensure that packets of data are correctly received.

The OSI model was based on 'virtual circuits' and was favoured by 'monopoly' service operators and computer equipment manufacturers. In this approach data transmission followed a packet-switched system, however an 'end-to-end' virtual connection was made over which packets travelled. The TCP/IP approach was 'connectionless' and was favoured by a number of computer scientists and researchers. The latter formed the International Networking Working Group (INWG) in 1972 and developed a standard based on a connectionless datagram, 'with no relationship established between sender and receiver. Things just go separately, one by one...' [9]. A final agreement on specification was presented to the CCITT, a precursor of the International Telecommunications Union (ITU), an agency of the United Nations in 1975. The proposal was defeated by the CCITT whose members at the time were mostly telecommunications engineers.

A committee comprising telecommunications engineers and computer companies set about to standardise packet-switched communications under the auspices of the International Organisation for Standardisation (ISO). The new committee set out to define an open working approach to gather input from all stakeholders, in the process formulating a seven-layer architecture into which different computer communication tasked were divided. Their approach connected computers with computers as peers allowing communications between devices made by different manufacturers. However, the involvement of many players, from manufacturers to network operators, meant that the standard was complex and difficult to implement.

A number of the INWG members left and went to work on the Arpanet and further developed the connectionless philosophy. With funding from the US government the Internet architecture was developed using four distinct layers. With the slow pace of OSI standardisation many companies started using the TCP/IP approach for their immediate connectivity needs.

Andrew Russel in his article 'OSI: The Internet That Wasn't' summarises the history of how OSI promise was never realised and how the IPapproach came to dominate the packet-switched telecommunications landscape. He writes about the major flaw of the OSI approach as follows:

[The OSI's] fatal flaw, ironically, grew from its commitment to openness. The formal rules for international standardization gave any interested party the right to participate in the design process, thereby inviting structural tensions, incompatible visions, and disruptive tactics [9].

Case Study Questions

- What do you see as the relative merits of the 'virtual circuit' and 'connectionless' approaches to packet switching?
- Why do you think the network operators preferred the 'virtual circuit' approach?
- What are the merits of 'layered' design in designing the packet network?
- What were the reason for overall adoption of TCP/IP over OSI?
- How should an international standardisation be carried out?

Protocols

Packaging data as described above provides a mechanism to send and receive information between two machines. However, a number of processes must be in place before actual communication of information may start. The receiver needs to know for example when a data transfer starts, which 'language' it uses, and at which 'port' it should be received. Within the received signal, the destination machine needs to know when a packet starts and when it ends; and detect whether there are errors and how to deal with them. Precise communications protocols are needed for two often heterogeneous machines to send and correctly receive data.

Open Systems Interconnection Model

As discussed above packet-switched architectures divided the processes necessary for establishing and transmitting of data into 'layers'. The OSI model defined seven functional layers as shown in Figure 8.4.

The seven layers of OSI architecture are designed to create a virtual end-to-end connection between a transmitter and a receiver over which data may be communicated. Each layer covers a certain number of functions and communicates with layers immediately above and below to transfer data. The inter-layer interface is standardised, which ensures equipment designed for different layers can seamlessly connect. This structure simplifies design of complex networks, and perhaps more importantly, facilitates competition in network equipment supply by enabling different vendors to build devices that can readily work together.

An example of how the OSI layer works is as follows. An 'email' program produces data at the application layer (L7). The letters of the alphabet produced by the email application may be represented by ASCII characters which may be transferred into an 8-bit Unicode Transformation Format at the presentation layer (L6). The session layer (L5) establishes an end-to-end session between the applications residing in two machines and ensures data flow. The data are packetised in the transport layer (L4) into parts of suitable size for transmission

7	Application	Provides a means for the user to access information on the network through an application
6	Presentation	Provides independence to the application processes from differences in data representation (syntax)
5	Session	Establishes, manages and terminates connections (sessions) between applications
4	Transport	Provides reliable data transport: e.g. end-to-end error recovery and flow control
3	Network	Provides means of transferring variable length data sequences from a source to a destination in network(s)
2	Data Link	Provides means to transfer data between network entities and to detect and possibly correct errors
1	Physical	Concerned with transmission of unstructured bits over a physical medium

Figure 8.4   OSI seven-layer functions

through the network. It also provides for end-to-end error control. The network layer (L3) routes data packets forward between network nodes, and manages data flow. The data-link layer (L2) is responsible for data transmission integrity between nodes, including error control. Finally the physical layer (L1) is concerned with physical modulation of data onto an analogue link. Note a physical-only connection exists at layer 1, through which physical signals carry data to the next node in the network. Each layer generally adds a header to the packet data which carries protocol information necessary for the receiver device and its layers to correctly extract the data contained within the packet.

All transmit-side layers are paired with their counterparts on the receiver side, physically at layer 1 and logically at layers 2–7. When packets pass through intermediate relay points, the relay device extracts network layer header information in order to determine the next node to which the packet is to be sent.

Packet flow from the application layer down to the physical layer and on to its destination through two intermediate nodes is shown in Figure 8.5. Note the logical pairing between layers across the network, and how an end-to-end virtual circuit is established. As discussed above, the OSI architecture defines the interfaces between layers and therefore different software and hardware vendors may design and develop products for data communications networking. The standardised interface facilitates interoperability between different vendors' products.

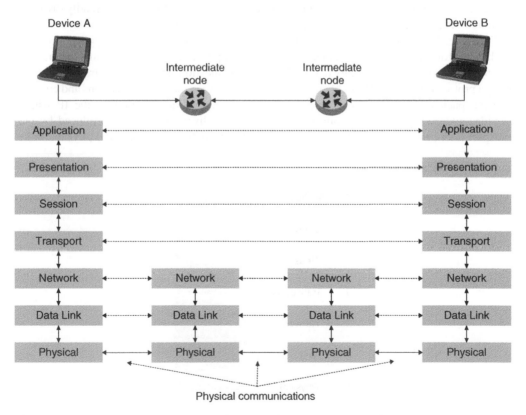

Figure 8.5 Data flow through different OSI layers (dotted lines are virtual peer-to-peer connections)

As discussed in the IP and OSI case study, the OSI seven-layer architecture is overtaken by the IP suite in carrying most packet-switched data traffic. However, the model is used widely in the design of telecommunications systems such as mobile telephones and wireless local area networks. In particular, layers 1–3 specifications are used for defining how a transmission medium may be accessed by multiple users. We will return to OSI layer definitions in future chapters, however packet-switched communications is exclusively discussed using the IP below.

Internet Protocol

The alternative architecture for data communications is the suite of protocols associated with the IP, including TCP, developed at the same time as the OSI protocol. The TCP/IP stack is built in a similar way to that of the OSI model, in that functions are divided across several layers. There are however a number of differences. The most important difference is the fact that TCP/IP is connectionless, in that no physical or virtual end-to-end connection is established between two applications. Packets are individually sent from a source to a local switch, which decides on where the packet should be forwarded based on the receiver's address. The end-to-end path which the packet takes remains open and transparent to the transmitter and intermediate nodes. Another difference between the internet and OSI protocols is how the layers are specified. The IP is made of fewer layers since many of the higher layer functions are included within the applications. A comparison of the two protocol stacks is shown in Figure 8.6, broadly dividing the functions included in respective layers of the protocols. Internet's application layer covers OSI's application, presentation and part of session layers' functionalities. The application layer produces information that is exchanged with a peer application, with a mutually understood presentation format, and capability to establish a communication. The internet's transport layer, using a protocol such as TCP, establishes a connection between the two applications and ensures received packets are error free. The network layer functions are covered by the IP which provides addressing as well as control mechanisms to deliver data packets produced by the transport later protocol. The network access (link) layer establishes logical and physical links

OSI Model		TCP/IP
7	Application	Application
6	Presentation	
5	Session	
4	Transport	Transport (Host to Host)
3	Network	Internet
2	Data Link	Network Access and Physical
1	Physical	

Figure 8.6   A comparison of the OSI and TCP/IP stacks

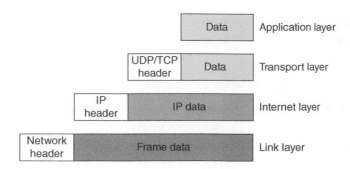

Figure 8.7   IP suite stack headers

between two nodes in the network. An example of such protocols is the Ethernet which includes the physical means of data transmission such as digital modulation, and includes the physical layer functions as part of the network access layer.

Each layer of the IP stack adds control information that enables the receiver to retrieve the embedded information. Each layer effectively creates an envelope that carries information necessary for extracting the data transmitted between two end applications. A typical internet stack and header construction is shown in Figure 8.7. A brief description of different layers and typical protocols are given below.

Application Layer Protocols

Countless applications use the IP to communicate with applications running in other devices. A Skype application may communicate with a Skype server in a remote telephone exchange to facilitate an international telephone conversation, a browser application may communicate with a Google server to retrieve search results and so on. Applications may encrypt the data they communicate (an email application), a function of the 'presentation layer', or a Skype call may be initiated between two physical end devices, an example of a 'session layer' function. Each application is specified through a well-defined protocol. It may be possible for an application to communicate with another based on the protocol. Some applications layer protocols are developed through the Internet Engineering Task Force (IETF) and/or the World Wide Web Consortium (W3C) and their structure is in the public domain. Examples are Simple Mail Transfer Protocol (SMTP), Hypertext Transfer Protocol (HTTP) and File Transfer Protocol (FTP). There also exist many proprietary protocols whose internal structure is not disclosed. Examples of these applications include Skype, Twitter and YouTube, all of which communicate using the IP suites.

Transport Layer Protocols

Transport layer protocols provide host-to-host connectivity to deliver information generated by an application in one device to an application in another device. The function carried out in this layer ensures data are passed on to the host's application reliably and in the correct order.

The prominent internet transport layer protocol is the TCP, which is often used as part of the TCP/IP suite. The TCP protocol enables communications between applications within two

Bits	0–15			16–31
0–31	Source port			Destination port
32–63	Sequence number			
64–95	Data offset	Reserved	Various	Window size
96–128	Checksum			Urgent pointer
127–159	Options			
160~	Data (up to 1500 bytes)			

Figure 8.8 TCP header and data fields

Bits	0–15	16–31
0–31	Source port	Destination port
32–63	Length	Checksum
64–65635 (max)	Data	

Figure 8.9 UDP header and data fields

remote devices, structures the data output from the application, and delivers it to the IP layer for transportation. TCP also facilitates flow control including adjustment of transmission speed in response to the physical link transmission capability. It further provides a mechanism to check whether the received data are correctly received and if so send an acknowledgement to the transmitter. Conversely it discards an erroneously received packet and asks for retransmission.

The TCP header and data field structure is shown in Figure 8.8. It specifies the port in a device from which the data are transmitted and the port which receives the data at the receiver device. The TCP header also includes a checksum for the header to inform a receiver of any errors in the header and data. The checksum is a modular addition of all 2-byte words in the header and data fields. A sequence number enables the receiver to reconstruct the full transmitted data from the parts it receives. The maximum payload of TCP (the data from the application layer) is 1500 bytes.

Another popular transport layer protocol is the User Datagram Protocol (UDP), which is commonly used for voice over IP (VoIP) and audio and video streaming applications. UDP is a simpler, and therefore faster to process, protocol and is used for applications which can tolerate lost packets. The UDP packet structure is similar to that of TCP but since UDP does not need to establish a host-to-host connection it does not include a number of fields such as sequence number. Furthermore, UDP is not designed for flow control as no retransmission feedback mechanisms are included. The UDP header and data field structure is shown in Figure 8.9.

Internet Layer Protocols

Similarly a number of network layer protocols exist which provide a logical connection between two end devices (as opposed to ports within devices which connect through transport layer protocols). The most prominent network layer protocol is the IP. Two versions exist: IP

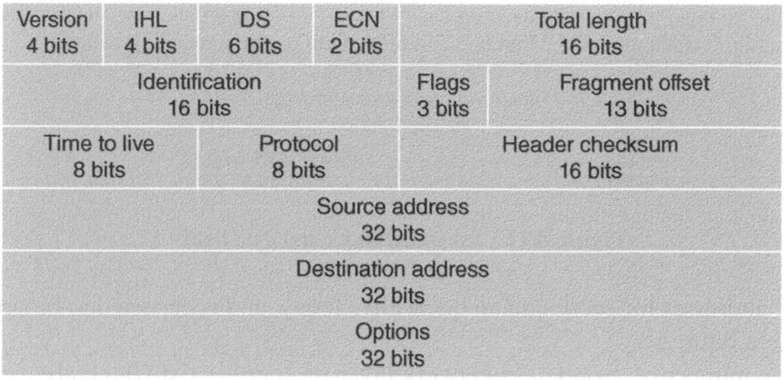

Version 4 bits	IHL 4 bits	DS 6 bits	ECN 2 bits	Total length 16 bits		
Identification 16 bits				Flags 3 bits	Fragment offset 13 bits	
Time to live 8 bits		Protocol 8 bits		Header checksum 16 bits		
Source address 32 bits						
Destination address 32 bits						
Options 32 bits						

Figure 8.10   IPv4 header

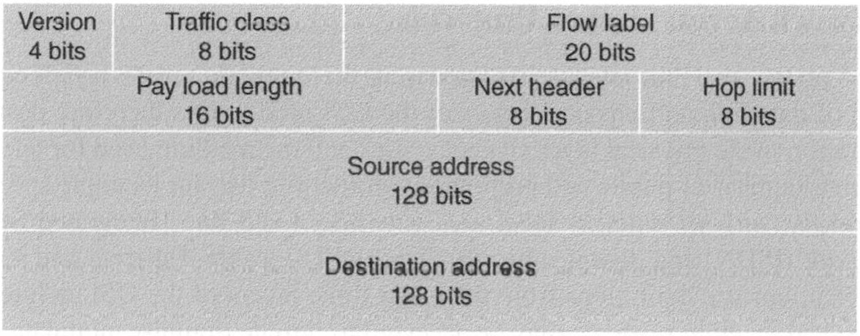

Version 4 bits	Traffic class 8 bits	Flow label 20 bits		
Pay load length 16 bits		Next header 8 bits	Hop limit 8 bits	
Source address 128 bits				
Destination address 128 bits				

Figure 8.11   IPv6 header

version 4 (IPv4) is an older version which is still commonly used, and IP version 6 (IPv6) which is gradually replacing it.

Figure 8.10 shows the IPv4 header structure. The header size is 20 bytes[4] and carries device address information, a checksum to ensure correct detection of header data, and other information such as the length of the payload and time to live.[5] The length of the two address fields of 32 bits is of special interest. These address fields provide for some 4.3×10^9 (4.3 billion) addresses, and were standardised in the early days of the internet when the number of devices connected to the internet were very few, and therefore the field size was expected to be sufficient. With the explosive growth of the internet in recent decades, this address field is now exhausted leading to a shortage of addresses for devices using IPv4. Furthermore, a number of IP header fields were proven to be redundant. These gave rise to the development of a new version of the IP approach in the early 1990s. The new version, IPv6, has been standardised with a much larger address field and provision for more efficient, more secure usage of header space. The IPv6 header structure is shown in Figure 8.11. It is 40 bytes long and provides 128-bit addresses, or 3.4×10^{38} addresses for source and destination. This provides

---

[4] The IP header size can be as long as 40 bytes.

[5] Time-to-live is used to ensure infinite transmission loops are avoided. Since IP protocol is connection-less, there exist the possibility that a packet is routed in a loop without finding its way to the final destination. This field ensures that a packet is discarded after a certain number of hops between 'relay' points.

Preamble 7 bytes of 10101010	Start of the frame 10101011	MAC destination 6 bytes	MAC source 6 bytes	802.1Q Header 4 bytes
Payload (data and padding) 46–1500 bytes				
CRC32 4 bytes	Inter-frame gap 12 bytes			

Figure 8.12   Ethernet header and data fields

some 5×10^{28} addresses for every human being alive today, and is expected to be sufficient for the foreseeable future.

Similarly a number of proprietary network layer protocols exist which are used for proprietary applications.

Link (Network Access) Layer Protocols

As discussed above, the link layer (also known as the network access layer) equates to a combination of data link and physical layers of the OSI model. It is therefore designed with due consideration to the physical layer characteristics and the medium used for telecommunications. A large number of public and proprietary protocols exist: for example specific protocols for wireless (such as wireless local area network) fixed line [Integrated Services for Digital Network (ISDN) and digital subscriber line (DSL)], cable, Ethernet and so on. Many of these applications use definitions from the lower three layers of the OSI architecture such as how a physical medium resource may be accessed to initiate digitally modulated signals.

For example, the Ethernet protocol enables two devices to send and transmit a sequence of 1's and 0's. This is facilitated through a specific header structure as shown in Figure 8.12. A 7-byte preamble consisting of a predetermined sequence of 1's and 0's is utilised to alert a receiver of an incoming data packet before data transmission starts. A receiver continuously monitors the transmission medium and detects signals as a random series of 1's and 0's. The probability that random noise may generate this specific preamble is small, and therefore a monitoring device is unlikely to falsely detect a reception. The receiver then detects the network layer address of a device [usually a media access control (MAC) address] and proceeds to extract the Ethernet payload. Ethernet protocol employs a strong error detection mechanism to ensure incorrectly received data are quickly discarded. Since a transmitter using a TCP expects to receive an acknowledgement for a correctly received packet, all discarded packets will automatically be retransmitted. A UDP communication will simply disregard the lost packet.

Example 8.1: Communication Establishment

A data communication using the internet goes through the following processes. An application in device A (e.g. a client) initiates a data transmission to another device B (e.g. a server) by generating a request to the transport layer. A transport layer protocol such as TCP attempts to establish a logical connection with the server by sending a connection request to the IP layer, indicating the application ports of both server and client. The IP layer adds the IP address of the client and server devices with further control information, forwarding the packet to the link layer which will forward the message to the next node in the local network. Each node is

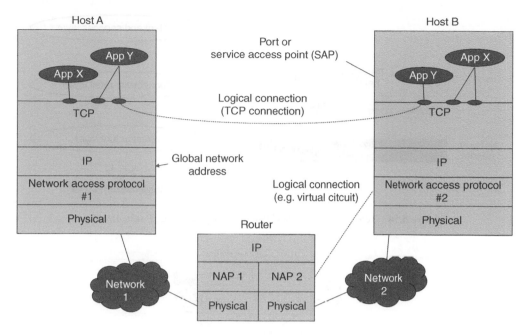

Figure 8.13 Internet packet flow

identified by its static MAC address, a 48-bit address which uniquely identifies a device (a computer, a printer, a switch, etc.) A MAC layer packet is usually referred to as a 'frame', and carries the data and headers of all higher layers. Routers forward the message towards its destination from node to node, based on a locally determined path. The destination switch then receives the link layer packet, extracts the IP layer packet, then transport layer packet and finally the application data. Through this process an application-to-application connectivity, and layer-to-layer logical (and not physical) connectivity is established over which data flows. This is illustrated in Figure 8.13.

Telecommunications Ecosystem and the Internet Protocol

Different players of the telecommunications business ecosystem may be divided according to the different layers of the IP suite. The infrastructure layer businesses are active in the network access/physical layer, providing a mechanism for packets to be sent to the retail layer. The retail layer business serves customers using the TCP/IP internet and transport layers and connects them to the application and content providers in the application layer, as illustrated in Figure 8.14.

Switching and Routing

As discussed above internet packets are individually and independently forwarded at each network node. Two modes exist for forwarding packets at a network node: switching; and routing.

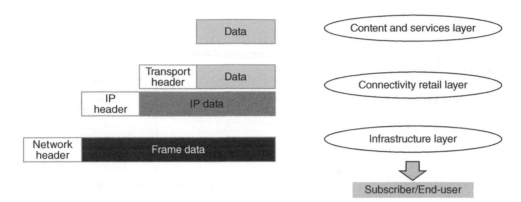

Figure 8.14 Telecommunications business ecosystem and IP suite layers

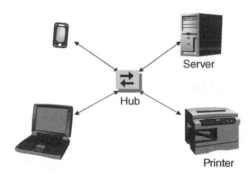

Figure 8.15 Local switching through a hub

In practice, any packet from a local device is sent to a local switch. If the receiver is also within the local network, the switch knows its network (MAC) address and therefore can switch the message directly to the recipient. Switches therefore operate at the link layer. An example of a switching operation could be a file sent to the local network printer. At times the switch, sometimes called a hub, broadcasts the received message to all its devices including the destination device. This is illustrated in Figure 8.15.

If the packet is intended for an address outside the local network, the destination MAC address is a gateway device to which the packet is sent. The local gateway then forwards it to the edge router of the internet service provider (ISP). A router receives packets from a number of nodes and forwards them to another node depending on the destination IP address. This is illustrated in Figure 8.16.

Each router maintains a routing table which it updates regularly to determine a suitable direction for forwarding a packet. A packet sent from Sydney to Los Angeles may be sent in a different direction compared with a packet destined for London. The routing table maintains the number of nodes a packet needs to pass through before it reaches a certain IP address.

To summarise, a receiving node examines a packet's MAC layer header information to determine who the intended recipient is, and therefore to where the packet should be sent. If

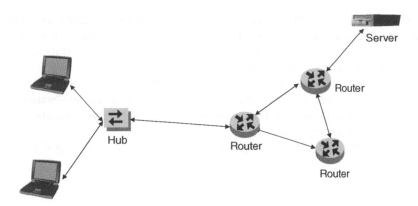

Figure 8.16   A simple routing network

the link layer header indicates that the packet is intended for the local area network, a 'switching' function is carried out. If the MAC layer indicates an external gateway address, the packet is forwarded to the gateway. At the gateway, the IP layer header is examined to find the destination IP address. A routing table is then consulted to find to which direction the packet is to be sent. This process is known as 'routing'.

Packet Switching Performance Metrics

As discussed in Chapter 3, circuit-switched voice communications systems performance is measured using the grade of service metric, which shows the probability that a call may not find a free circuit and may therefore be blocked. This is because the number of concurrent circuit-switched calls is limited by the number of circuits. When all circuits are busy any new incoming call is dropped as no end-to-end physical link can be established.

Because packet-switched data communications systems are (relatively) delay insensitive, incoming packets may be stored in a local buffer if all links are in use. The packet may then be placed in a queue and transmitted when the link becomes free. The only restriction is the size of the memory buffer: if an infinite buffer can be installed, no packet may ever be lost. The circuit switching grade of service metric is therefore not applicable to packet-switched systems.

The metrics used for packet switched systems are based on probabilities a packet may be queued, that is a circuit is not immediately available, and the length of time it has to wait to be processed. This queuing process however happens at every single node through which the packet passes (e.g. Figure 8.2). The total delay is the sum of the delay experienced by the packet as it traverses the packet-switched network. The metrics of interest are therefore the overall delay, and the percentage of packets which will wait more than a certain length of time.

As a packet traverses through network nodes, each nodes takes a finite length of time to 'process' the packet. This processing time is mainly for a switch or a router to open the appropriate address field and determine to which port the packet is to be sent. A received packet at a router for example is stored in a buffer, the IP header is read and then the packet is copied to a port buffer as determined by the routing table. The outgoing buffer is then emptied gradually depending on the speed of the outgoing link. For example a 1 megabit per second (mbps) link requires 10 ms to transmit a 10 000 bit (1250 bytes) packet. If 1 ms is necessary for the router to read the IP header and copy the packet form input to the output buffer, then a total of 11 ms is necessary to process this packet. The processing time can be reduced if more servers and more/faster outgoing links can be installed.

If a new packet arrives while an earlier packet is being processed, it is stored in the input buffer until the router is free. If more servers are available, then the average 'waiting time' for packet processing can be decreased. The overall service time is therefore a function of servers' processing time and outgoing links.

The probability that a packet may be queued is calculated from the Erlang-C formula:

$$P = \frac{\dfrac{a^n}{n!}\dfrac{n}{n-a}}{\displaystyle\sum_{x=0}^{n-1}\dfrac{a^x}{x!} + \dfrac{a^n}{n!}\dfrac{n}{n-a}}$$

where a is the offered traffic in erlangs and n is the number of servers. From this formula several parameters such as average waiting time, required length of buffer, probability of buffer overflow and so on can be calculated [10]. Calculation of the queuing (pictured)[6] probability and average waiting times are outside the scope of this book, and generally complex to calculate. On-line tools  (e.g. http://www.math.vu.nl/~koole/ccmath/ErlangC/index.php) can be used to calculate the desired performance parameters of a packet-switched system.

Example 8.2

A router in a packet-switched system receives 360 000 packets every hour. Each packet takes on average 10 ms to be processed. The router has two servers to process the received packets.

• What is the probability that a packet has to wait?
• What percentage of packets wait more than 10 ms?

Answer

• The probability that a packet has to wait is 33.3%.
• The percentage of packets that wait more than 10 ms is 12.3%.

It should be noted that the Erlang-C formula assumes the behaviour of incoming packet traffic and processing time to be modelled by exponential probability distribution. In practice this is not a realistic assumption and more complex behaviour models have been developed to

---

[6] Alasdair McLellan (own work) [CC BY-SA 3.0 (https://commons.wikimedia.org/wiki/File%3AVolcano_causes_ queue_at_Bergen_Station.jpg)].

characterise packet traffic and processing. These do not lend themselves easily to closed-form formulas, and therefore the performances of complex networks are evaluated using computer simulations, for example OPNET [11].

Call Centre Application

The Erlang-C formula has found an application in calculating the number of operators at call centres. The principle behind the calculation of the number of operators is the same: how long it takes to serve one customer, how many customers call every hour, and how many operators there must be to provide a certain quality of service.

Example 8.3

Assume 200 customers call the Origin Energy help line every hour, and that each call lasts 5 min. How many operators are required to ensure no-one waits more than 1 min? How many operators are required so only 10% of customers wait more than 1 min?

Answer

- The number of operators for 100% waiting less than 1 min = 612
- The number of operators for 90% waiting less than 1 min = 18

Messaging using self-contained packets, be it clay tablets or letters, is perhaps the first type of information *tele*-communication. In contrast to telephony, packetised telecommunications does not require the establishment of a predefined, static end-to-end link. This flexibility allows for a more robust system design with diversity of channel link capacity. Moreover, data retransmission and queuing techniques permit virtually error-free information transfer. These advantages exist at the expense of system complexity as precise protocols are needed to ensure devices can communicate with each other through a host of diverse intermediate nodes. Nevertheless, advantages are significant enough that packet-switched information transfer is now the dominant mode of telecommunications.

Review Questions

1. What are the advantages of packet switching compared with circuit switching?
2. What are the relative merits of the OSI and IP stacks?
3. What is the role of the transport layer protocols?
4. What role do the internet layer protocols play?
5. How is the OSI architecture useful in the design of systems operating using the IP approach?
6. What is the role of a link layer protocol such as Ethernet? Can you describe the processes that the Ethernet header supports?
7. Name two factors which determine the size of a buffer at a router.
8. What are the metrics of importance to packet-switched networks?
9. How do switches and routers differ?

10. What role does a routing table play? Do you think such a table should be static? Why?
11. A call centre employs 10 operators. On average calls arrive every 15 s and last 120 s. Visit http://www.math.vu.nl/~koole/ccmath/ErlangC/index.php and answer the following questions:
    (a) What is the average waiting time?
    (b) What is the probability that a received call is queued?
    (c) What is the probability that a received call waits more than 60 s?
12. Carnegie Mellon University Australia decides to migrate to a VoIP system which uses an 8 kbps voice coding technique. What outgoing transmission link size is required if 20 staff and faculty are constantly on the telephone at peak time?
13. What is the business value of communication protocols?
14. Why has IP been successful whereas the OSI model has not?
15. How have the intellectual properties of the IP developers been protected? Speculate how a patent protection regime would have succeeded.

References

[1] https://commons.wikimedia.org/wiki/File:Magnetic-tape_hg.jpg, accessed 23 August 2015.
[2] https://www.flickr.com/photos/elonuniversity/7105615519, accessed 23 August 2015.
[3] Kleinrock, L. (2010) An Early History of the Internet. IEEE Communications Magazine.
[4] Mathison, S.L., Roberts, L.G. and Walker, P.M. (2012) The History of Telenet and the Commercialization of Packet Switching in the U.S. IEEE Communications Magazine (May).
[5] http://www.printerworks.com/Catalogs/CX-Catalog/CX-Fed-Ex.html, accessed 23 August 2015.
[6] http://www.entrepreneur.com/article/219445, accessed 23 August 2015.
[7] Hochman, D. (1962) Digital Systems in Space Communications. Proceedings of the Third International Congress of Aeronautical Sciences.
[8] Costigan, D.M. (1971) *Fax: The Principles and Practice of Facsimile Communication,* Chilton Book Company.
[9] Russel, A.L. (2013) OSI: The Internet That Wasn't. IEEE Spectrum Magazine.
[10] Kleinrock, L. (1975) *Queueing Systems,* Wiley Interscience.
[11] http://gb.riverbed.com/products/performance-management-control/opnet.html, accessed 23 August 2015.

9

Fixed Broadband Communications Systems

Preview Questions

- How can telephone lines be used for data transmission?
- Why does ADSL2 perform better than ADSL over short distances but similarly over long distances?
- What is Vectored VDSL?
- How did cable TV operators become internet service providers?
- How do optical fibre cables augment DSL systems?

Learning Objectives

- Characteristics of telephone line medium
- Principles of xDSL operation and how xDSL transmission rates are calculated
- Cable and hybrid fibre cable systems characteristics
- Principles of optical fibre communications
- Combination of fixed line systems for higher transmission rates

Broadband Telecommunications Technologies and Management, First Edition. Riaz Esmailzadeh.
© 2016 Riaz Esmailzadeh. Published 2016 by John Wiley & Sons, Ltd.
Companion Website: www.wiley.com/go/BTTM

Historical Note

Packet-switched computer data communications bet-ween a computer terminal (such as the DEC VT100 pictured [1])[1] and a mainframe computer emerged in the 1960s. The subsequent growth in importance and value in the 1970s and 1980s led to major changes in the telecommunications industry. These included theoretical work in packet switching and queuing the-ories as well as development of network equipment to facilitate transmission of digital traffic over wide areas. As a result, a new 'digital' infrastructure became necessary to support these 'wide-area networks' and connect customer premises with exchanges and com-

puter centres, and with each other. The new digital infrastructure used fixed lines media, that is wired as opposed to wireless, and a number of standards emerged to specify how telecommunications infrastructure could be built on this fixed media.

The traditional voice telecommunications network as discussed in Chapter 3 comprises a number of local, regional and international exchanges and the connecting infrastructure is gen-erally made of copper wires. A subscriber's premises is commonly connected to a local exchange using twisted pair copper wires, whereas exchanges are connected by high capacity links: copper wires, coaxial cables and more recently optical fibres. These systems use fre-quency division multiplexing (FDM) and time division multiplexing (TDM) technologies to simultaneously transport multiple voice calls.

Digital modulation of voice necessitated changes to the network equipment at both sub-scriber premises and exchanges. New digital phones with analogue to digital conversion (and reverse) capabilities at subscriber premises, as well as digital switches at exchanges were needed. On the other hand, mainframe computer-terminal data communications also required special links between computer centres and the telecommunications network nodes. However, construction and roll-out of a totally new infrastructure was impractical and therefore new sys-tems had to be developed to utilise and augment the existing system of twisted pair copper

wires and coaxial cables. These led to development of new tech-nologies and standards, and has led to the modification and extension of the telephony infrastructure.

Three fixed-line media are of main interest to broadband tele-communications: twisted pair copper wires traditionally used for the telephony system; coaxial copper cables used for

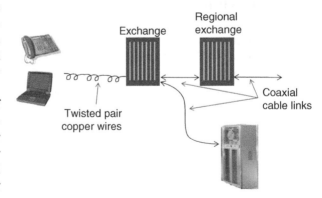

---

[1](FN) DEC VT100 terminal at the Living Computer Museum (apparently connected to the museum's DEC PDP-11/70). Creative Commons Attribution 2.0 Generic license.

inter-exchange links and cable TV broadcasting; and optical fibre systems used for international and long-distance telephony and telecommunications backhaul. The following is a chronological account of major milestones in fixed-line data and digital voice systems.

Copper wires trace their history to the early analogue telephone system in the 19th century and continue to be used. These were also used for discrete data transmissions such as automated telegraph, and analogue inter-exchange lines. In the early 1960s Bell Laboratories developed a system to communicate 32 digital voice calls over twisted pair copper wires between exchanges. The links were configured with a capacity of 1.5 megabits per second (Mbps) over a 2 km distance. This led to the standardisation of an inter-exchange digital voice system by the ITU in 1972, known as the Integrated Digital Network (IDN). This was further extended to create the subscriber-to-subscriber, end-to-end, digital voice and data network known as the Integrated Systems Digital Network (ISDN) in 1988 [2, 3].

Initial computer-terminal communications were carried out using telephony lines and in-band transmissions in the order of 300 baud (symbol) rates. These in-band (4 kHz) modulator/demodulators (modems) were further enhanced to support up to 56 kbps over two pairs of copper wires. Data communications from subscriber premises remained at these transmission rates until the turn of the 21st century, when digital subscriber line (DSL) standards provided multi megabit per second rates on twisted pair copper wires.

Coaxial cables invented in the 19th century were first used to connect telephone exchanges in the 20th century. This includes the first underwater telephony cable to connect Tasmania and mainland Australia, and a long-distance telephone cable between New York and Philadelphia in 1936. Cable TVs using coaxial cables were introduced soon after in the late 1940s as televisions became popular to provide reception to underserved areas. Coaxial cables were also used as a medium for local area networks data communications in the 1980s. As the popularity of the internet and data connectivity at home grew, cable TV operators started to move into internet service provision using data over cable service interface specification (DOCSIS) standards in the late 1990s.

The invention of optical fibres dates to the late 1960s. However, their very high capacity ensured that they were soon being rolled out in most inter-exchange links, with the first links being rolled out in the early 1980s. Optical fibres were initially used for international telephony links replacing coaxial undersea cables: for example the first transatlantic cable TAT-8 was laid in 1988. Optical fibre cable applications were soon extended to long-distance telephone and later on to data communications aimed at business subscribers. The usage of optical fibres to connect subscriber premises to the internet also dates to the turn of the 21st century. More commonly, a combination of fibre and copper is offered to provide fast internet connectivity to a subscriber premise.

Twisted Pair Copper Wires

The use of twisted pair copper wires as a communication medium is as old as telephony. These were developed for the transmission of voice over long distances as determined by the gauge or thickness of the wires. As discussed in Chapter 3, voice signals have frequency components in a range of 300–3400 Hz, which can be supported for telecommunications over distances of tens of kilometres over typical 0.4 or 0.5 mm gauge twisted pair copper wires with sufficiently low loss.

Figure 9.1 A Novation CAT acoustically coupled modem [4]. Source: Lorax on en.wikipedia (public domain): https://commons.wikimedia.org/wiki/File%3AAcoustic_coupler_20041015_175456_1.jpg

Table 9.1 Evolution of modem rates

Date	Modem	Transmission rate
1981	Hayes Smartmodem	300 bps
Late 1980s		2.4 kbps
1991	Superfax 14400	14.4 kbps
Late 1990s		56 kbps

Twisted pair copper wire telephone lines have also been used for low rate data communications since the 1920s. Applications have included teleprompters, automated telegraphs and radar-to-air defence systems communications. A later application was in telephony systems using dual-tone multi-frequency signalling for call set-up since the early 1960s. In the 1970s acoustically coupled modems (Figure 9.1) used different audio tones to represent 0 and 1 and transmit data between nascent desktop computers over the voice bandwidth of telephone lines (i.e. 300–3400 Hz). The maximum transmission rates achieved were in the order of 300 bits per second (bps).

Over the next two decades modem technology using the telephone improved and by the end of 1990s dial-up modems were capable of supporting 56 kbps in the downstream. A list of modems and their transmission rates is shown in Table 9.1. All these devices used the same frequency band as that of voice. The full potential of twisted pair copper wire was generally not used.

The bandwidth of the copper wire has capacity for data communications at much faster rates than 56 kbps. As noted above the medium was already being used for inter-exchange transmission rates of 1.5 Mbps, as well as ISDN voice communications which operated in the 80–120 kHz range of the spectrum. New systems needed to be developed to take advantage of this latent capacity.

Twisted Pair Copper Wire Characteristics

Twisted pair copper wire has a specific frequency response bandwidth. The path loss of 26 AWG (0.6 mm) twisted pair wire as a function of frequency is drawn in Figure 9.2. As shown, the path loss for signals at low frequencies, such as that of voice, is very small, and is

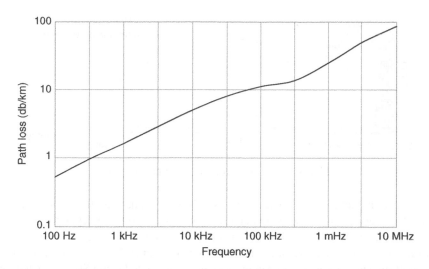

Figure 9.2 Twisted pair wire path loss per kilometre for 26 AWG medium as a function of frequency. Source: Ref. [5]

in the order of a few decibels per kilometre. However as carrier frequency increases, path loss increases such that a signal at 10 MHz loses power by a factor of approximately 70 dB every kilometre. Clearly twisted pair copper wire is more suitable to low frequency, low speed data communication applications. Nevertheless, the path loss is small enough in the range of up to several megahertz to support high data rate communications over short distances. The maximum length for such communication is calculated from path loss, line distortions, interference from other lines and frequency of operation.

Digital Subscriber Line

A number of technologies were developed to utilise the twisted pair copper wire spectrum at frequencies higher than that used for voice. The most popular were T-Carrier and E-Carrier technologies developed in the US and in Europe for multiplexing several voice channels over a high capacity inter-exchange link. Also known as T-1 and E-1, these technologies used twisted pair copper wires to transmit 24 (32 for E-1) digital voice channels and later on data at a rate of 1.544 Mbps (2.048 Mbps for E-1) between operator exchanges and later on to mostly business subscribers' premises [6]. These technologies utilised a much wider frequency spectrum of twisted pair copper wires and beyond the 4 kHz used for voice. T-1 and E-1 were marketed to business subscribers on a lease basis as data backhaul lines for inter-office and office–computer centre connectivity. The cost of leasing a T-1 was quite high and in the order of $2000 per month because repeaters and other electrical equipment were required every 2 km. This was because symmetrical data transmission rates in downstream and upstream for voice communications had to be maintained.

DSL technologies grew from T-1/E-1. The principle was the fact that the twisted pair copper wire spectrum has several useful megahertz of bandwidth over short distances and therefore higher data rates may be supported on telephone subscriber lines using this bandwidth.

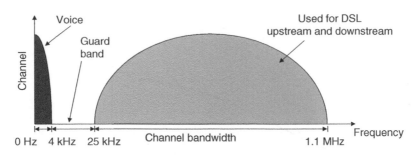

Figure 9.3   ADSL frequency band allocation

If symmetrical communication is not required, as is the case for data applications, then repeaters are not needed every 2 km and therefore many subscribers served by a telephone exchange may be able to connect to the internet at high transmission rates without a need for new infrastructure roll-out. Given the path loss characteristics shown in Figure 9.2, possible transmission rates over different distances for a system bandwidth can be calculated using Shannon's theorem.

Several DSL standards have been developed and standardised under the ITU G.99x series of standards since the early 1990s. The most common standards have been asymmetric DSL (ADSL) and its evolutions. Asymmetric refers to the fact that upstream and downstream transmission rates are unequal. The difference between these standards is mainly the overall range of frequencies used and the spectrum allocated to upstream and downstream transmissions. Recent ADSL systems are known as very-high-bit-rate DSL (VDSL) systems, which utilise a wider bandwidth and provide corresponding higher transmission rates.

DSL technologies are standardised under the auspices of the ITU-T. G.991 standard specifies ADSL transceiver operation including modulation, channel coding, system bandwidth and transmission power levels. Other G99x series standards specify call set-up procedures (G.994), testing (G.996) and physical layer management (G.997). In particular G.992 defines how ADSL systems share the copper wire bandwidth with voice in a frequency division multiplexing (FDM) manner. The 0–4 kHz frequency band is reserved for voice, and 25–1104 kHz for upstream and downstream transmissions as illustrated in Figure 9.3. A guard band is allocated between 4 kHz and 25 kHz to minimise mutual interference between voice and data transmissions.

Digital Subscriber Line System Architecture

The DSL system structure is built with a star topology centred at an operator's exchange, also known as the central office (CO). A CO is connected to a customer premises equipment (CPE) device using twisted pair copper wires used hitherto for telephony services. In some configurations a CO is connected to a remote terminal (RT) with an optical fibre link and an RT and a CPE device are connected using twisted pair copper wires. Because the same medium is used for both DSL and telephony, a device is needed to ensure the two services do not interfere with each other's operation. This is made possible through a 'splitter', a passive device consisting of a low-pass filter and a high-pass filter. The low-pass filter passes the voice signal up to 4 kHz and suppresses all frequency energy above it. The high-pass filter suppresses all frequency energy below 25 kHz, and passes all frequencies above it as used for DSL. Two splitters are needed for each subscriber, one at the customer's premises for downstream communications

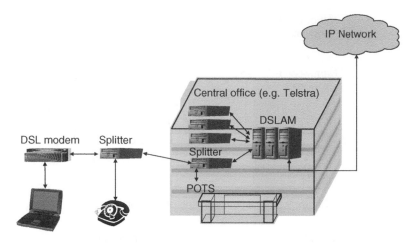

Figure 9.4 DSL system components

and one at the CO (or RT if one is used) for upstream communications. A DSL modem converts the digital bits into analogue signals using a discrete multi-tone (DMT) modulator on the transmitter side and demodulates analogue signals into data bits on the receiver side. The CO serves a large number of DSL subscribers' CPE devices. Signals from CPE devices in the upstream are multiplexed together at the CO, and are forwarded to the internet. In the other direction, incoming signals from the internet need to be switched to their final destination. These functions are carried out by a device known as a DSL access multiplexer (DSLAM). DSLAM is in effect an edge router, which switches internet protocol (IP) packets onwards in both upstream and downstream. A typical DSL system is illustrated in Figure 9.4.

Discrete Multi-tone

The transmission spectrum in the ADSL standards is divided into many narrow-band subcarriers, and data transmission is carried out in parallel over these subcarriers. Each subcarrier is known as a tone, and the data transmission is called DMT. In this process, the input data are first converted from serial-to-parallel, and mapped into a number of symbols. Each symbol is then modulated onto a subcarrier which are then frequency division multiplexed together and transmitted as illustrated in Figure 9.5.

Note that the subcarriers overlap and are orthogonal with respect to each other. The multiplexing and demultiplexing of DMT signals is carried out using discrete Fourier transform (DFT) techniques. Here data are modulated in the frequency domain, and each data symbol [e.g. a 16-QAM symbol] is represented by the amplitude and phase of a frequency carrier, which is orthogonal to its neighbouring frequency carriers. Multiplexing of DMT signals and their transmission is shown in Figure 9.6.

The receiver operation is the reverse of that of the transmitter and is shown in Figure 9.7.

In ADSL, each subcarrier occupies a bandwidth of 4.3125 kHz over which a 4.0 kilo symbol per second transmission is transmitted. Each subcarrier is individually evaluated for its signal to interference plus noise ratio (SINR) to decide the multiple QAM modulation level. Channel

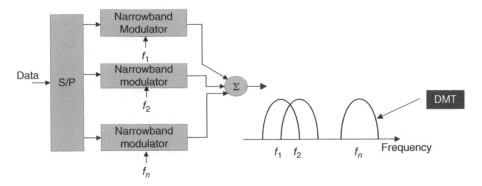

Figure 9.5 Data transmission using subcarriers

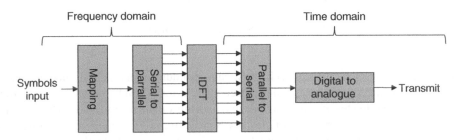

Figure 9.6 An inverse discrete Fourier transform (IDFT) DMT transmitter

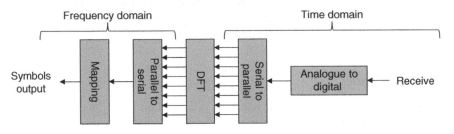

Figure 9.7 A DFT DMT receiver

capacity is calculated by the number of subcarriers allocated to the upstream and downstream as well as the system noise and distortions from other wires in the same cable bundle (crosstalk) as a function of the length of the medium (Figure 9.8).

As shown in Figure 9.2, twisted pair copper wire has a low pass frequency response, that is, the lower frequencies experience a lower path loss compared with higher frequencies. It follows that signals transmitted at lower frequencies will arrive at a higher signal power level at the receiver compared with higher frequencies and therefore with a higher SINR as illustrated in Figure 9.9.

Shannon's theorem shows that channels with higher SINR can support higher transmission rates, that is, more bits per second can be transported for each hertz of spectrum as shown in Figure 9.10 (note the new term I in the denominator, representing interference power.) This means that lower frequency ADSL carriers can support higher transmission rates. Multi-level QAM techniques are employed to take advantage of good channel conditions. Figure 9.11

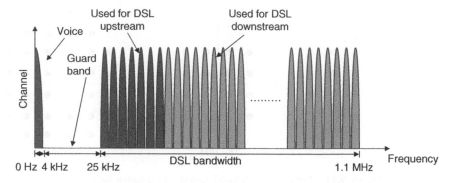

Figure 9.8   ADSL frequency band allocation with DMT modulation

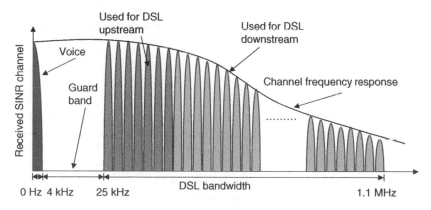

Figure 9.9   ADSL carrier received SINR level

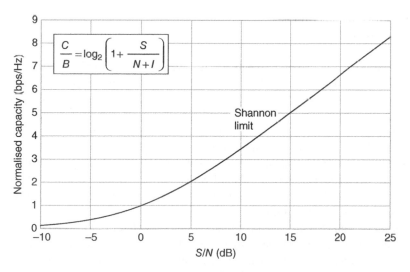

Figure 9.10   Normalised Shannon limit

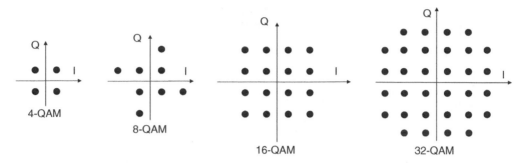

Figure 9.11   Some common xDSL QAM signal constellations

shows 4-, 8-, 16- and 32-QAM constellations. A symbol in these constellations carries 2, 3, 4 and 5 bits of data, respectively. Higher order modulations (64-QAM, 128-QAM, and up to 32, 768-QAM) levels which carry 6, 7 and up to 15 bits of data per symbol, respectively, have been specified in the xDSL standards.

CrossTalk

The performance of twisted pair copper wire systems is affected by interference from other telephone connections. In general a home telephone is individually connected to a telephone exchange. However, all telephone lines from a neighbourhood are bundled together in an underground duct and are further bundled with wires from nearby areas before they reach the exchange as shown in a typical configuration in Figure 9.12.

Although these wires are insulated from each other, they still act as electromagnetic antennas transmitting their signal to, and receiving signals from, other wires in the ducts. This interference is generally known as crosstalk, and exists in two main types:

- near end crosstalk (NEXT)
- far end crosstalk (FEXT)

NEXT occurs when interfering signals are received at nearby devices. For example, two ADSL modems may be sending and receiving data over their respective lines to/from a CO. The transmitted signal of one ADSL modem may induce interference over the other modem, impacting the received signal at the near end. FEXT occurs when the signal from one modem induces interference in the upstream, and impacts on the signal of another subscriber at a CO. The same may happen in downstream transmissions. In general NEXT power level can be quite high if the path length of the interfering signal travel is shorter compared with the desired signal (coming from an opposite end). The FEXT power level is of the same order of magnitude compared with the desired signal as both travel relatively equal distances to opposite ends. NEXT and FEXT are shown in Figure 9.13.

The impacts of both NEXT and FEXT are cumulative. Each two users induce a certain amount of interference on each other's signals as a function of signal power level, proximity, line gauge and insulation, and frequency of operation. As many lines are bundled together in the same physical cable each user signal induces interference on all its neighbours to varying

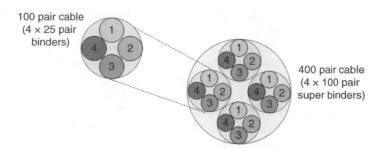

Figure 9.12 A typical copper binder

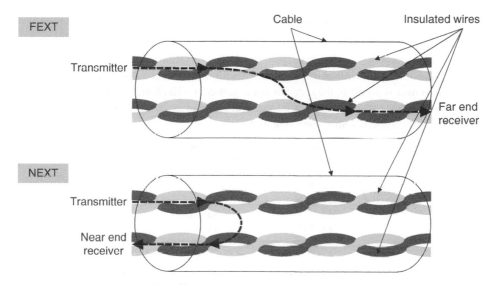

Figure 9.13 FEXT and NEXT

degrees. The cumulative crosstalk interference adds a significant amount of noise to the ADSL system, and is a major cause of reduced capacity.

Since ADSL and VDSL standards specify the use of different frequency bands for upstream and downstream, interference from upstream will not impact downstream reception significantly. As a result the NEXT process does not contribute much to ADSL performance and capacity calculation. The FEXT process on the other hand contributes significantly to the total noise and interference in ADSL/VDSL systems, and is a major reason for capacity degradation.

Digital Subscriber Line Capacity Calculations

Maximum possible transmission rates of ADSL/VDSL systems, or systems capacity, are similarly calculated from Shannon's theorem based on available bandwidth and SINR. DSL systems experience noise and interference from a number of sources. Thermal additive white Gaussian noise (AWGN) is generated by electrical devices and has a constant power spectral density. Impulse noise is generated by devices active in the vicinity of communications and have various sources such as electrical switching. Amplitude modulated (AM) radio, DSL signals of other operatorstravelling through the same cable bundle and so on are other sources of interference. Same operator FEXT also adds to the system noise.

The received signal power facilitates data communications in the presence of this noise and interference. The higher the SINR, or the cleaner the signal as compared with noise and interference, the higher the transmission rates that can be supported. The interference frequency response behaves similar to that of signal: it is lower at higher frequencies. As path loss increases at higher frequencies both signal and interference become weaker. AWGN however remains at a constant power. This results in transmission rate limitation being a function of mainly interference at lower frequencies but also of noise at higher frequencies. Since a DMT method of modulation is used, different subcarriers experience different capacities as a function of signal, noise and interference power levels at that particular frequency.

Example 9.1

What is the maximum downlink capacity of an ADSL2+ system which uses 80% of all available 512 subcarriers, each with an effective bandwidth of 4 kHz, at a distance of 2 km. Assume transmission power is 30 dBm, the receiver noise power is –70 dBm, and path loss follows the table shown. Further assume that FEXT interference level is 20 dB lower than the received power. What is the system's Shannon capacity?

Carriers	Path loss (dB/km)
1~128	12
129~256	24
257~384	36
385~512	48

Answer

Carriers	Path Loss (dB/km)	Rx Signal Power (dBm)	Rx FEXT Power (dBm)	Rx Noise Power (dBm)	Rx SINR (dB)	Number of Carriers	Shannon Capacity (Mbps)
1~128	12	6	–14	–70	20.0	26	0.7
129~256	24	–18	–38	–70	20.0	128	3.4
257~384	36	–42	–62	–70	19.4	128	3.3
385~512	48	–66	–86	–70	3.9	128	0.9
Total							**8.3 Mbps**

Note that carriers 1–384 capacity is mostly limited by FEXT interference, whereas carriers 385–512 capacity is limited by noise. Clearly interference removal can lead to significant capacity improvement.

Vectored DSL

While much of additive noise is random and therefore difficult to remove, the crosstalk interference signals are deterministic and therefore it is possible to calculate and remove them from the received signal. Such crosstalk interference cancellation can result in higher SINR and therefore higher transmission capacity.

Removal of interference at a CO is relatively easy. A CO receives and detects the upstream signals of all users within a link bundle. It may then recalculate the crosstalk value associated with one user and subtract it from the received signal of another and thereby produce a less noisy signal. Alternatively it may jointly detect the signals of multiple users given the amount of crosstalk as calculated from all received signals. Through this process, the received signal is made free of crosstalk and therefore signal detection is done without any FEXT.

While the CO receiver detects all upstream signals, in downstream an end user only detects its desired signal with FEXT present from all other active users. However it cannot calculate the crosstalk values since it does not know about the channel parameters of other users. However the transmitter of the CO can pre-decorrelate the transmitted signal using the upstream channel vectors. The transmitter multiplies the individual signals by the inverse of the vector matrix. This cancels the transmission channel vector matrix and as a result the received signal at any user is FEXT-free. This method is known as Vectored VDSL because it is generally used in VDSL.

In cases when two or more operators use the same copper bundle then Vectored VDSL may not be possible as one operator's transmitter may not know about another operator's downstream data. To implement Vectored operation, multiple operators need to collaborate.

ADSL/VDSL Standards Enhancements

ADSL/VDSL standards have evolved to deliver higher transmission rates in both upstream and downstream. DMT technology used in ADSL/VDSL standards facilitates a simple frequency spectrum allocation. Generally the first 4 kHz (or in the case of ISDN legacy systems the first 80–120 kHz) are reserved for plain old telephone service (POTS) voice communication. A guard band is then specified to ensure minimal voice/DSL mutual interference. Next is the upstream spectrum of 25 to 138–250 kHz. The remainder of the spectrum is used for downstream transmissions. The total bandwidth has evolved through different versions of the DSL standard and its annexes.

ADSL2

ADSL2 incorporates better channel coding, reduced frame overhead, and supports multiple wire pairs, thereby multiplying the capacity.

ADSL2+

ADSL2+ extends overall bandwidth from 1.1 to 2.2 MHz, thereby doubling the maximum possible transmission rate in the downstream. The upstream spectrum remains the same in a number of variants of the standards, or doubled in some others.

Figure 9.14 shows the signal to noise ratio (SNR) for the downstream of an ADSL system and possible bits that each carrier supports at that possible SNR level.

ADSL transmission rates are a function of received SINR. Received signal power diminishes as the distance between CO and CPE increases. As discussed above copper wire path loss is a function of carrier frequency, and higher frequencies experience higher path loss. This means that at short distances, all subcarriers experience low path loss, and therefore SINR is high across all subcarriers. At long distances, higher subcarriers experience large path loss, and therefore can contribute little to overall ADSL capacity.

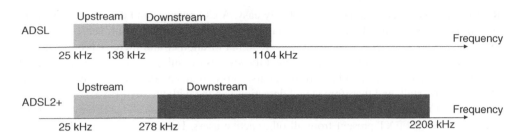

Figure 9.14   Variant of frequency allocation for ADSL and ADSL2+ standards

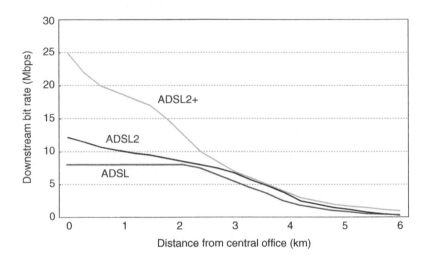

Figure 9.15   ADSL, ADSL2 and ADSL2+ downstream transmission rates as a function of distance

Downstream transmission rates are shown for ADSL, ADSL2 and ADSL2+ in Figure 9.15. At distances less than 2 km from a CO, downstream transmission rates of ADSL2+ are significantly higher compared with ADSL2 and ADSL. This is because extra subcarriers associated with increased bandwidth from 1.1 to 2.2 MHz contribute significantly to overall transmission rates. However at distances of more than 3 km, these subcarriers have very low SINRs and therefore do not contribute much extra transmission rate capability.

ADSL2++

A further evolution of ADSL standard is ADSL2++ (also known as ADSL4) which further extends transmission bandwidth from 2.2 to 3.75 MHz.

VDSL

VDSL further extends transmission bandwidth to 12 MHz and incorporates a new frequency plan with a mixture of downstream and upstream frequency allocations. The transmission rates can be as high as 55 Mbps in the downstream and 3 Mbps in the upstream.

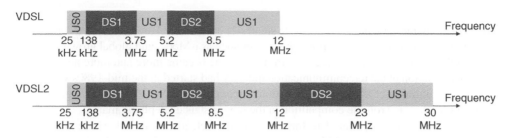

Figure 9.16 Frequency plan and possible transmission rates for downstream (DS) and upstream (US) for VDSL and VDSL2

VDSL2

VDSL2 extends system bandwidth to 30 MHz with a mixture of downstream and upstream frequency allocations. The transmission rates can be as high as 100 Mbps in the downstream and 100 Mbps in the upstream. VDSL and VDSL2 frequency plans and transmission rates are shown in Figure 9.16.

The SINR for the downstream of a VDSL system with a DS1, DS2, and US1 and US2 configuration can reach as much as 60 dB. This supports a maximum of 15 bits per symbol/carrier, which combined with a 4 kilo symbol/sc/carrier, results in a 60 kbps per carrier transmission rate.

Latest Transmission Rates

ADSL and VDSL maximum transmission rates are a function of available bandwidth and SINR. Higher bandwidths of up to 300 MHz have been considered for VDSL systems for short range communications, and theoretical calculations of capacity have been reported by Ericsson with possible transmission rates of as high 3500 Mbps [7]. Using a 30 MHz bandwidth VDSL2, Ikanos Communications demonstrated a 300 Mbps VDSL system using a single twisted pair copper wire in July 2013 [8].

These rates are limited by the length of copper medium and FEXT level between users within the same cable bundle. However, given these limitations it is possible to provide very high transmission rates to customer premises using a hybrid fibre–copper topology.

Case Study 9.1: Yahoo Broadband

Yahoo Broadband (BB) is a joint venture company between Yahoo! Inc. and Softbank K.K. a software development company (and an early investor in Yahoo). Yahoo Japan, also a joint venture of Yahoo! Inc. and Softbank K.K., is an e-commerce company which offers services including search engines, auctions, news portals and so on. Yahoo BB offers broadband ADSL and more recently optical fibre to the home services in Japan. It started its services in September of 2001 and successfully grew to become one of the largest internet service providers (ISPs) in the Japanese market.

The internet service market in Japan was small compared with many OECD countries such as the United States. One reason was the comparatively lower personal computer ownership by Japanese households due to small size apartments. However, with the growth in popularity of mobile internet through services such as i-Mode, and global growth of the World Wide Web, internet connectivity from the home became more and more necessary. The deregulation of the telecommunication industry had started in the mid-1980s with the partial privatisation of the incumbent national operator NTT. In 1996 further legislation was put in place to promote competition in the long-distance telephony market.

A number of ISPs appeared in the Japanese market, including Tokyo Metallic and eAccess and soon flat-rate 64 kbps ISDN data services were offered by NTT at the cost of around $80 per month. Faster DSL services were offered at a range of rates of up to $350 per month with 1.6 Mbps downstream and 288 kbps upstream transmission rates. Despite these, the take up of ADSL services in Japan was quite slow due to the small market size of households with PCs, low recognition of the value of the services broadband could offer and the relative high prices for these services. Moreover, the cost of ADSL modems was relatively high and in the order of $500.

Yahoo BB management realised these obstacles and set out on an aggressive marketing campaign. Their new ADSL service offered transmission rates of 12 Mbps in the downstream, and was offered at $25 a month. Moreover, modems were given away for free with 2-year contract sign-ups. It was estimated that the cost of each new subscriber acquisition was in the order of $600.

The marketing campaign was greatly successful. Within 3 months Yahoo BB was the market leader, and within 14 months revenues overtook costs. The success of Yahoo BB empowered Softbank K.K. to fully enter the telecommunications market and acquire Vodafone Japan. Softbank has successfully grown these businesses and in 2013 acquired Sprint, the third largest mobile operator in the United States [9, 10, 11].

Case Study Questions

- What were the obstacles to the growth of ADSL services in Japan?
- What were the government policies in lowering these obstacles?
- Softbank was a software development company. Why do you think it entered the ISP market (through Yahoo BB)?
- Which layers do you see Yahoo BB play in the broadband telecommunications ecosystem?
- In 2005 Softbank bought a baseball franchise in Japan. How do you see this move fits with its overall telecommunications strategy?

Coaxial Cable Systems

Usage of coaxial cable to deliver TV signals to houses dates back to the late 1940s in the United States. TV signal can be broadcast through TV tower antennas and received over the air. However, TV reception in some areas can be weak due to physical obstacles blocking the signals. Cables were initially used to fill in TV signal coverage, and by entrepreneurial operators in bringing signals to areas where coverage was yet to be achieved. Cable TV service later on grew because the number of terrestrial broadcast channels was limited due to frequency spectrum restrictions and new channels could only be received via cable. By 2006 it is reported that some 58% of American household had basic access to cable TVs, while cables passed about 96% of house units [12]. TV cables and telephone lines constituted the two most common fixed-line telecommunications services into an American home. The cable TV architecture consisted of a cable headend, which receives the TV programmes from terrestrial as well as satellite broadcasts, uses a frequency division multiplexing to transmit up to 500 TV channels on a single medium and feeds it to cables (coaxial or optical fibre) which carry the signals to distribution hubs (Figure 9.17). Each hub can connect to 500–2000 subscribers' homes located within a radius of several hundred meters, using a coaxial cable. A set-top box at the receiving home demodulates the desired channel to the baseband and shows it on TV.

The useful frequency spectrum of a coaxial cable is divided into downstream (50 MHz–1 GHz and above) for TV signals and upstream (2–45 MHz) for signalling purposes such as facilitating subscriber pay-per-view channel selection. The frequency allocation is shown in Figure 9.18. The upstream channels can be as narrow as 200 kHz to 3.2 MHz and 6.4 MHz (in later standard releases.) The downstream bandwidth corresponds to the TV channel bandwidth and is 6 MHz in North America and 8 MHz in Europe.

With the growth of the internet in the late 1990s, cable TV operators also moved into internet service provision over their infrastructure. The data communications standards were developed through CableLabs, a research and development body established by a global consortium of cable operating companies. The standards are known as Data Over Cable Service Interface Specification (DOCSIS).

DOCSIS standards work within the cable TV upstream and downstream frequencies and use one or more of these carriers for data communications. In the downstream the data packets are time division multiplexed and sent to all cable modems connected to the headend. Since the packets are intended for one IP address, only the intended receiver is capable of detecting

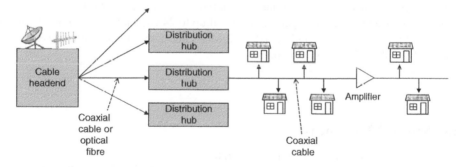

Figure 9.17   Cable TV architecture

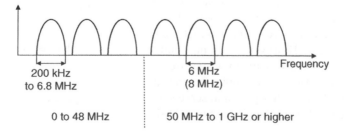

Figure 9.18   Cable TV frequency allocation

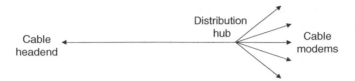

Figure 9.19   Cable TV frequency allocation

the data. In the upstream, packets are scheduled by the cable headend and are transmitted in a time division multiple access fashion (Figure 9.19).

Downstream and upstream transmissions use quadrature phase shift keying (QPSK), 8-QAM, 16-QAM, 32-QAM, 64-QAM, 128-QAM, 256-QAM and in the latest release of the standard 4096-QAM modulation techniques. The maximum transmission rates can be calculated from the available bandwidth and the modulation technique.

Example 9.2

Calculate the downstream transmission rate for a North American cable system which uses 256-QAM modulation. (Note this is not Shannon capacity.)

Answer

Bandwidth = 6 MHz, supporting a 5.36 mega symbol/s transmission rate.
Modulation = 256-QAM or 8 bits/symbol
Transmission rate = 8 × 5.35 = 42.88 Mbps. (This is the channel level transmission rate. If overheads are considered, the data transmission rate is reduced to 38 Mbps.)
The DOCSIS 3.1 standard release specifies usage of up to 24 channels in the downstream and 8 channels in the upstream, yielding 1029.12 Mbps and 245.76 Mbps transmission rates, respectively, with a 256-QAM modulation.

Case Study 9.2: Comcast

Comcast is the largest cable broadband operator in the United States by coverage area, offering a range of connectivity plans [13]. The entry of Comcast into the broadband market dates to 1996 when @Home its joint venture with several other cable TV companies offered home internet and IP telephony services. However, the joint venture did not fare well, and was finally taken over by Comcast in 2002. Meanwhile Comcast started an independent ISP which has grown to serve some 20 million subscribers in the United States [14].

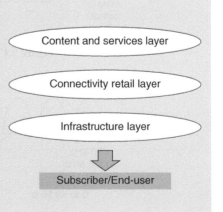

Starting out as a small cable TV company in 1963, Comcast has grown to become the biggest Cable TV operator in the world by revenue. It has interests in TV and radio broadcasting, film studios, theme parks and telecommunications. It is a highly vertically integrated company, producing, retailing and delivering content to its many subscribers.

Comcast entered the broadband telecommunications market as an ISP using its cable connections. Comcast and other cable TV operators are in a strong position as they have a fixed-line medium entering or passing many households in the United States. In this way Comcast competes in the same field as telephone operators which connect to homes with twisted copper wires. The development of DOCSIS standard 3.1 theoretically enables Comcast to offer broadband service at the gigabits per second transmission rate. This is a highly competitive rate and can protect Comcast from competition from optical fibre-based service providers such as Verizon FiOS and Google Fibre. The physical cable connection has afforded Comcast and other operators a golden business opportunity.

Comcast's highly vertically integrated business affords it great competitive advantage. However, it is also risky from a policy point of view. Comcast has been accused of slowing video content from its competitors such as Netflix. This has given rise to a long policy discussion on net neutrality [15]. A structural separation of the company across the content delivery function may be unlikely, but a firmer competitive parity may be enforced.

Case Study Questions

- How did cable TV operators first enter the ISP market?
- What were their subsequent moves?
- What are Comcast's strengths and competitive advantages?
- How do you see a competitor to Comcast emerging?
- Compare the competitive landscape in the United States with Australia which has a relatively small cable subscriber base.

Optical Fibre Systems

Information transmission using light signals is perhaps the most ancient form of telecommunications. A more modern form of light communications was studied by Alexander Graham Bell and he invented the photophone, where voice signals were sent using a light beam. The limitation of this and other forms of light communications has been the dispersion of light in the air and the requirement of a line-of-sight between the transmitter and the receiver. A number of scientists contributed to the development of an optical cable within which light could be confined and transmitted to remote locations. These were based on the fact that light will be reflected within a medium if the incident angle is smaller than a certain value as shown in Figure 9.20.

Early optical fibres however were hampered by impurities within the glass medium which absorbed or reflected light, leading to large signal loss. Scientists at Corning, a glass manufacturing company, developed techniques that reduced signal loss in fibres to 17 dB/km in 1970. Further improvements have reduced path loss to less than 0.5 dB/km.

A glass fibre structure is shown in Figure 9.21. The glass core is very thin and in the order of several micrometres. It is covered by several layers of protection made of plastic to ensure that the core is not broken. The resulting strand of fibre is less than 0.5 mm thick.

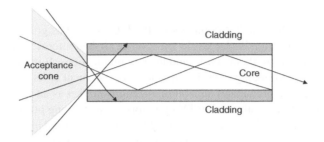

Figure 9.20   Diagram of how light is confined within a glass fibre

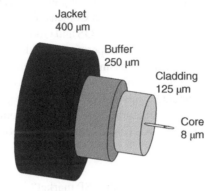

Figure 9.21   Structure of a glass fibre [16]

Because of its very small diameter, special devices are needed to launch coherent, information modulated light into the fibre core. The light source needs to be monochromatic, and in many cases also coherent to ensure best transmission performance is obtained.

A typical silica glass fibre exhibits different path losses at different wavelengths. Path loss is minimum in three regions of light 'colour' with wavelengths of 1.2, 1.3 and 1.5 µm with losses of less than 1 dB/km. The equivalent bandwidth for signal transmission at these wavelengths is very wide: for example a system operating around the 1500 nm wavelength can use the following channel bandwidth:

- Assume upper band = 1530 nm, or 196078 GHz
- Assume lower band = 1550 nm, or 193548 GHz

and therefore a bandwidth of 196078 − 193548 = 2530 GHz.

Example 9.3

If the SNR at an optical fibre receiver is 9 dB, what is the Shannon capacity if the bandwidth is 2530 GHz?

Answer

Capacity = 2530 x $\log_2$ (1+8) = 8.0 Tbps.
Clearly this is a very large transmission capacity. In general, optical fibre systems can be developed to satisfy any broadband connectivity transmission rates.

Table 9.2   Several frequency bandwidths and wavelengths

Nominal carrier frequency (GHz)	Nominal central wavelength (nm)
195 650	1532.3
195 700	1531.9
195 750	1531.5
195 800	1531.1
195 850	1530.7

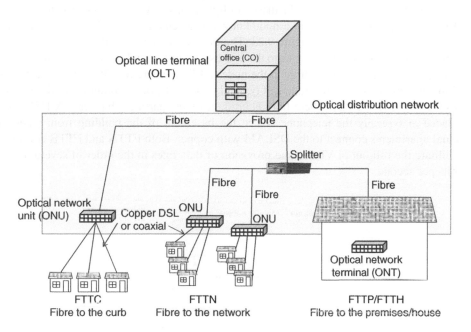

Figure 9.22   A passive optical network architecture

In practice only a portion of the bandwidth is used as the light sources can only use a limited bandwidth. ITU-T specifies typical carrier frequencies in the 1500 nm range, with bandwidths of 12.5, 25, 50 and 100 GHz. Table 9.2 shows several central carrier frequencies for 50 GHz bandwidth channels [17]. At times an optical fibre operator may use several carrier frequencies by emitting at a number of light sources with different wavelength to multiplex more data streams and transmit at higher speeds. This technique is known as wavelength division multiplexing (WDM).

In the context of broadband communications, optical fibres may be used in isolation or in combination with copper-based xDSL or coaxial cable systems. These together are referred to as a passive optical network (PON), and is illustrated in Figure 9.22. The system is called passive since data are broadcast to all users in the downstream, and only detected by intended users based on IP addressing and encryption. An optical line terminal (OLT) transmits signals to all optical network units (ONUs), which forward the downstream packets to all subscribers within their networks. A CO is connected to a subscriber's premises optical network terminal

(ONT) either directly by an optical fibre through a splitter, or via a DSL/coaxial cable. The upstream signals are communicated in a TDM fashion through scheduling by the ONT, which allocates a time window to each ONU. An ONU acts as a DSLAM, buffering and aggregating the traffic from subscriber premises.

Fibre-to-the-x

As noted above, DSL technologies are effective only for distances of up to 4 km due to copper medium characteristics. One way to increase the range of DSL coverage is by incorporating optical fibre into the DSL system. In these configurations, a CO is connected to a remote DSLAM using optical fibre lines in a node known as a remote terminal (RT). A number of configurations exist as shown in Figure 9.23. The difference between all these configurations is essentially how close the fibre comes to the end-user premises and remote terminal. The two most common configurations are fibre to the node (FTTN), where the DSLAM is located at a 'node' in a neighbourhood with a radius of typically 1 km. The other common configuration is fibre to the building (FTTB), where fibre is brought to an apartment building. A DSLAM is then placed at typically the telephone box at the basement of the building from where all individual apartments connect to the DSLAM with copper. Both FTTN and FTTB configuration facilitate the roll-out of VDSL, and provision of data rates in the order of several tens of megabits per second.

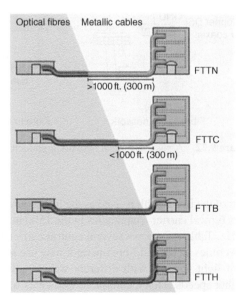

Figure 9.23   Fibre-to-the-x configurations. FTTN, fibre to the node; FTTC, fibre to the cabinet; FTTB, fibre to the building; FTTH, fibre to the home [18]. Source: Riick (own work) [GFDL (http://www.gnu.org/copyleft/fdl.html) or CC-BY-SA-3.0 (https://commons.wikimedia.org/wiki/File%3AFTTX.svg)]

Copper-based media, whether twisted wires or coaxial cables, have long been used to connect customer premises to sources of information. The rise of the internet and a need for connectivity from home and office spurred research on telecommunications systems which use this existing copper infrastructure. DSL and hybrid fibre cable systems use a combination of optical fibre and copper media to provide connectivity rates in the order of 100 Mbps or more. Additionally all-fibre systems promise 1 Gbps connectivity. Fixed broadband telecommunications system are capable of satisfying typical connectivity needs.

Review Questions

1. In ADSL:
   a. What medium is used to transmit signals in ADSL?
   b. What is the frequency spectrum range?
   c. What is the role of the splitter?
   d. What is the role of the DSLAM?
2. What is the maximum downlink throughput of an ADSL system assuming the system uses 256 subcarriers, each with an effective bandwidth of 4 kHz, with received SNR = 12 dB, 10 dB, 8 dB, and 6 dB for 64 subcarriers each?
3. If 10% of resources are used for uplink transmissions, which carriers are used? What would be the uplink maximum capacity?
4. What is the difference between ADSL2+ and VDSL systems?
5. What are FEXT and NEXT?
6. How does 'vectoring' reduce the FEXT? What is the resulting benefit?
7. How does the discrete multi-tone (DMT) technique help with the frequency response model of the twisted copper wires?
8. Draw an architecture model for a hybrid fibre cable system. Why is optical fibre used here?
9. What is WDM and how is it valuable?
10. What determines the capacity of an optical fibre strand?
11. Optical fibre communications use a special form of binary ASK, where the presence of a light pulse in the transmission period represents a 1 and lack thereof a 0. Why do you think multi-level ASK is not used?
12. List four different fibre-to-the-x systems. How do optical fibre systems complement copper-based DSL systems?
13. What are the typical nodes in a fibre-to-the-premises (FTTP) network?
14. What are the main costs in rolling out an FTTP network?
15. What are the advantages of an FTTP or an FTTN system?
16. When can the roll out of a national fixed broadband network be left to the private sector? When should it be taken over by the government?
17. What are the policy considerations in broadband service provision?
18. What are the implications of monopoly broadband telecommunications infrastructure roll-out? What are the relative policy and business advantages and disadvantages?
19. How can two hybrid fibre cable (HFC) service providers compete?
20. Compare FTTH, FTTN and HFC using the technology–business–policy framework.

References

[1] https://commons.wikimedia.org/wiki/File:DEC_VT100_terminal.jpg, accessed 21 August 2015.

[2] Flood, J.E. (ed.) (1997) *Telecommunication Networks*, IEE Publishers.

[3] Lechleider, J.W. (1991) High bit rate digital subscriber lines: a review of HDSL progress. *IEEE Journal on Selected Areas in Communications*.

[4] http://en.wikipedia.org/wiki/Acoustic_coupler#mediaviewer/File:Acoustic_coupler_20041015_175456_1.jpg, accessed 23 August 2015.

[5] http://www.sce.carleton.ca/courses/sysc-4700/w14/SYSC4700-W14-LectureNotes/10-TransMedia-HYanikomeroglu-06Feb2014.ppt, accessed 23 August 2014.

[6] https://www.princeton.edu/~achaney/tmve/wiki100k/docs/T-carrier.html, accessed 23 August 2015.

[7] Almeida, H. (2009) VDSL2: Taking the Wire to the Limit. Ericsson Review.

[8] http://www.ikanos.com/press-releases/ikanos-demonstrates-300mbps-throughput-at-200-meters-on-a-single-pair-copper-line-further-raising-the-vdsl-rate-reach-performance-bar/, accessed 23 August 2015.

[9] http://www.intercomms.net/AUG03/content/utstarcom.php, accessed 23 August 2015.

[10] http://www.itmedia.co.jp/broadband/0303/31/lp22.html, accessed 23 August 2015.

[11] http://bb.watch.impress.co.jp/cda/news/2213.html, accessed 23 August 2015.

[12] http://www.marketingcharts.com/television/snl-kagan-cable-subscription-data-contradicts-fcc-chairman-kevin-martin-2634/snl-kagan-cable-summary-data-2006jpg/, accessed 23 August 2015.

[13] http://broadbandnow.com/Comcast, accessed 23 August 2015.

[14] http://cmcsk.com/releasedetail.cfm?ReleaseID=781496, accessed 23 August 2015.

[15] http://www.cnet.com/news/netflixs-hastings-makes-the-case-for-net-neutrality/, accessed 23 August 2015.

[16] https://commons.wikimedia.org/wiki/File:Singlemode_fibre_structure.svg, accessed 23 August 2015.

[17] ITU. ITU-T G.694.1 Standard Specification: Spectral Grids for WDM Applications: DWDM Frequency Grid.

[18] https://commons.wikimedia.org/wiki/File:FTTX.svg, accessed 23 August 2015.

10

Terrestrial Broadband Wireless Telecommunications

Preview Questions

- How can a limited frequency resource be shared by a large subscriber base?
- What has characterised 1G → 2G → 3G → 4G transitions?
- What is the role of a base station in a packet-switched network
- How do multiple mobile users connect to a base station using OFDMA?
- How does MIMO technology work?

Learning Objectives

- Characteristics of terrestrial wireless channels
- Frequencies allocated for broadband services and reusing of this spectrum
- Multiple access systems
- Different generation of mobile communications standards
- Circuit and packet switching in wireless networks

Historical Note

Whereas wireline communications followed the invention of electricity by Alessandro Volta, wireless communications are based on the discoveries of a number of scientists including Hans Christian Ørsted, Michael Faraday, James Maxwell, and Heinrich Hertz

Broadband Telecommunications Technologies and Management, First Edition. Riaz Esmailzadeh.
© 2016 Riaz Esmailzadeh. Published 2016 by John Wiley & Sons, Ltd.
Companion Website: www.wiley.com/go/BTTM

(pictured)[1] on the propagation of electromagnetic waves in the air. In particular Hertz' experiments, first reported in 1888, with equipment to generate, transmit and receive electromagnetic waves were essential to the invention of radio telegraphy in the late 19th century. He developed a device to produce an electromagnetic wave, means to transmit and receive the wave, and a means to detect the received wave, as shown in Figure 10.1. This equipment essentially facilitated the same processes of present day wireless telecommunications as used for example in mobile phones.

Besides Hertz' experiments a number of further developments were necessary before 'information' could be transmitted over electromagnetic waves. These included devices to generate signals with sufficient energy for transmission over long distances, means to modulate these signals with information, efficient antennas for transmission and reception, and means for detection of the information.

The clear candidate system for information transfer was the telegraph. However suitable antennas and transceivers for long-distance telecommunications were still to be developed. A number of scientists, including Ernest Rutherford (the father of nuclear physics) and Nikola Tesla (the inventor of alternating current and the electric motor), worked on wireless telegraphy based on Hertz' experiments. However the first successful demonstration of transmission and reception of Morse codes was carried out by a 20-year-old inventor, Guglielmo Marconi (pictured),[2] in 1894 in Bologna, Italy.

Figure 10.1 Transmission equipment based on Hertz' experiment. Source: Physikalische Apparate, Preisliste No. 18 (1904) Ferdinand Ernecke, Berlin, Germany, p.304, fig. 8800 in Instruments for Science collection, Smithsonian Institution. https://commons.wikimedia.org/wiki/File%3AEarly_Lecher_line.png

---

[1] Robert Krewaldt [public domain or CC BY 4.0 (https://commons.wikimedia.org/wiki/File%3AHeinrich_Rudolf_Hertz.jpg)].

[2] Published on LIFE. https://commons.wikimedia.org/wiki/File%3AGuglielmo_Marconi_1901_wireless_signal.jpg.

Marconi learned about electromagnetic wave propagation from a teacher/neighbour, Augusto Righi, who replicated Hertz' experiments. Marconi enhanced the experimental set-up using an improved antenna and a more sensitive electromagnetic wave receiver, and using these, managed to transmit Morse signals over a distance of 2.4 km in 1896.

Similar to Morse and Bell, Marconi's status as the inventor of 'radio' owes much to his acumen as a businessman. His invention was not well received in his native, Italy: the Italian Ministry of Post apparently did not consider his invention significant and capable of replacing the electrical telegraph. Perhaps the ministry was concerned about the then small range of transmission of Marconi's wireless telegraph, or comparatively satisfied with their present wired telegraph. Regardless, the ministry's refusal did not daunt Marconi, and he quickly took his invention to the UK. (He may not have been the first to leave his native land in search of better economic prospects elsewhere, but he is one of the most prominent.) It was in the UK that he established his company to commercialise radio telegraphy. His Wireless Telegraph & Signal Company, later known as the Marconi Company, established in 1897, was so successful that by 1900 his devices were being used on ships, and a permanent wireless telegraph system was installed linking two Hawaiian Islands. While the ownership of the Marconi Company has changed many times since its inception, it existed as an on-going firm more than a century later until it was acquired in 2006 [1].

A Wireless Telecommunications System

A general block diagram of a telecommunications system is shown in Figure 10.2. A large degree of commonality exists in technologies developed to realise fixed and wireless telecommunications. There also exist a number of differences due to the medium over which electromagnetic signals are sent and received. These differences mainly concern the modulation techniques and means for sending and receiving electromagnetic waves. Some modulation techniques are more suited to transmission over copper wires, optical fibres or air. While a

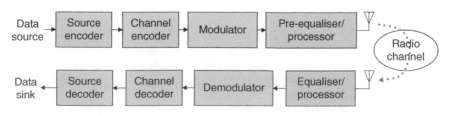

Figure 10.2   A simple block diagram of a wireless communications system

modulation technique may be more superior, cost and complexity considerations may result in selection of a slightly inferior technique for a particular application. The decision generally depends on the required degree of efficiency and cost of implementation.

The physical means for launching an electromagnetic wave into a communications medium differ considerably for wires, air or optical fibres. For example, a twisted pair of copper wires may be directly connected to a telephone; a coaxial cable requires a special connector; an optical signal needs a very accurate set-up for a laser light source to be connected to a fibre. Wireless systems use a number of different types of antennas depending on the application and the frequency of operation. In general an antenna's size and shape are dependent on the required gain and the frequency of operation, and the characteristics of the transmission medium. A number of antennas are shown in Figure 10.3.

Antennas

Telecommunications antennas are devices that convert electrical energy to electromagnetic energy and radiate them and vice versa. An antenna may be designed to radiate energy equally in all directions or to concentrate radiation towards a designed direction. Conversely, a receiver antenna converts radiated electromagnetic energy into an electrical current which may then be processed by the receiver electronics. The size and shape of antennas depend on the applications for which they are used. A directional antenna concentrates energy in a particular direction and therefore has a higher 'gain' compared with an omnidirectional antenna. The antennas in Figure 10.3 are (from left to right) for satellite, TV reception and a mobile communications base station. All these antennas are directional. Antennas are generally characterised by their degree of directionality and the amount of electromagnetic energy they can radiate or collect. The term antenna gain refers to how much energy is radiated, and how concentrated it is in a desired direction compared with an omnidirectional, simple antenna element. For example a typical TV antenna gain is 10 dB and a base station antenna gain is 18 dB. The satellite antenna gain can be in the order of 60 dB or more.

 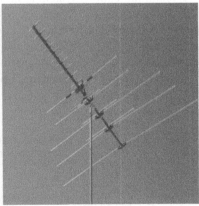 

Figure 10.3 Different antennas (from left to right) for satellite, TV reception and a mobile communications base station. Source: NASA (public domain): https://commons.wikimedia.org/wiki/File%3AC-band_Radar-dish_Antenna.jpg

The signal loss of the medium is a function of frequency, and dependent on whether a line-of-sight exists between the transmitter and the receiver. For example, signal path loss in free space where nothing blocks the path between transmitter and receiver is much smaller compared with a system operating over a terrestrial channel where a number of physical objects may block or reflect a signal.

The block diagram of a wireless digital communications system in Figure 10.2 also shows components necessary for source coding, channel coding of source information. A pre-equaliser/processor block is necessary to compensate for variations of signal as it propagates through the communication medium. The signal energy in the form of electromagnetic waves is then radiated using antennas. The receiver antenna absorbs a portion of the radiated energy and delivers it to equipment which further processes the signal. Processing is required to reduce signal distortions introduced by the channel, before demodulation and decoding processes are performed to extract the transmitted information.

Wireless communications may be point-to-point, such as between the Earth station and a satellite, or point-to-multipoint such as from a satellite to many Earth stations. Communication may also take place in a single direction, called simplex or half duplex (when communication is in one direction at any one particular time). Examples are broadcasting (radio, TV) or walkie-talkies. Alternatively communication may be bidirectional, such as communication between cellular base stations and mobile phones. Generally point to point communication is carried out using directional antennas, and point to multi-point using omnidirectional or sector-directional antennas.

Wireless communications may be carried out using different media and technologies: in the air (such as radio, TV, mobile communications, etc.), space (satellite communications), and even water (sonar, visible light). Usually the frequency of operation for each of these media is different and a function of the special characteristics of the particular medium. For example, communication within water must be at very low frequencies because of the conductive character of water. Similarly space communication can use very high frequencies, since line-of-sight can usually be established between the transmitter and receiver. In the context of broadband telecommunications services, this book focuses on terrestrial fixed-wireless and mobile communications. These systems usually operate within a frequency range of 450 MHz to 5.8 GHz as will be discussed below.

Mobile Terrestrial Telecommunications Systems

Mobile terrestrial telecommunications systems, used in public and private broadband service provision, are generally characterised as a star network, shown in Figure 10.4, where a central fixed device communicates with many moving subscriber devices. The central device may be called a base station or an access point. The subscriber device may be called user equipment or mobile phone/device (cell phone) depending on the wireless standard and functionality.

The central device communicates with subscribers using an allocated frequency spectrum. Some standards operate within a licensed band, where the operating frequency band is allocated by the government through a specific process, such as an auction. Some systems operate within an unlicensed band, where all devices need to comply with government regulations which may limit transmission power levels.

In contrast to fixed (wired) communications where usually a dedicated link exists between a transmitter and a receiver, terrestrial wireless communications are carried out in the 'open'.

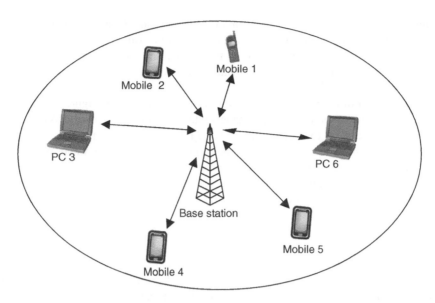

Figure 10.4   A wireless star network

Any such link, whether a frequency channel or a time slot is a 'public' resource and is shared between a large number of end-users. A big challenge of wireless communications is how to use this limited resource in an efficient way.

Duplex Modes

Most wireless telecommunications systems operate in a full duplex mode. That is, a channel carries information from a central node to an end-user, and another channel is used for the reverse communications. In this way, both central and subscriber devices transmit and receive data. Usually communications from the central node to the end-user are called downlink, and the reverse uplink, transmissions.

Duplex operation is possible using orthogonality in either frequency or time domains. In the frequency division duplex (FDD) mode two different frequency channels are used for the downlink and uplink transmissions. In contrast, time division duplex (TDD) mode transmissions in the downlink and uplink are carried within the same frequency band, but during different time frame channels. FDD and TDD modes of operation are illustrated for a terrestrial mobile telecommunications system in Figure 10.5.

Frequency Bands

FDD and TDD uplink and downlink transmissions use a number of frequency bands in the 450 MHz to 5.85 GHz range. Many of these frequency bands are internationally used to allow the same devices to be operable in many countries. There also exist frequencies bands which are only used in a particular country/region. Moreover, some frequencies are specified for FDD and some for TDD operation. A list of frequencies used for mobile communications and the region where they are used is given Figure 10.6. In total more

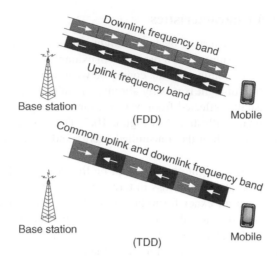

Figure 10.5 FDD and TDD modes

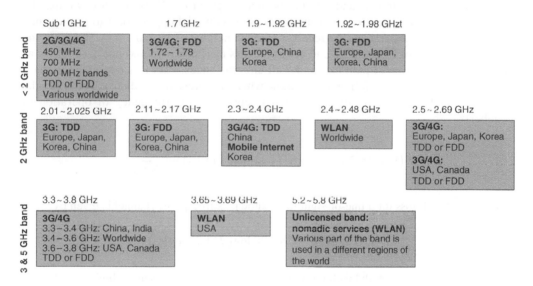

Figure 10.6 Several mobile communications system frequency bands

than 1 GHz of frequency bandwidth has now been allocated to mobile (broadband) communications to satisfy the demand for broadband wireless services at the time of writing (2015).

The frequency bands allocated to mobile and broadband services are suitable as they allow for large possible bandwidths, small size antennas as well as low path loss. In general very little capacity is free in frequency bands below 450 MHz, and frequencies above 3–5 GHz experience very large path loss, making them unsuitable for long-distance terrestrial communications as will be explained below. Frequency allocation to different mobile operators therefore follows strict business and policy guidelines to ensure this scarce resource is efficiently utilised.

Terrestrial Channel Characteristics

Some radio channels are characterised by line-of-sight signal propagation, that is, nothing blocks the path between a transmitter and a receiver. An example of such a channel is satellite communications links where nothing but air comes between the satellite and Earth station antennas. In contrast, a line-of-sight is usually absent in mobile terrestrial communications: transmitted signals are usually reflected from or travel through physical objects before they arrive at a receiver antenna, as illustrated in Figure 10.7. The physical objects blocking or reflecting the signal absorb much of the transmitted energy and therefore the received signal is significantly weakened.

While signal loss for line-of-sight channels can be theoretically derived from the distance and frequency of operation, path loss in terrestrial channels is a function of terrain and building topology, and an exact formula cannot be derived. Instead, a number of empirical models have been developed based on path loss measurements in a real physical environment. These models calculate path loss as a function of frequency of operation, the height of transmitter and receiver antennas and the type of environment (dense urban, urban, suburban or rural). A broad summary of these models can be found in electromagnetic wave propagation textbooks and references such as given in [2]. In this text we use the simplified extended Hata model for urban areas. This model applies best to a frequency range of 2–3 GHz, and transmitter/receiver antenna heights of 30 and 1.5 m. We apply the model for a range of frequencies and environments as illustrative and not necessarily as an accurate calculation of path loss under all circumstances. Path loss expressed in decibels is calculated from the following formula, where d represents the distance between the transmitter and receiver in kilometres, and f represents the frequency of operation in megahertz.

$$L\left(\text{Hata}\right) \cong 26 + 35 \log\left(d\right) + 34 \log\left(f\right) + 10 \log\left(\frac{f}{2000}\right)$$

In contrast path loss for a line-of-sight (also known as freespace) channel is given by:

$$L\left(\text{FreeSpace}\right) = 32.4 + 20 \log\left(d\right) + 20 \log\left(f\right)$$

As might be expected, path loss for a similar propagation distance is much higher in a typical mobile terrestrial compared with line-of-sight channels. Example 10.1 illustrates this.

Figure 10.7   Signal propagation in terrestrial channels

Signal Fading

In contrast to line-of-sight satellite channels which are largely time-invariant, terrestrial communications channels are time-variant. That is, the received power level changes rapidly over time. This phenomenon is referred to as signal fading, as illustrated below.

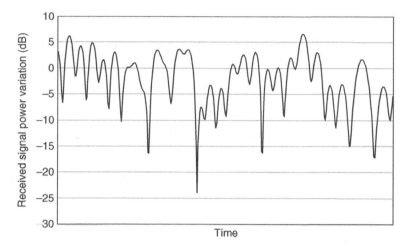

Fading occurs because radio communications channels are characterised by signal arrival by multiple paths. A received signal is usually composed of many reflections of the transmitted signal – reflections from natural and man-made objects such as mountains, buildings, cars, and the ground, as illustrated below.

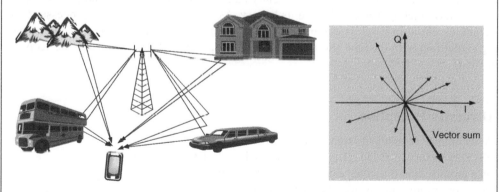

Each received signal ray travels through a different path and has a different propagation delay time, amplitude and phase. As a result each arrives with a certain amplitude and phase as a function of distance travelled and efficiency of reflection. When combined, the sum of the received signal rays (vectors) may combine constructively or destructively, resulting in the signal power characteristics shown above. The range of received power variation can be as much as 60 dB.

Fading can cause significant degradation of system performance: for example a system operating in a fading environment requires nearly 100 times more transmission power to achieve the same quality of communications compared with a non-fading environment.

Example 10.1

Compare the path loss between a line-of-sight and terrestrial channel over a distance of 1 km at the operating frequency band of 2.3 GHz.

Answer

- Free space:

$$L(\text{FreeSpace}) = 32.4 + 20\log(d) + 20\log(f)$$

- $L = 99.6$ dB
- Urban (Hata):

$$L(\text{Hata}) \cong 26 + 35\log(d) + 34\log(f) + 10\log\left(\frac{f}{2000}\right)$$

- $L = 140.9$ dB (higher than line-of-sight by a factor of more than 13 000 times, over a distance of only 1 km)

The large path loss presents two problems to the telecommunications technologist. One is the required strong transmission power for compensating the path loss, which may not be practical or even possible in small hand-held devices. This limits the transmission range and network coverage. The other is the variability of path loss as a function of distance. A device must adjust its transmission power as it moves nearer to or farther from a receiver, to ensure that (1) the received power level is sufficiently large for correct detection and (2) it does not transmit too much power and deplete its battery. As a result, power control processes have become essential in wireless communications. Power control is also important in combatting signal fading in the time and frequency domains, topics which are outside the scope of this text.

Diversity Reception

A technique to mitigate signal fading is through diversity combining. Diversity combining is based on a premise that if a signal can be received through two (or more) independently faded paths the combined signal is less faded as a trough in one path is compensated by a possible peak in another. Independent reception of signals may be accomplished through several means, including employing 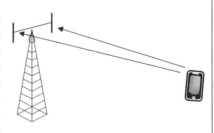 multiple receiver antennas. Because the probability that all paths may fade at the same time is significantly less than the probability that a single path fades, the overall combined signal experiences less fading. This is illustrated below, where the stronger of two fading paths is selected, and as a result the combined signal has fewer instances of signal fading.

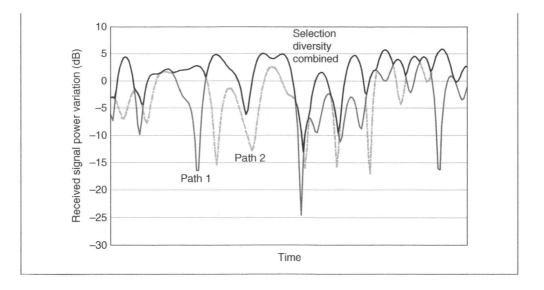

Case Study 10.1: Indoor Office Wireless Communications Systems

The 1980s witnessed the growth of office automation and usage of desktop computers, on-line printers and databases. However most offices were not wired for the necessary local area networking infrastructure and equipment. While most offices had been wired for internal and external telephony, the new data communications infrastructure required significant investment in cables and wiring and reconfiguration of office layout. This motivated research into wireless local area networking.

The wireless medium of indoor offices is characterised by signal fading because of the presence of many objects such as chairs, desks, partitions, furniture, false walls and so on. Moreover, office workers both reflect and absorb electromagnetic waves. The resulting signal fading required very high received signal to noise ratio to maintain an acceptable communications quality, making it impractical to use existing conventional wireless data communication techniques.

A number of systems were developed towards the end of the 1980s to facilitate wireless local area networking. WaveLAN, an early system developed by NCR and later on Lucent, used a bandwidth of 26 MHz in a 900 MHz unlicensed band for a 2 megabits per second (Mbps) maximum transmission rate in air. As the fading channel causes the carrier phase to vary during transmission, a differential modulation technique was used. Another system was Motorola's Altair which operated in the 18 GHz band. Both systems used the channel sense multiple access (CSMA) technique to facilitate sharing of communications resources. A test by PC Magazine in 1992 showed that both systems could provide throughput rates of approximately 500 kbps with several active users, which compared well with faster Ethernet transmission rates of approximately three times. Test reports from 1999 by Network World show WaveLAN transmission rates of 1.15 Mbps, an improvement over the 1992 results [3]. While these systems realised wireless indoor data communications, clearly usage of the available spectrum was inefficient (Table 10.1).

Table 10.1   Wired and wireless local area throughput figures (in megabits per second) in 1992 [4]

System	1 User	2 Users	3 Users	4 Users
Ethernet	1.44	1.29	1.06	0.9
WaveLAN	0.51	0.46	0.43	0.39
Altair	0.47	0.41	0.34	0.26

    Wireless local area network (LAN) standards have been developed under the auspices of the IEEE since the early 1990s and under the 802.11 series. Many technological advancements such as higher processing power devices, more advanced antennas and modulation techniques have helped increase transmission rates to several hundred megabits per second. While Ethernet (wired local area networking) is still common, wireless office communications are largely replacing the wires in offices.

Case Study Questions

- What were the business drivers behind the research and development of wireless systems for local area networking?
- What characterises the indoor office transmission medium? Is it conducive to high quality communications?
- The frequencies band used for wireless LAN systems are generally unlicensed? Why?
- Why is the frequency usage efficiency so low in early wireless LAN systems?
- Why is CSMA used for resource sharing and media access?

Generations of Mobile Communications Systems

Mobile communications technologies have been broadly classified into a number of 'generations'. Four generations are widely recognised in the industry and broadly defined as follows:

- First generation (1G) systems are analogue circuit-switched systems, and were used primarily for voice telecommunications, although minimal circuit-switched data communications capability also existed.
- Second generation (2G) systems are digital circuit-switched systems used primarily for voice telecommunications. Significant data communications capability, both on circuit-switched and minimal packet-switched systems was realised. Later generations of 2G systems standardised the capability of fully packet-switched communications (e.g. GPRS and EDGE).
- Third generation (3G) systems are digital circuit and packet-switched systems used for voice and data communications with variable transmission rate capabilities.
- Fourth generation (4G) systems are broadly, all-internet protocol (all-IP) end-to-end systems from a wireless mobile device all the way through the base station and core network to the internet. They are fully packet-switched systems with voice communications also carried out using packet-switching protocols. The network architecture and nodes which support end-to-end connectivity are also based on IP systems.

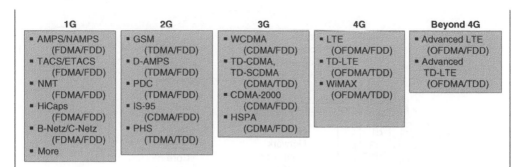

Figure 10.8 Successive generations of mobile telecommunications systems

The generational transitions are generally characterised as follows:

- 1G to 2G evolution with the shift from analogue to digital
- 2G to 3G evolution with the introduction of data communications, higher transmission rates, and packet switching
- 3G to 4G evolution with the provision of mobile broadband services, full packet-switching technologies, and advanced network architecture.

A number of standards broadly associated with these four generations are shown in Figure 10.8. Some of these standards did not succeed in the marketplace and were not widely rolled out. Some standards evolved across generations and remain in operation, while evolution of some standards failed to succeed. All 1G, most of the 2G and a number of 3G systems have been phased out, and 4G systems are now being rolled out globally. A number of duplexing and multiple access technologies were used in the four generations to date and beyond. These will be discussed in detail below. The process of standardisation for the fifth generation of mobile communications systems (5G) has started, with a target roll-out data of 2020 [5].

Nodes in a Mobile Telecommunications System

A number of networking devices are necessary to connect a mobile device to the public telecommunications system and to other mobile devices. The network generally consists of the mobile device (or user equipment – UE as known in some 3G/4G standards), base stations (or 'node-B' as known in some 3G/4G standards) and a central network, also known as the core network, which controls the base stations and interfaces the fixed telecommunications network as shown in Figure 10.9.

The mobile device (user equipment) is essentially a device that transmits and receives signals through the air using electromagnetic waves. A mobile phone is made of electronic circuitry, control systems, a battery, a transmitter/receiver (transceiver) which performs radio signal coding and decoding, a radio antenna system, and a microphone and speaker. Later generations of mobile phones have visual interfaces and data communication capabilities (means to input and view data). Mobile devices communicate with base stations, which are a part of a mobile telecommunications operator network.

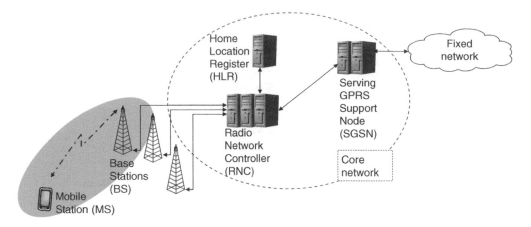

Figure 10.9   A mobile telecommunications network

Figure 10.10   Base station antennas

Base stations are devices that communicate with mobiles and relay the communications to a core network and through that to a fixed network or another base station and another mobile device. Base stations contain transceivers, which send and receive signals to and from mobile phones, and modems that connect to the core network. Base station antennas are usually located on transmission towers or on top of tall buildings, and are a feature of urban and rural landscapes. Figure 10.10 shows some base station antennas. Base station equipment may be placed near the antenna or in housing at the base of the building or tower.

Core networks connect to the mobile telecommunications carrier's fixed telecommunications infrastructure, which allow communications with the mobile phone to be relayed to other parts of the network. A base station connects to a number of nodes that control different

Figure 10.11   Cellular coverage

functions associated with mobile communications. These functions may include allocation of resources, registering the physical location of a mobile device and billing. Different 1G–4G standards perform these functions at network nodes referred to as the core network. Different terminologies are used in different standards, from base station controller (BSC) and mobile switching centre (MSC) in a 1G standard, to radio network controller (RNC) in several 2G and 3G standards to enhanced Node-B in a 4G standard. The functionalities are very similar: a BSC determined which mobile phones were allowed to communicate with a base station and an MSC relayed the voice traffic onward to the fixed telecommunications network. Similar functions are carried out by RNCs or enhanced Node-Bs.

A base station transmit signals to (and receives from) mobile devices within a coverage area known as a 'cell'. The radius of this area is determined by the maximum range of transmission as found through a link budget analysis as well as the amount of traffic generated to/from mobile devices. Based on this a 'cellular' network is constructed to cover entire populated areas. The size of a cell is smaller in densely populated urban areas compared with suburban and rural areas because the offered traffic amount, and path loss is larger. A typical metropolitan area may be covered by cells with a radius of ~500 m in a central business district area, and 3–4 km in outer suburbs. A typical configuration may appear as shown in Figure 10.11. Each cell has a base station which serves the mobiles that roam within its borders.

Resource Allocation

The frequency spectrum is a limited resource and is used for a number of services, from public emergency, police and military wireless communications applications, to radio, TV, and amateur radio and so on. Governments control the allocation of the spectrum within their jurisdiction and ensure that different services do not interfere with each other. Each mobile operator is allocated a certain portion of the spectrum, within which it provides mobile

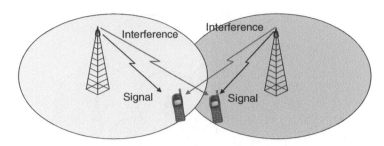

Figure 10.12   Interference to a neighbouring cell by mobiles reusing the same carrier frequency

telecommunications services to its subscribers. This allocated spectrum is a precious resource, for which an operator may need to spend billions of dollars. Frequency usage efficiency is very important.

For example assume a 10+10 MHz of spectrum (downlink and uplink) is allocated to a typical operator to operate a 1G mobile telecommunications system. Further assume each voice channel requires 25~30 kHz to provide an acceptable quality, and therefore only 350~400 channels are possible in a 10 MHz spectrum. These channels have to be reusable if a large portion of the population of a region is to be served.

If two mobile phones were simultaneously to use the same uplink channels (i.e. the same carrier frequencies) for the purpose of conducting a voice call to a base station on a mobile network, the radio signals would interfere with each other. Therefore the same frequency carrier may not be used in the same cell. Similarly the same carrier frequency may not be used in a neighbouring cell as the level of interference may still be too large. An example is shown in Figure 10.12 where two users at cell borders communicate with one base station while interfering with the other. This problem was especially acute in the 1G standards which used analogue frequency modulation radio technology for communications between mobiles and base stations. Interference in these systems manifested itself as crosstalk: a conversation could be overheard by anyone using the same carrier frequency. This meant that 1G analogue mobile telecommunications systems had to be designed to ensure that interference between users was below a designed level.

It follows that a certain distance needs to be kept between two base stations which use the same carrier frequency. In the 1G, some 2G and to some extent 4G standards each cell is allocated a subset of all available carriers. A neighbouring cell is provided with a different (and mutually exclusive) set of carriers, and so on before the cell sites are sufficiently apart so the same subset of carriers can be reused. Frequency reuse factor is defined as the number of cells which do not use the same carrier frequency. A frequency reuse factor of 7 was used in 1G standards as illustrated in Figure 10.13. This reuse regime resulted in a signal to interference ratio of approximately 30 dB, which was deemed sufficient for analogue voice communications.

As noted above, cell size is determined by two factors. One factor is the maximum distance a signal may travel before its power falls below a certain threshold. Such calculations use path loss characteristics of the channel, which depend on the geographical as well as the built structure of the cell area. A densely urban cell has a higher path loss as a function of distance, whereas a rural cell may be mostly characterised by line-of-sight communications. The other factor is the amount of expected traffic a user may generate and the user density in an area. Since the number of communications channels is limited, the number of users that may be served at any particular time is also limited. The cell area therefore must be designed with this

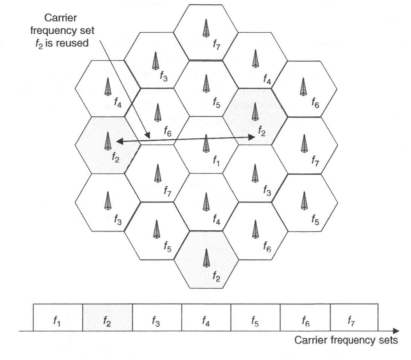

Figure 10.13   Frequency reuse factor of 7 for 1G systems

limitation in mind. If a cell size is too large, the number of subscribers will be too many, and therefore connectivity demand on the base station will be larger than the capacity of the base station leading to unacceptable grades of service.

Frequency Division Multiple Access

Given a certain channel capacity in each cell a number of users may be connected to a base station. Several multiplexing technologies were developed over the first to fourth generations of mobile communications standards to connect multiple devices to a base station.

The first generation of mobile telecommunications systems used a multiplexing technology known as Frequency Division Multiple Access (FDMA). In this technology individual user resource allocation is achieved by using different carrier frequency channels. In FDMA a mobile network assigns to a mobile phone a 'channel' within an allocated frequency band for the duration of its voice call. For example, a 1G mobile network may use its allocated 10 MHz of radio spectrum and offer 400 voice channels each with a 25 kHz bandwidth. With a frequency reuse factor of 7, 57 voice channels will be available in each cell in this system. An FDMA system, serving these 57 mobiles in one cell is illustrated in Figure 10.14.

Using FDMA technology, each mobile phone active on the network in a particular geographic area is assigned a downlink frequency channel and an uplink frequency channel which it exclusively uses for the duration of its call. After the call is ended, the channel is released and may then be assigned to another mobile for a new call.

Example 10.2

An example of an FDMA system is the Advanced Mobile Phone System (AMPS) standard. In AMPS, an operator's available bandwidth is divided into several parts for frequency reuse, and then each part is subdivided into frequency bands of 30 kHz, each of which is then used for one connection between a mobile user and a base station. One such AMPS operator has been allocated 8.4 MHz of spectrum (4.2 MHz for downlink and 4.2 MHz for uplink). How many voice channels can be offered in each cell if the frequency reuse factor is 7?

Answer

The maximum number of users which can be simultaneously connected to an AMPS system in each cell is calculated to be 20. AMPS system's frequency utilisation efficiency in terms of voice channels per megahertz of bandwidth per cell is 20/4.2 ≈ 4.8 voice channels/MHz/cell. This is illustrated in Figure 10.15.

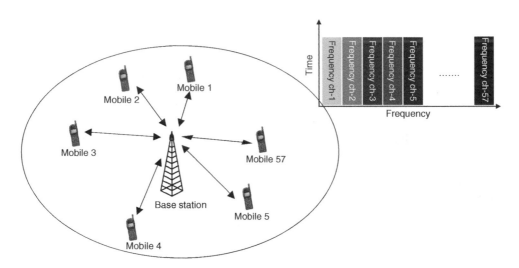

Figure 10.14   FDMA as used in a typical 1G system

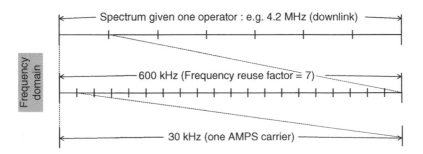

Figure 10.15   Resource allocation in an AMPS system

Table 10.2 1G standards

Standard	Countries/region
NMT	Nordic countries, Switzerland, Holland, Eastern Europe
AMPS	North and South America, Australia
TACS	UK
C-450	West Germany, Portugal, South Africa
Radiocom 2000	France
RTMI	Italy
TZ-801, 802, 803	Japan (NTT)
JTACS	Japan (DDI)

Circuit-switching technology was used in 1G mobile systems. An exclusive end-to-end connection between a mobile and a base station and beyond to the telephony network was established and maintained for each call. That is, the radio wave spectrum resources required for a call (e.g. a 25 kHz voice channel) and the public telephony fixed network links, were assigned at the beginning of the call and maintained until the call ended.

A number of 1G FDMA system were developed in the late 1970s and early 1980s. A list of these standards is given in Table 10.2. Most major telecommunications manufacturers independently developed analogue FDMA systems, and marketed them to their operator clients. Because of this independent development, little to no interoperability existed between different standards, and therefore no roaming was possible between different operators and countries. However mobile phone services were deemed a success, and policy makers, operators and manufacturers worked together to develop a new generation of mobile communications standards with a more universal reach. Digital technologies were deemed as essential in achieving a higher degree of efficiency.

Time Division Multiple Access

The growing popularity of mobile telephony, and an increasing need for higher efficiency and quality of wireless communications services by using digital technologies led to the development of a number of new standards in the 1980s. Led by national and regional regulators, many manufacturers and operators were brought together to develop more universal standards. These standards are generally classified as the second generation or 2G.

Two of these standards were based on Time Division Multiple Access (TDMA) technology. One is the Global System for Mobile communications (GSM), and the other is the Digital Advanced Mobile Phone System (D-AMPS). GSM was developed by the European Telecommunications Standards Institute (ETSI) under the auspices of the European Community as a pan-European standard. However, global prominence of several European manufacturers such as Ericsson, Siemens and Alcatel meant that GSM was marketed to operators outside Europe as well and as a result grew to become a global standard. The GSM standardisation process commenced in the early 1980s, the first set of GSM standards was published in 1990, and the first network commenced operation in 1991.

The other TDMA standard, D-AMPS, was developed in North America by the Electronic Industries Alliance (EIA) and the Telecommunications Industry Association (TIA), two

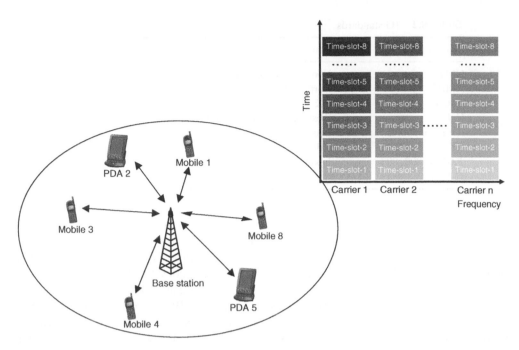

Figure 10.16   A 2G GSM resource allocation

US-based organisations – and is also known by its standard numbers IS-54 and IS-154. The D-AMPS standard was closely based on the Analogue AMPS standard which used 30 kHz for each voice channel. The change from analogue to digital allowed using the 30 kHz channel for 3 users on a TDMA basis, thereby increasing system efficiency by three times. The standards development occurred in the late 1980s, and the relatively minimal evolution meant that a smooth roll-out was carried out in the early 1990s.

As noted above, both GSM and D-AMPS use TDMA technology to enable connectivity between several mobile devices and a base station. In TDMA, the same frequency channel is shared among several mobile devices on a time slot basis. That is, mobile devices take turns to use a frequency resource one at a time. In this way, the mobile devices do not interfere with each other despite using the same carrier frequency. Note that FDMA operation still exists as the allocated bandwidth is divided into a number of carriers as specified by the standard. The operation of a GSM system based on TDMA and FDMA is illustrated in Figure 10.16.

As noted, GSM uses TDMA in combination with FDMA in that several voice channels are supported on the same frequency carrier (TDMA) and a number of carriers supporting different voice channel clusters. The GSM standard specifies that each carrier frequency has a bandwidth of 200 kHz, which is then shared among 8 users, each of which is assigned one of eight independent time slots every 577 μs. Figure 10.17 shows how a voice channel resource is allocated to an active subscriber. Effectively a voice channel is given a one-eighth share of a 200 kHz carrier on a time-shared basis. Each cell may be assigned several such carrier frequencies, the number of which determines the total number of available voice channels in the cell. To increase the number of users that can simultaneously communicate with the base station in the cell, it is necessary to increase the number of assigned carrier frequencies.

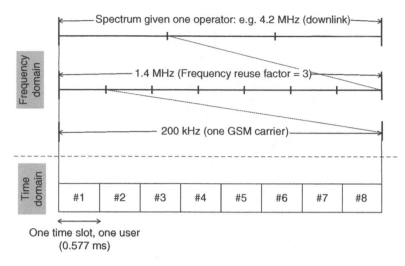

Figure 10.17   Resource allocation in a GSM system

An important feature of 2G systems was usage of digital technologies. Voice was digitised using several different technologies. GSM uses a voice coder with a transmission rate of 12.2 kbps. The data bits thus produced were packed in frames at intervals of 20 ms. GSM was designed to tolerate a 1% frame error rate, that is, of every 100 voice frames with duration of 20 ms that are transmitted, one frame may be lost without significant impact on the voice quality. To achieve this 1% frame error rate, a typical signal to interference plus noise ratio (SINR) in the order of 7 dB is needed. This compared well with analogue transmission, which necessitated a SINR in the order of 30 dB.

A result of this reduction of required SINR was that the same carrier frequency could be reused in cells that were geographically closer to each other compared with analogue communications. In other words, GSM is much more tolerant of inter-cell interference compared with analogue communications. Whereas the same carrier frequency could be reused at every seventh cell in analogue 1G systems, the same carrier frequency was reused up to every third cell in GSM communications, as shown in Figure 10.18. This increases the number of voice channels that can be accommodated within a cell on a mobile network using the same bandwidth.

Example 10.3

A GSM operator has been allocated 8.4 MHz of spectrum (4.2 MHz for downlink and 4.2 MHz for uplink). How many voice channels can be offered in each cell if the frequency reuse factor is 3?

Total available bandwidth	8.4 MHz (4.2+4.2 up & down)
Re-use factor	3
Single carrier bandwidth	200 kHz
Voice channels per carrier	8

Answer

The maximum number of users which can be simultaneously connected to a GSM system in each cell is calculated to be 56. The frequency utilisation efficiency is equal to 13.33 voice channels per megahertz of bandwidth per cell. GSM resource allocation is illustrated in Figure 10.17.

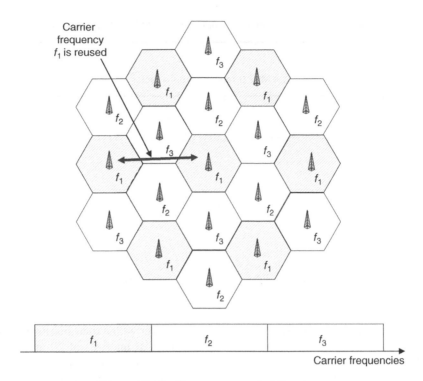

Figure 10.18   Frequency reuse in 2G systems

Table 10.3   2G standards

Standard	Countries/region
GSM	Pan-Euopean and then worldwide
D-AMPS	North and South America, Australia
PDC	Japan
PHS	Japan
CDMA	Worldwide

GSM was by far the most successful among the TDMA-based 2G technologies, and was rolled out in 212 countries and territories as reported by GSM World Statistics [6]. In comparison the D-AMPS standard was not so successful and was primarily rolled out in North and South America. Most D-AMPS operators completely migrated to GSM and its 3G evolution in the late 2000s.

In addition to GSM and D-AMPS two TDMA standards were developed and commercialised in Japan: Personal Digital Cellular (PDC) and Personal Handy-phone System (PHS). These remained mostly confined to the Japanese market and have also been mostly phased out. Table 10.3 lists the 2G TDMA technologies.

Code Division Multiple Access

Another digital voice, and later data, mobile communications system is considered to be a 2G standard. It was developed and standardised in the United States under the auspices of the TIA. It is called Interim standard 95 (IS-95) and used a novel technology known as Code Division Multiple Access (CDMA) and was developed by Qualcomm, Inc., a US company. IS-95 (or cdmaOne) became a successful global system and a direct competitor to other 2G standards such as GSM, D-AMPS and PDC [7].

Similar to FDMA and TDMA, CDMA also enables a base station and multiple number of mobile devices to communicate with each other at the same time. The difference is that in CDMA all users communicate using the same carrier frequency and at the same time. Inter-user interference is however mitigated through usage of a unique code in spreading narrow-band signal to a wideband spectrum. A receiver then detects its desired signal in the presence of the interfering signals using that unique code. Figure 10.19 illustrates how several users are connected using CDMA technology.

A practical analogy to explain CDMA is as follows. With FDMA, one can think of two pairs of users having separate conversations, each pair conducting their separate conversations in different rooms. With TDMA the two pairs of users conduct their separate conversations in the same room but at different times. With CDMA the two pairs of users conduct their separate conversations in the same room at the same time but each pair speak using a different language. As a result, although CDMA users interfere with each other's conversations, each conversation remains comprehensible while the number of users remains sufficiently low.

Because the frequency reuse factor is 1 – the same frequency may be used in all cells – the signal to interference ratio calculation must take all active users into account. That is, in the downlink the interference comes from all active base stations. Similarly in the uplink the

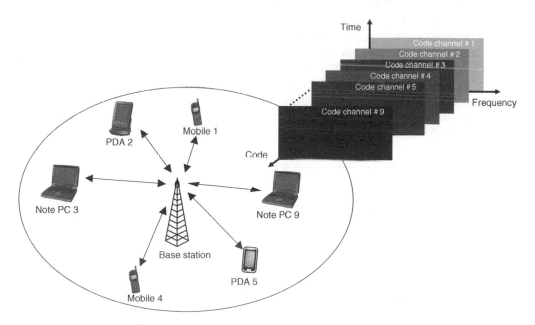

Figure 10.19   Resource allocation in a CDMA system

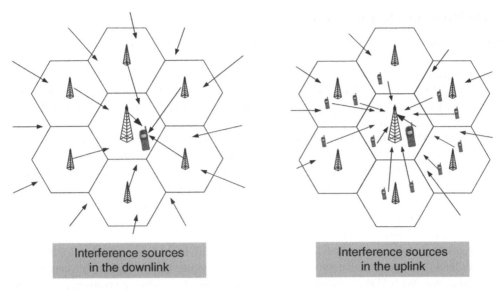

Figure 10.20   Interference sources in the downlink and the uplink of a mobile telecommunications system

interference comes from all active mobile devices. These are illustrated in Figure 10.20. The total sum of these interferences and noise power compared with the desired signal power yield the SINR.

The capacity of an IS-95 system in terms of the number of voice channels is calculated based on the required SINR for a voice channel. Given the length of code, the fact that a user is only actively speaking around 40% of the time, and a required SINR of 7 dB, the total number of simultaneous voice channels may be estimated as around a third of the total code space. For IS-95 standard the number of users per carrier can be estimated at 20–25 per cell. This is a rule of thumb that a third of the code space can be used in a CDMA system considering all the interference from uplink and downlink users in cells near the nearby cells, fading and noise. It follows that the same frequency may be used in all cells (i.e. frequency reuse factor is equal to 1). This simplifies frequency planning significantly compared with both FDMA and TDMA systems [8, 9].

Example 10.4

An IS-95 operator has been allocated 10 MHz of spectrum (5 MHz for downlink and 5 MHz for uplink). How many voice channels can be offered in each cell if the frequency reuse factor is 1?

Answer

Assuming that the allocated 5 MHz spectrum can accommodate 3 IS-95 system carriers, and that 23 active voice users per carrier may be supported, the total number of users is $3 \times 23 = 69$ users, resulting in a 13.8 voice channel/MHz capacity. Note this is similar to capacity figure for the GSM system.

Third Generation: Digital Mobile Communications Systems

The mobile communications standards landscape in the middle of the 1990s was a mix of regional and proprietary standards. GSM had made large in-roads in becoming a world-wide standard. However it was not widely adopted in a number of regions and countries such as the Americas, Japan and Korea. The two Japanese 2G standards had remained confined to Japan and Japanese handsets and handset manufacturers were largely constrained in developing devices for other markets. The IS-95 standard had been adopted by several major operators including those in the US, Korean and Japanese market. However the handsets were not internationally compatible and international roaming was limited. There was a need for a truly non-proprietary international standard to create a global market. Moreover, the emerging growth of the internet and computer communications, as well as nascent electronic commerce required fast wireless data transmission rates. These all gave rise to a global collaboration between a core group of manufacturers, operators and regulatory organisations from several countries to develop a new generation of mobile communications standards.

Wideband CDMA

The core group initially worked on several standard tracks. One was based on TDMA technology as an evolution of GSM to provide faster transmission rates. Another track was based on a combination of TDD and CDMA communication, an idea based on the PhD work of this author and his advisor in the early 1990s [10]. Another track was based on a combination of FDD and CDMA and similar to IS-95 but used a wider bandwidth to facilitate comparatively faster transmission rates. This latter technology came to be known as Wideband CDMA (W-CDMA), and became the prominent standard.

ETSI and ARIB joint efforts led to the development of a W-CDMA standard which was largely based on the GSM core network. This choice of network architecture meant a smooth transition to W-CDMA for GSM operators as a period of co-existence with GSM was possible. In 1998, other standards-setting bodies from Korea, China and the US joined ETSI and ARIB to form the Third Generation Partnership Project (3GPP) and worked to develop a truly global mobile communications standard. The first W-CDMA standard was released in 1999, and the first commercial network was by NTT Docomo in late 2001 (before the start of the 2002 FIFA World Cup).

W-CDMA is based on CDMA technology, and similarly users are distinguished by unique codes. As noted above one requirement for the W-CDMA standard was to have a wider bandwidth (compared with IS-95) to facilitate the required transmission rate of 2 Mbps. This resulted in allocation of 5 MHz bandwidth for W-CDMA frequency carrier, compared with the approximately 1.66 MHz bandwidth of IS-95. Another feature of W-CDMA was usage of different length codes, which facilitated multi-rate data transmissions. This meant that higher transmission rates could be supported by W-CDMA as compared with IS-95. The first release of W-CDMA standard specified transmission rates for uplink and downlink as listed in Table 10.4.

Two initial releases of the W-CDMA standard (Release 99 in 2000 and Release 4 in 2001) were based fully on circuit switching technology. While data communications were supported in the standard, its implementation was inefficient as the data traffic behaviour best fits with packet switching. Subsequent releases of the W-CDMA standard in 2002 and later specified a

Table 10.4   W-CDMA transmission rates for uplink and downlink

Uplink		Downlink	
Spreading factor	Transmission rate (kbps)	Spreading factor	Transmission rate (kbps)
2	1920	2	1920
4	960	4	960
8	480	8	480
16	240	16	240
32	120	32	120
64	60	64	60
128	30	128	30
256	15	256	15
512	7.5		

packet-switched air interface which heralded the age of broadband wireless communications based on the IP.

The W-CDMA standard has been updated regularly through the issue of successive 'releases'. The system of ongoing releases allows for the addition of new features. It also enables mobile communications equipment manufacturers to develop devices that are interoperable with devices made by other manufacturers for previous releases of the standard. Furthermore, each release is amended with consecutive new 'versions' containing relatively minor enhancements and changes to that release of the WCDMA standard. While the 4G standards have been released and rolled out, the 3G standards still form a major part of operators' networks worldwide.

Case Study 10.2: Time Division Synchronous CDMA (TD-SCDMA)

The original CDMA IS-95 standard used the FDD mode for its uplink and downlink channels. In 1991, my PhD supervisor Masao Nakagawa and I proposed a time division duplex mode of a CDMA system in order to improve the system's performance especially with respect to transmission power control to combat channel fading. Our first papers were presented at a national seminar in Japan in August 1991 and the IEEE Vehicular Technology Conference in May 1992. My PhD thesis focused on TDD-CDMA.

As the 3G standardisation process gathered pace in Japan and Europe, both FDD and TDD modes of CDMA were considered. The FDD mode was supported by a large number of contributors including Ericsson and NTT Docomo. The TDD mode on the other hand was supported principally by Panasonic and Siemens in the development of the TD-CDMA standard. In 1997, the China Wireless Telecommunication Standards (CWTS) group in collaboration with Siemens proposed a new CDMA system using the TDD mode using a narrower bandwidth (one-third of wide band CDMA) and therefore lower transmission rate. The standard's main feature was a synchronous uplink in order reduce the uplink co-channel interference through orthogonal code usage. The technology was called Time Division Synchronous CDMA (TD-SCDMA), and was mainly aimed to provide voice communications services and therefore high transmission rates for data communications was not considered a priority. TD-SCDMA was included in the initial release of the 3G standards in 1999.

Over the ensuing decade a number of Chinese operators announced plans to commercialise the TD-SCDMA network and a number of experimental networks were rolled out in several Chinese cities in 2006–2009. China Mobile finally announced a RMB 600 million investment in handset and equipment development for the technology in 2009. In 2013 China Mobile announced that they had invested RMB 21 billion in network roll-out and handset subsidy. The announcement also showed that this investment was also intended for future roll-outs of TD-LTE, a 4G standard.

Meanwhile, the rollout of the 4G mobile telecommunications systems and standardisation of the new TDD technology called TD-LTE has meant that most TD-SCDMA investment is being reconsidered. Both TD-CDMA and TD-SCDMA have failed to gain much traction in the market place. Despite strong support from governments and major companies, the worldwide mobile telecommunications industry is converging to one main set of standards, and it is uneconomical to roll out technologies which do not gain a worldwide following [11, 12].

Case Study Questions

- How is TDD superior to FDD in power control?
- How is an uplink synchronous system valuable?
- Why do you think the focus of TD-SCDMA was voice communications?
- How did the government influence the choice of mobile telecommunications technology in China and why?
- How do you see the importance of Chinese government support in the emergence of the TD LTE standard?

W-CDMA Release 5 – Introduction of HSDPA

The first version of Release 5 of the W-CDMA standard was published in 2002. Release 5 introduced a new downlink data transmission scheme called High Speed Downlink Packet Access (HSDPA). HSDPA enabled increased data transmission speeds, from the base station to the mobile device, to a speed of several megabits per second. This meant that it became practical for mobile devices to perform functions such as downloading large files or streaming video, for example by using YouTube.

If channel conditions are good (e.g. if the SINR is high) then according to Shannon's theorem the channel capacity is high. This means that higher transmission rates may be supported within the same bandwidth. The mechanism used to realise higher transmission rates is the application of higher order modulation techniques, such as 64-QAM (quadrature amplitude modulation) or 16-QAM instead of quadrature phase shift keying (QPSK). As discussed in Chapter 7, higher order modulation techniques can enable more bits per signal constellation point/symbol.

Release 5 of the W-CDMA standard incorporated 16-QAM, which enabled higher data rates for users that have high SINR. A mobile device measures the strength of the signal received from a base station, produces a parameter known as the signal quality indicator (SQI), and reports it back to the base station. Based on this parameter, the base station then decides the appropriate transmission rate for that mobile device. A different way of controlling the transmission rate is through the use of different channel coding schemes. The lower the coding rate (e.g. 1/3 or 1/2), the more protection the coding scheme provides to the transmitted data. However this comes at the expense of lower transmission rates, as more redundant bits are added for the purpose of error correction. If the channel is of high quality, then higher coding rates (e.g. 3/4 or 7/8) are sufficient. This results in higher effective data rates.

The HSDPA W-CDMA standard specifies an adaptive modulation and coding (AMC) rate. Each AMC rate includes a spreading factor, coding rate and modulation method, the combination of which results in a unique transmission rate. The AMC rate is controlled based on information conveyed through the SQI sent by a mobile device.

Most mobile communication technologies until Release 5 of the W-CDMA standard were primarily designed for voice communication. They were based on circuit-switched technology in that a dedicated end-to-end channel was established for a voice call. This was logical because a voice call is a continuous process which fully utilises a circuit when the call is active. In contrast, data communications use packet switching, which is characterised by periods of active and inactive usage. HSDPA uses packet-switching technology as it is better suited to supporting data communications.

In voice communication, the capacity parameter of interest is the maximum number of users that can be accommodated simultaneously. In packet-switched data communications the parameters of interest are the peak transmission rate, the average transmission rate experienced by a large number of users geographically dispersed throughout an area served by a base station, and the minimum transmission rate before a user is cut off.

The peak transmission rate is of interest to a mobile network operator as it is useful for marketing purposes. It is calculated based on the best or the highest AMC scheme and is available to those users located 'very close' to a base station, that is where the SINR is very high. At these locations, maximum transmission rates may be supported.

The average transmission rate is the average rate experienced by a user travelling through a cell area. When the user is 'close' to a base station with high SINR, its mobile device can experience high transmission rates. When a user is 'far' from a base station, its mobile device will experience a lower transmission rate. The average transmission rate for a user is calculated by averaging the possible transmission rates over a geographic area covered by a base station.

The minimum transmission rate is the rate at which a base station cuts off transmission to a mobile device. While theoretically (according to Shannon's theorem) transmission may be possible at very low rates, a mobile network operator stops transmitting to a mobile device to avoid situations of very low transmission rates and poor transmission quality. The minimum transmission rate is decided by a mobile network operator.

W-CDMA Release 6 – Introduction of HSUPA

Release 6 of the W-CDMA standard, which was first published in 2005, introduced High Speed Uplink Packet Access (HSUPA). HSUPA enables a mobile device located 'close' to a base station, that is where SINR is high, to transmit at high rates in the uplink. Until Release 6 of the W-CDMA standard, uplink transmission speeds had remained the same since Release 99. Introduction of HSUPA meant that it became practical for mobile devices to, for example,

Table 10.5   Transmission rates for different releases of 3G standards

Release	Technology	Downlink (Mbps)	Uplink (Mbps)
5	HSDPA	14.4	0.384
6	HSUPA	14.4	5.7
7	2 × data	28	11
8	Multi-carrier	42	11
9	2 × 2 MIMO DL (10)	84	23
10	2 × 2 MIMO UL (20)	168	23
11	4 × 4 MIMO (40)	336–672	70

upload large files or stream data including images and video through applications such as Facebook. Table 10.5 shows transmission rate specifications for HSDPA and HSUPA (together known as HSPA) systems for different releases of 3GPP standards.

MIMO Antennas

Multiple-input and multiple-output (MIMO) in general refers to installation of multiple antennas at both transmitter and receiver sides as illustrated. MIMO systems provide two main advantages: one, they create diversity transmission and reception, thereby mitigating the fading channel characteristics. More importantly they create virtual parallel channels, thereby increasing the system's telecommunications capacity. For example the 2×2 MIMO system in the figure creates two parallel channels and therefore doubles transmission capacity. In general an $M \times M$ MIMO system is capable of providing M-times capacity.

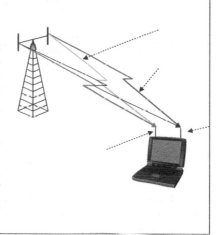

Packet-switched Telecommunications in Mobile Networks

HSPA systems transmit data using packet-switched technology. In a wireless configuration the RNC acting together with a base station operate as an edge router which sends packets to, or receives them from, an end-user. The network nodes may have several queues which give priority to different traffic. For example, a voice over IP call, which has delay constraints can be switched faster than email traffic. The operation of a base station in a packet-switched system is illustrated in Figure 10.21.

Orthogonal Frequency Division Multiplexing

The fourth generation of mobile telecommunication standards are based on a novel modulation and multiple access technology, called Orthogonal Frequency Division Multiplexing (OFDM). OFDM is a similar technology to Frequency Division Multiplexing (FDM) used in

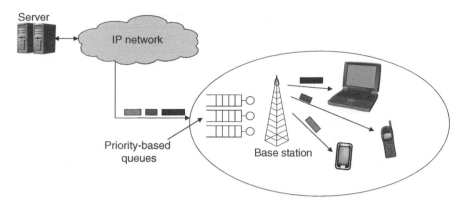

Figure 10.21 Base station in a packet-switched mobile communications system

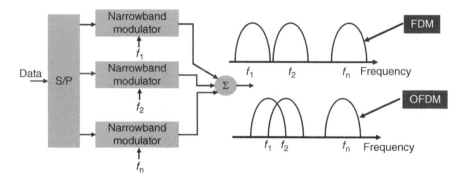

Figure 10.22 FDM and OFDM transmitters

the first generation of standards described above. Both OFDM and FDM systems assign a number of frequency carriers for transmitting data in parallel, as shown in Figure 10.22. The difference is that in FDM systems, filters and guard band are used to ensure neighbouring carriers do not interfere with each other. Carrier frequencies in an OFDM system overlap, and as a result do generally interfere with each other. However OFDM works in a way that at certain frequency sampling points, the interference is zero. If the received signal is sampled at these points, then a clear interference-free detection can be made. This 'orthogonality'" is very significant in interference-limited multiuser mobile telecommunications systems. The history of OFDM traces to the mid-1960s [13] and efforts to better utilise long-distance telephony lines. It was first used in the wireless local area network standards of IEEE, namely 802.11a and 802.11g (more details in Chapter 12).

Frequency orthogonality arises from the fact that the frequency domain representation of a time-domain square waveform is a sinc function. For example a square waveform with a duration of T_s and amplitude of 1 has a frequency representation of sinc(f_s) where $f_s = 1/T_s$ as represented in Figure 10.23. The square waveform may take any one of different values of amplitude and phase as specified by a signal constellation. For example the square waveform may be one of 4 points in a QPSK signal constellation. As a result the frequency-domain representation will have corresponding different amplitude and phase value.

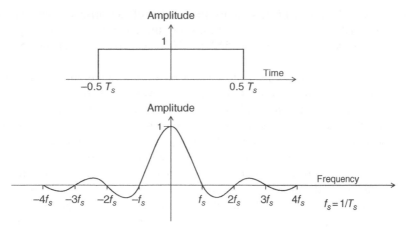

Figure 10.23   A square waveform in the time domain and its frequency-domain representation

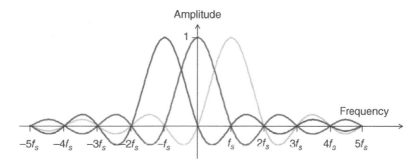

Figure 10.24   Three frequency domain orthogonal waveforms

As can be seen from Figure 10.23, the sinc function periodically passes a zero value at frequencies equal to integer multiples of f_s. It follows that if carriers are separated by exactly such integer multiples of f_s, then they will be orthogonal to each other if sampled at exactly these frequencies. This is illustrated in Figure 10.24, where three waveforms are transmitted together. It can be seen that at frequencies which are integer multiples of f_s, the interference from neighbouring carriers is zero. It means that carriers can be received orthogonally, that is without cross-carrier interference.

In practice the available bandwidth to transmit a signal is divided into narrow bandwidth subcarriers. The subcarrier bandwidth is generally specified by standards: for example 4G LTE standards specify a subcarrier bandwidth of 15 kHz. Different standards have different subcarrier bandwidth as determined by their specific applications and available total bandwidth.

An advantage of OFDM systems is the relatively inexpensive (in terms of processing requirement and device cost) discrete Fourier transform (DFT) technology it uses. DFT and inverse DFT (IDFT) techniques convert signals from the time domain to the frequency domain equivalent and back. In most OFDM systems, the number of subcarriers is generally a power of 2 as DFT and IDFT sample sizes work best using these numbers. For example 4G LTE standards specify systems with 128, 256, 512 and 1024 subcarriers of narrow 15 kHz bandwidth [14].

An OFDM transmitter adjusts the transmission phase and amplitude of each individual subcarrier based on the signal constellation (e.g. 16-QAM symbols have different phase and amplitude). It can further adjust the amplitude and phase in response to variations of transmission channel as described with respect to DSL systems in Chapter 9. For example the three subcarriers shown in Figure 10.24 may be individually controlled in response to varying signal constellation or channel conditions. This is demonstrated in Figure 10.25.

It follows that each subcarrier will carry a data symbol, which is then transformed into the time domain and transmitted as shown in Figure 10.26. Each symbol may carry one or more bits of data as a multiple PSK or multiple QAM signal constellation is utilised. The constellation chosen and therefore the number of bits a symbol may carry is found from the capacity of the channel as determined from the subcarrier bandwidth and the received SINR. Power control and channel equalisation may adjust the received SINR level to ensure an overall channel achieves maximum capacity as effects of channel variations are minimised.

The receiver operation is the reverse of that of transmitter and is shown in Figure 10.27.

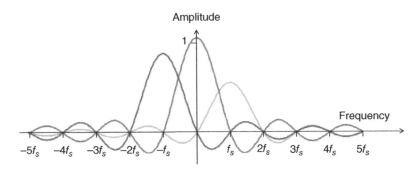

Figure 10.25   Three orthogonal waveforms individually power controlled in the frequency domain

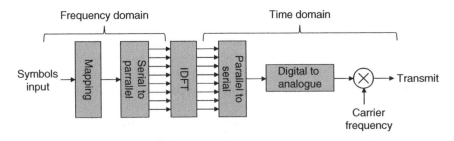

Figure 10.26   An IDFT OFDM transmitter

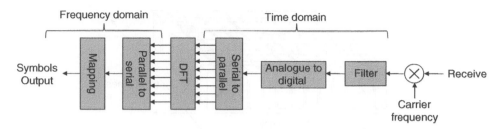

Figure 10.27 A DFT OFDM receiver

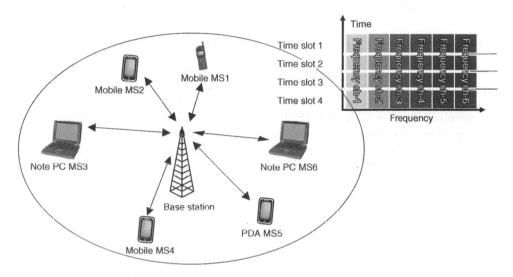

Figure 10.28 Resource allocation in an OFDMA system

Orthogonal Frequency Division Multiple Access

Multiple users need to access the same network in a mobile telecommunications system. In OFDM-based systems, multiple users may send and receive data to/from a base station using different orthogonal frequency carriers, and at different times. This access mode is known as Orthogonal Frequency Division Multiple Access (OFDMA). Here each OFDM subcarrier is assigned to one user at a time, over which the subscriber data are transmitted. Subcarriers are used on a TDMA basis, that is, multiple users may use the same subcarrier on a time-slot by time-slot basis. This is illustrated in Figure 10.28. Furthermore, the telecommunications resource (frequency subcarrier and time slot) is allocated according to a user's need. For example voice communications may need a relatively small number of time slots in one carrier. In contrast a video streaming service may need to reserve multiple subcarriers on a continuous time basis. This is illustrated in Figure 10.29, where different users (shades) are allocated different amounts of telecommunications resources. Such granularity and flexibility of resource provision is of great importance to data communication applications which have time–space variant transmission rate requirements. Because of these and a number of other reasons, OFDMA is the technology of choice for the 4G standards.

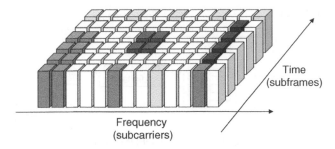

Figure 10.29   Diverse capacity resource allocation to users in an OFDMA system

Fourth Generation: LTE and LTE Advanced

With the growth of the internet and a growing need for wireless connectivity, ITU started the process of a new generation of mobile system standards in 2004. Initially a list of requirements were approved for these new technologies. These included transmission rates and delay requirements as shown in Table 10.6 [15]. Furthermore, the new standards were designed from ground up to be a packet-switched, all-IP end-to-end system: from a wireless mobile device all the way through the core network to the internet. All transmissions were to use IP to packetise data and to transport the packets over the air interface as well as the core network. For example, voice communication was to be designed using voice over IP (VoIP) technology. The main emerging technology with this characteristic became known as Long Term Evolution (LTE) and LTE Advanced. The first set of standards was released in 2008 and the first operational system rolled out in 2010.

LTE is defined in Release 8 of the 36 series of the 3GPP UMTS standard [14, 16]. As noted above, the first release of the LTE standard was published in 2008. This standard is characterised by a number of new features, including the choice of OFDMA. The LTE standards specify an OFDM system with a narrow subcarrier set of 15 kHz bandwidth in the downlink. This allows for smooth transition of a 3G operator to 4G. For example, a W-CDMA system operator which has been allocated a 5 MHz bandwidth as a single carrier to transmit data, will operate an OFDMA system with 300 narrowband subcarriers as shown in Figure 10.30. A CDMA-2000 operator which has a 1.25 MHz bandwidth allocation can use 75 subcarriers of 15 kHz in an OFDMA 4G system. The resource is also allocated on a TDMA basis. An LTE frame is 10 ms, comprising ten 1 ms subframes, which in turn comprises two 0.5 s time slots to facilitate seamless FDD and TDD modes of operation as illustrated in Figure 10.31. Resource allocation algorithm at the base station grants users capacity in multiples of 'resource blocks'. One resource block is defined as a cluster of 12 subcarriers and one 0.5 s time slot, or 7 OFDM symbols, as shown in Figure 10.32 [16]. Some important LTE standard parameters are listed in Table 10.7. A single-carrier FDMA (SC-FDMA) system is used in the uplink.

LTE OFDMA also facilitates device operation over a number of different possible bandwidths. The LTE standard specifies device operations using a range of bandwidths: 1.25, 5, 10 and 20 MHz. This allows for flexible bandwidth allocation to different operators. The specification for transmission rates and delay satisfy the requirements listed in Table 10.6. A typical calculation shows the peak throughput for a 5 MHz bandwidth allocation to be equal to 37.8 Mbps as shown in Table 10.8.

Table 10.6   LTE specification for transmission rates and delay

Parameter	Minimum requirement set by operators
Peak transmission rate (downlink)	100 Mbps
Peak transmission rate (uplink)	50 Mbps
Downlink peak user throughput	5 b/s/Hz
Uplink peak user throughput	2.5 b/s/Hz
Spectral efficiency – Downlink	1.6–2.1 b/s/Hz/cell
Spectral efficiency – Uplink	0.66–1.0 b/s/Hz/cell
Cell edge rate – Downlink	0.04–0.06 b/s/Hz/cell
Cell edge rate – Uplink	0.02–0.03 b/s/Hz/cell
Delay two way	<10 ms
Connection set up	<100 ms
Channel bandwidths (MHz)	1.4~20
Duplex modes	FDD and TDD

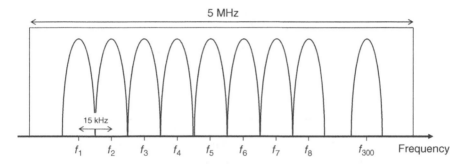

Figure 10.30   Resource allocation in an OFDMA-based LTE standard downlink

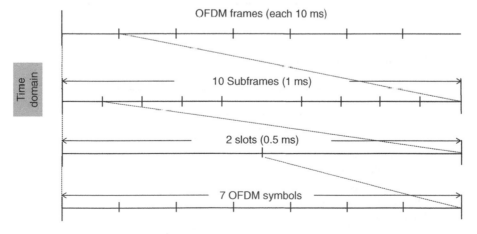

Figure 10.31   Division of LTE telecommunications resources in the time domain

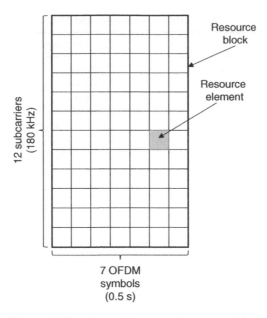

Figure 10.32   A resource unit and a resource block

Table 10.7   LTE specification for transmission rates and delay

Bandwidth	1.4, 3, 5, 10, 20 MHz
Duplex mode	FDD, TDD, half duplex FDD
Channel coding	Turbo 1/3
Modulation	QPSK, 16-QAM, 64-QAM
Multiple access scheme	DL: OFDMA, UL: SC-FDMA
Multiple antenna	MIMO

Table 10.8   Peak throughput calculation for an LTE system with 5 MHz downlink bandwidth allocation

Parameter	Value
Number of bits / symbol (64-QAM)	6
Number of OFDM symbols/slot	7
Number of slots/subframe	2
Number of subframes/frame	10
Number of frames/s	100
Number of subcarriers/5 MHz	300
Total bits/s	25.4 Mbps
After MIMO 2×2	50.4 Mbps
Assume overhead of 25%	37.8 Mbps

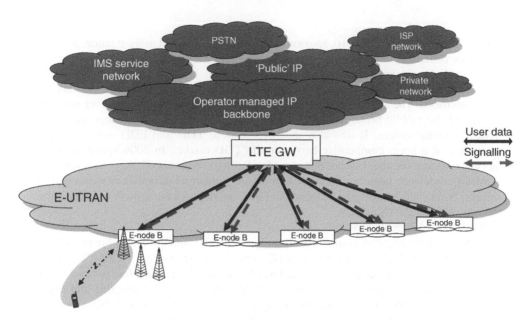

Figure 10.33 Resource allocation in the OFDMA-based LTE standard

The use of the more efficient access scheme in OFDMA, as well as a wider bandwidth, enables LTE to achieve higher transmission rates compared with W-CDMA. Furthermore, LTE uses an all-IP end-to-end network, which results in a simplified core network architecture and a less costly network roll-out. This is illustrated in Figure 10.33.

Case Study 10.3: IEEE 802.16 (WiMAX) Standard

With the growth of data communications and the internet, wireless broadband became a desired alternative to fixed xDSL and cable systems. There were a number of reasons to choose wireless: some regions did not have a wired infrastructure and the cost of new roll-out was excessive. Moreover, wireless systems could provide blanket coverage much more quickly compared with the time required to build a fixed infrastructure. Accordingly in the late 1990s 2G and 3G standards moved to incorporate fast data connectivity in addition to voice services. These systems initially specified transmission rates of several hundred megabits per second, for example the first 3G release supported a 384 kbps downlink transmission link.

At the same time several companies developed proprietary systems based on the OFDM technology. An ecosystem grew over these systems under the auspices of two IEEE standardisation groups: 802.16 for wireless metropolitan area networking (MAN) and 802.20, known as Mobile Broadband Wireless Access (MBWA). In essence the two systems were targeting the same market, provision of wireless broadband connectivity using OFDM technology.

Of the two standards, 802.16, also known as WiMAX, gained more traction. Several countries rolled out commercial systems to provide broadband access in underserved areas, including Australia, Pakistan and Russia. The first set of 802.16 standards provided a point-to-multipoint star topology architecture through fixed antennas at customer premises. The 802.16 standard did not support mobility at first, however with the growth of the mobile broadband market, the specifications evolved to support a full-fledged mobile system. It was also specified in both TDD and FDD modes to take advantage of a lower competitive barrier in the TDD market. In 2006 Sprint, a major mobile operator in the United States and Clearwire announced they had selected WiMAX to provide broadband mobile service. A number of manufacturers, including Lucent, Nokia and Samsung, developed handsets and network equipment to support the roll out of the WiMAX system.

While WiMAX created a significant buzz in the mobile market in the mid- to late 2000s, it eventually wilted against competition from the 4G LTE standard. Several reasons have been given, including the fact that as LTE grew from the matrix of GSM/WCDMA standards, the network equipment was backwards compatible and therefore posed a lower risk/cost to incumbent operators. Further reasons include slow call set-up speed, and insufficient support at higher layers for call set-up and data packet transfer. A bigger reason perhaps was the unwillingness of major manufacturers to support multiple standards. Costs of research and development, intellectual property protection and management become too large, and a single global standard became the most efficient way forward.

By 2012, the main supporters of WiMAX conceded its unviability. Two modes of LTE standards are now set to become universal technologies for provision of wireless broadband [17].

Case Study Questions

- What were the motivations behind the development of wireless broadband systems?
- What was the initial focus of 2G and 3G systems?
- Why was the TDD market more attractive to WiMAX?
- What are the costs associated with supporting two standards for operators and manufacturers?
- Why was LTE successful in defeating WiMAX as a competitor, and was this inevitable?

Frequency Reuse for LTE Systems

LTE systems are based on OFDMA technology and therefore a combination of FDMA and TDMA. As a result a frequency reuse factor of 1 cannot be possible since a user may be impacted by interference from a neighbouring cell as illustrated in Figure 10.12. Therefore a reuse factor of 3, as shown in Figure 10.18, is a starting operation.

In practice, a lower reuse factor may be devised. This is illustrated in Figure 10.34 for a WiMax system [18]. The frequency reuse is 3, however, all subcarriers may be used in the cell centre as these are sufficiently separated from the sources of interference in other cells. In this way the frequency reuse factor can be employed. Moreover, dynamic resource allocation techniques can be employed to use subcarriers in a way that minimises inter-cell interference, thereby further reducing the reuse factor.

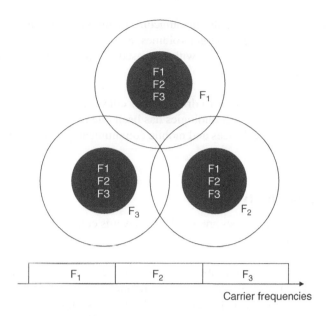

Figure 10.34 Frequency reuse in LTE systems

LTE Advanced

Release 10 (2010) of the 36 series of the LTE standard defines further improvements. This standard is known as LTE Advanced and it includes a number of new technologies, including technologies supporting advanced services and applications, and technologies supporting transmission rates of up to 100 Mbps for high mobility and 1 Gbps (downlink) for low mobility. Standardisation of the next generation (5G) has already started for systems which are expected to be rolled out around 2020.

Convergence in Mobile Communications Standards

There has been significant convergence in the mobile communications standards used around the world over the past three decades. For example, many different 1G analogue mobile communications technologies existed. In contrast, there were significantly fewer 2G technologies. By the arrival of 3G systems, there were two principal technologies: W-CDMA and CDMA2000. It appears very likely that in 4G systems there will be one principal technology used around the world, namely LTE.

Originally there were many 1G technologies because initial developments in mobile telecommunications occurred independently by different manufacturers in different countries. Manufacturers initially made technology choices based on their own expertise and marketed their devices to telecommunications network operators with little regard for interoperability, either internationally or with devices made by other manufacturers. By the time 2G standards were being developed, a number of forces drove convergence and these included the desire for:

- interoperability, to enable roaming of mobile devices onto networks operated by different mobile network operators;

- economies of scale by enabling the manufacture of mobile communications equipment using standardised technology in larger volumes; and
- 'open' telecommunications industries with increased competition in the supply of network equipment and handsets.

These same forces have continued to drive further convergence in the context of 3G and 4G technologies. Convergence of technologies has been one of the main reasons driving the reduction in the cost of mobile devices and mobile communications tariffs over the past three decades.

Backwards Compatibility

Mobile networks and mobile devices are usually backwards compatible with earlier standards and/or technologies because:

- a network operator is obliged to continue to serve mobile devices using older technologies;
- network roll-out is a process that may take several months or years and a network operator needs to support multiple technologies during this process; and
- mobile devices need to operate with many older technologies because mobile networks are not homogeneous in many parts of the world. A mobile device may need to roam to a network that uses an older technology, for example because there is no network coverage by the newer technology, including temporarily when there is insufficient capacity on the newer technology network.

In practice, a base station and a mobile device communicate with each other and inform each other of the standards they support. This allows both to communicate using the standards they share.

Terrestrial wireless telecommunications systems resolve two challenges: one operating in channel conditions where signals are blocked by physical objects, and where multiple signal arrivals can lead to fading. Furthermore, multiple users transmit together and therefore their signals mutually interfere. The other is allocation of scarce bandwidth resources among large number of simultaneous users. Technology development over some four decades has not only made terrestrial wireless telecommunications possible, but highly competitive with fixed broadband.

Review Questions

1. What was the fundamental discovery of Heinrich Hertz that led to the invention of wireless telegraphy?
2. In which applications would directional antennas be used?
3. Why are omnidirectional antennas more useful in mobile communications?
4. What is a simplex communication system? How does it differ from a half-duplex system?
5. What is the difference between the characteristics of a satellite channel and a terrestrial mobile channel?

6. Calculate the path loss for a device transmitting over a 2 km path at 2 GHz frequency band in a Free Space system and a Hata system.
7. Name and describe two methods for compensating fading effects.
8. What is bit error rate (BER) and why is it of interest?
9. How can improvement in system performance due to diversity be quantified?
10. What is the main difference between FDD and TDD systems?
11. Give examples of TDD and FDD systems.
12. In what circumstances does TDD help improve frequency utilisation efficiency?
13. Give examples of TDD and FDD systems.
14. In what circumstances does TDD help improve frequency utilisation efficiency?
15. What is frequency reuse? Why is it necessary?
16. Calculate the number of voice channels for a 2G system. Assume an operator has been allocated 10 MHz bandwidth for downlink and 10 MHz bandwidth for uplink. The standard specifies 25 kHz carriers, each supporting 3 (or 6) voice channels. Assume frequency reuse factors of 7 and 4. How many voice channels exist in each cell? How much traffic (in erlangs) can be supported in each cell? [You need to calculate 4 sets of figures.]
17. What is spreading gain and why is it important?
18. How are users distinguished from each other in a CDMA system?
19. A CDMA base station connects to 8 mobiles simultaneously as shown in the figure.
   a. How is signal to interference ratio calculated at each mobile (uplink)?

   Assume interference from neighbouring cells is insignificant, and that the received signal power from each mobile at the base station is equal as power control is applied.
   b. What is the signal to interference ratio experience for each mobile's signal?
   c. What is the signal to interference ratio if the spreading factor is equal to 64?

   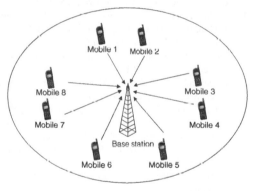

20. A source generates a bit stream with a 32 kHz bandwidth. It is transmitted using a CDMA system with a bandwidth of 1.25 MHz. What is the spreading gain?
21. Calculate the peak and average Shannon capacity for the following W-CDMA system. Assume the effective bandwidth is 3.84 MHz. (SIR, Signal to interference ratio.)

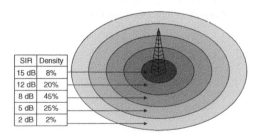

SIR	Density
15 dB	8%
12 dB	20%
8 dB	45%
5 dB	25%
2 dB	2%

22. Where is link budget calculation used in mobile communications? What business implications are there?

23. What are the policy concerns in allocating 4G spectrum?
24. What are the advantages of 4G standards over 3G? Compare the two systems using the technology–business–policy framework.
25. Should 3G and 4G technologies coexist or is it better that older technologies are phased out?
26. What are the business drivers in the convergence of standards from 1G to 4G? What were the policy drivers?
27. Why do mobile operators work with manufacturers in developing standards?
28. Most countries have several mobile operators. Considering the high cost of rolling out a wireless mobile network, would a monopoly operator not be more cost effective?
29. How have multimode, multifrequency mobile handsets helped the roll-out of new standards?

References

[1] http://www.ericsson.com/news/1017515, accessed 23 August 2015.
[2] http://morse.colorado.edu/~tlen5510/text/classwebch3.html, accessed 26 March 2015.
[3] Holt, A. (2010) *802.11 Wireless Networks: Security and Analysis*. Springer.
[4] Gunn, A. (1992), Wireless LANs. PC Magazine.
[5] http://www.totaltele.com/view.aspx?ID=483904, accessed 23 August 2015.
[6] GSM Association (2010) GSM World Statistics. gsmworld.com, accessed 5 November 2015.
[7] CDMA Development Group (2010) CDG Technology: Quick Market Statistics..
[8] Kudoh, E. (1993) On the Capacity of DS/CDMA Cellular Mobile Radios under Imperfect Transmitter Power Control. *IEICE Transactions on Communications*.
[9] Gilhausen, K.S., Jacobs, I.M., Padovani, R., Viterbi, A.J., Wevaer Jr, L.A. and Wheatley III, C.E. (1991) On the capacity of a cellular CDMA system. *IEEE Transactions on Vehicular Technology*.
[10] Esmailzadeh, R., Nakagawa, M. and Kajiwara, A. (1992) Power Control in Packet Switched Time Division Duplex Direct Sequence Spread Spectrum Communications. Proceedings of VTC '92.
[11] Gao, X. and Li, J. (2014) Technological capability development in complex environment, the case of the Chinese telecom equipment industry, in *China's Evolving Industrial Policies and Economic Restructuring*, Routledge.
[12] http://news.zol.com.cn/360/3606207.html, accessed 23 August 2015 (in Chinese).
[13] Weinstein, S. B. (2009) The History of Orthogonal Frequency Division Multiplexing. IEEE Communications Magazine (November).
[14] Dahlman, E., Parkvall, S., Skold, J. and Beming, P. (2010) *3G Evolution: HSPA and LTE for Mobile Broadband*, Academic Press.
[15] Requirements for Evolved UTRA (E-UTRA) and Evolved UTRAN (E-UTRAN). 3GPP TR 25.913 V7.3.0.
[16] http://www.3gpp.org/dynareport/36211.htm, accessed 23 August 2015.
[17] Segan, S. (2012) WiMAX vs. LTE: Should You Switch? PC Magazine.
[18] A. Stolyar, A. and Viswanathan, H., (2008) "Self-organizing dynamic fractional frequency reuse in OFDMA systems," IEEE Conference on Computer Communications INFOCOM.

11

Satellite Communications

Preview Questions

- How do satellites remain in the Earth's orbit?
- What are geostationary satellites?
- How are satellites in low earth orbit used for telecommunications?
- What are VSATs?
- How does GPS work?

Learning Objectives

- Principles of satellites and satellite telecommunications
- Private and public satellite operators
- Mobile satellite services
- Provision of broadband connectivity using telecommunications satellites
- Global positioning systems and associated businesses

Historical Note

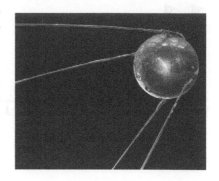

Artificial satellites, long confined to the domain of science fiction were first realised with the launch of Sputnik (pictured),[1] a 58 cm diameter sphere with 4 transmitting antennas, by the Soviet Union in 1957. While the principles behind artificial satellites and the physics of maintaining orbiting the Earth were well known, rocket science was not advanced enough to launch these satellites to sufficiently high altitudes where satellites could maintain their orbit.

The impact of the launch of Sputnik was greatly felt at the height of the Cold War: the United States rightly felt that it was behind in the 'space race' and started a number of programmes to catch up. In time these programmes have led to space probes landing on the Moon, Mars and soon a comet, and travelling to the edge of the Solar System and beyond. Many countries have started a space programme, and contributed to these developments. A permanent space station hosting scientists from many countries has been orbiting the Earth since 1998. Soon there will be regular space tourists travelling to the edge of the space for several minutes, and later on staying in 'space hotels' for longer periods. This is another example of the extent of technological advances of the second half of the 20th century in a large number of fields which have facilitated these achievements.

Sputnik's telecommunications system was a simple battery powered transmitter, sending beacons at two frequencies of 20 and 40 MHz, and its impact is perhaps the greatest on the field of telecommunications. Some by-products of the 'space race' were indirect: for example in order to catch up and regain lead in the race, the United States established the Advanced Research Projects Agency (ARPA) 4 months after the launch of Sputnik. Arpanet, the forerunner of the internet, was one of the programmes funded and managed through this agency. Many others were very direct: systems and equipment developed for space (and military) telecommunications found terrestrial and civil applications including special antennas and microwave frequency equipment, digital communications techniques such as source and channel coding and modulators and demodulators, and so on. While many of the initial satellite applications were in the military and scientific domain, soon these were used for civilian telecommunications.

The theoretical background of satellites communications had been well known before Sputnik. Arthur C. Clarke a radar specialist in the British Royal Air Force and a member of the British Interplanetary Society, and later a famous science fiction writer, in a letter to the Wireless World magazine in the dying days of the Second World War in February 1945 proposed 'Peaceful Uses for V2' rockets. He suggested these German rockets could be used to carry instruments to space and measure the characteristics of different layers of the ionosphere. At the end of his letter he showed how three geostationary satellites may be used for telecommunications purposes [1]:

---

[1] NSSDC, NASA: https://commons.wikimedia.org/wiki/File%3ASputnik_asm.jpg.

An 'artificial satellite' at the correct distance from the earth would make one revolution every 24 hours; i.e., it would remain stationary above the same spot and would be within optical range of nearly half the earth's surface. Three repeater stations, 120 degrees apart in the correct orbit, could give television and microwave coverage to the entire planet.

A mere 6 years after the launch of Sputnik, the first geostationary satellite was placed in the Earth's orbit, and was later used to directly broadcast television from the 1964 Tokyo Olympics. This milestone can be considered to mark the birth of our 'global village' where faraway events can be witnessed in real-time on our TV screens. The establishment of the Intelsat in 1964 as a commercial intergovernmental organisation (IGO) to operate a geostationary tele-communications satellite system is a global policy milestone [2]. Intelsat-1 was placed over the Atlantic Ocean and provided 24 telephone lines or one TV channel relay capability between Europe and the Americas. This was the first telecommunications satellite for international telephony and TV broadcasting, a network that now provides Earth coverage many times over.

The development of optical fibres lessened the value of telecommunications satellites. These have now found a new lease on life in broadband telecommunications service provision to underserved, sparsely populated locations, for maritime uses and regions with poor infra-structure. Satellite telecommunications systems now provide connectivity to nearly half of the world population who otherwise would not be connected to the internet.

A Relay Station in the Sky

Relay stations are used to deliver communications messages from one point to another. Their main function is strengthening the 'message' and/or its 'carrier', as these become weak the further they travel. For example, the Royal Highway stations (see Historical Note in Chapter 1) housed fresh horses and couriers who carried the message forward to the next station. The distance between stations was calculated to maximise speed of message delivery and maintain a low operational cost. A closer example of a modern relay station is a Chappe telegraph tower. Again, the distance between these towers was calculated based on how far the signals could be visible by the telescopes utilised in the system – given a line-of-sight. Similarly, the distance between Morse stations depended on the required received power of the electrical signal for it to be detected, the signal loss as it traversed the copper wires, as well as the oper-ational cost of having a station. In essence relay stations are needed to receive, strengthen and repeat the signal towards its final destination.

As noted above, the idea of having a relay station 'fixed' in space to connect two points on the Earth was first proposed by Arthur C. Clarke in 1945. Such a space station can receive signal from a point on the Earth, strengthen the signal and then transmit it to another point. The distance between these two points can be tens of thousands of kilometres, as such a satellite can have a line-of-sight to nearly half of the surface of the Earth.

How a satellite maintains an orbit around the Earth can be explained by a centrifugal force it feels as it rotates around the Earth which balances the Earth's gravitational pull as illustrated in Figure 11.1.

The satellite's orbital velocity determines the centrifugal force: the faster the speed, the more the centrifugal force. On the other hand, the gravitational pull is calculated from the dis-tance of the satellite from the centre of the Earth. The further a satellite is from the Earth, the lower the gravity pull. Therefore a particular application may choose a satellite orbit with a

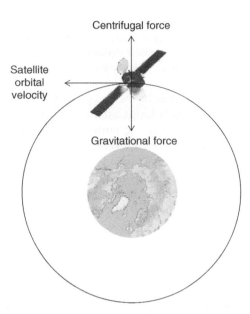

Figure 11.1 Balancing of centrifugal and gravitational forces

particular altitude and an associated orbital velocity and therefore orbital period. The lower the orbit, the higher the orbital velocity and the shorter the orbital period, and the higher the orbit, the lower the orbital velocity and the longer the orbital period. The orbital period, T, is calculated from the following classic physics formula:

$$T = \sqrt{\frac{4\pi^2 r^2}{\mu}}$$

where T is orbiting time in seconds, r is the radius from the centre of the Earth in kilometres and constant $\mu = 398\ 600\ \text{km}^3/\text{s}^2$.

Satellite Telecommunications System Structure

As noted above, the first international telecommunications satellite system, Intelsat Early Bird, was launched in 1964. This system was designed to provide international telephony service and TV broadcasting to consortium member nations. A number of systems nodes had to be developed for the satellite network to operate: a block diagram of the system's equipment is shown in Figure 11.2. It consists of the following major parts:

- Ground networks: satellites usually act as inter-regional or international 'relay' links. Ground networks connect and switch telephone calls between the satellite system and the regional/national telephony networks.
- Ground antennas (earth stations): because of the large distance between a satellite and the Earth, a signal travelling between the two loses a large portion of its power. This signal loss is compensated by high-gain antennas on the ground.

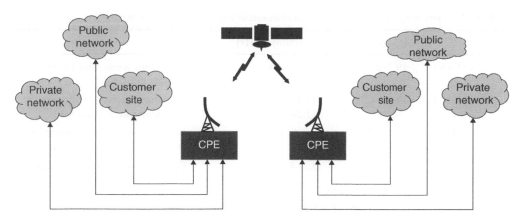

Figure 11.2 A satellite communications system (source Intelsat). CPE, customer premises equipment

- Satellite antennas: satellites transmission and reception coverage can be designed for a particular country or geographic region. Specially designed antennas on-board a satellite can focus their communications for a particular area.
- Transponder: a signal received from the Earth needs to be amplified and relayed back using a different frequency in order that uplink and downlink signals do not interfere with each other. The devices that carry these functions are called 'transponders'. A satellite typically has several tens of transponders, each of which may be assigned to a region/country.

Operating Frequencies

Because of the international nature of satellite communications, a number of frequency bands have been internationally assigned for satellite communications. In order to avoid interference, ITU has divided the world into three regions within which frequency spectrum allocation is harmonised. Table 11.1 lists the S, C, X, K_u and K_a bands in region 2 used for satellite telephony and data telecommunications [3, 4, 5, 6]. Until recently the K_a band had mostly been used for military applications. However with the growth of broadband services and the wide available spectrum bandwidth, many satellite operators are using this band for civilian applications. For a satellite operator, the choice of frequency band depends on what bands are available in a region to ensure little mutual interference with other satellite operators. Other important factors are the amount of available bandwidth, and the path loss the signal experiences.

As Table 11.1 shows, the higher the frequency band, the greater the available bandwidth. However, bandwidths allocated to civilian applications have been small. Only the C and K_u band had been allocated a substantial bandwidth of 500 MHz. With the growing need for global internet coverage using satellites, 3500 MHz of spectrum is being made available in the K_a band for broadband telecommunications systems such as those employed in the Australian National Broadband Network (NBN).

As for path loss, the higher the frequency band, the more the path loss becomes. Among the satellite bands, the L band exhibits the lowest path loss, and is least susceptible to signal fading due to rain. However it is more susceptible to interference from terrestrial communications

Table 11.1 Typical spectrum allocated to civilian satellite services in various bands

Band	Uplink (GHz)	Downlink (GHz)	Usage
K_a (18~40 GHz)	19.7~20.2	18.3~18.8	Communications satellite service (mainly military)
K_u (12~18 GHz)	14.0~14.5	11.45~12.2	Communications satellite service
		12.2~12.7	Broadcast satellite service
X (8~12 GHz)	7.15~7.24	7.2~8.6	Military satellite communications Deep space communications
C (4~8 GHz)	5.9~6.4	3.7~4.2	Communications satellite service
S (2~4 GHz)	2.31~2.37		Digital Audio Radio Service
	2.0~2.2 MSS		MSS

MSS, mobile satellite services.

and devices active in radio frequency ranges. Signals in the K_a bands exhibit a higher path loss and fading in the presence of rain. However, K_a band systems can operate with smaller size antennas. Most present day satellites are designed to operate in both C and K_u bands, where the C band is used for wide geographic coverage and the K_u band for regional coverage. The K_a band satellites are being rolled out to provide broadband internet coverage on a global basis.

Satellite Communications Antennas

A range of different antennas are used for satellite communications. Early systems required the signal power to be much larger than noise power as the quality of analogue voice and video communications reception would be greatly impacted. Powerful antennas were used on the ground at earth stations to communicate with satellites. These antennas have gains in the order of 60 dB, that is, they can boost the received power by 10^6 times. Because of their size, and also to reduce the amount of background noise, these antennas were usually constructed far away from cities and generally in rural areas. The earth station antenna, used by OTC (one of the companies which formed Telstra Ltd) to communicate with an Intelsat satellite is shown in Figure 11.3. It is reported that in 1984 half of Australia's international telephone traffic passed through this antenna [7].

Antennas on board the satellite were generally of a smaller size and gain because of launch logistic limitations. One or more reflectors receive and transmit signals to Earth via a feed array which enables the satellite to focus communications on a particular spot on Earth. A typical footprint of a satellite operating over the Indian Ocean is shown in Figure 11.4. Usually the broader coverage uses the C band and the more localised, country-level coverage operation uses the K_u band.

The transmitted power from the satellite transponder is amplified through the antenna and is directed towards a footprint on the ground. Generally the power received in the centre of a footprint is larger than towards the edge. Equivalent isotropically radiated power (EIRP) is a measure of the effective power transmitted from a satellite in the direction of a specific region within the footprint. EIRP can be calculated by combining the transmission power from a

Figure 11.3   The earth station antenna at Ceduna, South Australia [8]. Source: Nachoman-au. GFDL (http://www.gnu.org/copyleft/fdl.html), CC-BY-SA-3.0

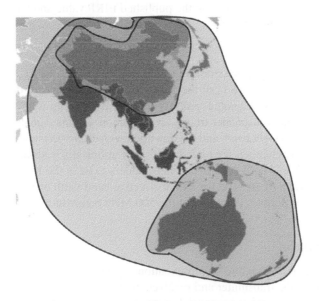

Figure 11.4   A satellite's spot beam footprint

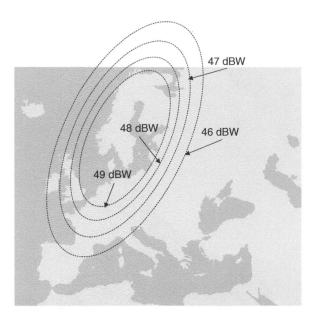

Figure 11.5   Spot beam footprint EIRP map

transponder and the antenna gain aimed at a particular geographical area. Figure 11.5 shows a typical EIRP for a regional footprint of an Intelsat satellite. It shows that the radiated power from a satellite directed towards Sweden is larger than 49 dBW (79 dBm). Similarly signal energy directed towards Portugal is 42–44 dBW (72–74 dBm). Ground station antennas are chosen to deliver a desired signal to noise ratio (SNR) for a designed reception quality. These can calculate the received SNR based on the published EIRP value and the path loss.

Transponder

Transponders are frequency convertors and amplifiers and are placed on-board satellites. Their function is shown in Figure 11.6: they receive a signal at one end, down-convert it to a different frequency by a fixed amount, amplify the signal and send it downlink. Figure 11.6 shows a K_u-band satellite transponder frequency allocation. Uplink signals are received at the satellite in the 14.0–14.5 GHz band, and the downlink signals are transmitted in the 11.7–12.2 GHz band. Transponder channels are paired, and every uplink slot is separated from its downlink pair by exactly 2.3 GHz. A total of 24 transponders exist in the K_u band, and each transponder has a bandwidth of 36 MHz for a total effective bandwidth of 864 MHz (vertical and horizontal polarisation doubles the usage of the 500 MHz bandwidth) [9].

Path Loss in Space

As discussed above, most satellite communications systems are designed to operate with a line-of-sight between the transmitter and receiver. Natural structures generally will not block this line-of-sight, in contrast to terrestrial communications where natural and man-made objects may block a link.

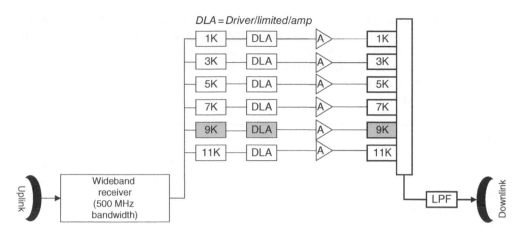

Figure 11.6   Transponder functions

However, there exists another limitation which is faced by all telecommunications systems. This is signal weakening as a signal traverses a carrier medium. The distance of a satellite from the Earth can be from hundreds to tens of thousands of kilometres, and the path loss over this distance can be very large. Therefore the signal arriving at a receiver is very weak compared with locally added noise. Signal amplification through strong antennas on the ground and on-board the satellite is necessary. Such antennas must face a point in space where the satellite is – meaning that the satellite needs to be stationary with respect to an observer on the Earth.

Although a line-of-sight exists between a transmitter on the Earth and a satellite in space, the signals transmitted to and from the satellite do travel a very long distance. Only a small portion of the signal energy transmitted through an antenna arrives at a receiver, as the transmitted energy is dispersed in space. The ratio of received signal power compared with transmitted signal power is a function of the distance between the two and the frequency of operation, as well as characteristics of the medium. For line-of-sight communications through space the path loss (L) in decibels is calculated from the following equation:

$$L = 32.4 + 20\log(d) + 20\log(f)$$

where d is the distance in kilometres and f is the carrier frequency in MHz.

Example 11.1

What is the path loss for a signal in the C band (4.0 GHz) from a satellite positioned 36 000 km from the receiver antenna on the ground?

Answer

$$L = 32.4 + 20 * \log(36\,000) + 20 * \log(4000) = 195.5\,dB.$$

This means the signal power diminishes by a factor of nearly 10^{20} as it travels the distance between this satellite and the earth station.

Without powerful antennas, a signal transmitted to/from a satellite will be received at a power significantly lower than noise power, and cannot therefore support reliable, high quality communications. The path loss is therefore compensated through boosting of signal power on both transmitter and receiver sides through powerful antennas on both the satellite and earth station. The required antenna gain is found through a link budget calculation process as described in Chapter 4.

Example 11.2

For a geostationary Earth orbit (GEO) satellite communication system, the downlink transmit signal power $P_t = 40$ W or 46 dBm; the satellite transmitter antenna gain, $G_t = 15$ dB; the earth station antenna gain, $G_r = 30$ dB. If frequency of operation is 12 GHz, and noise power level at the receiver $N_r = -120$ dBm:

- What is the EIRP?
- What is the path loss?
- What is the received SNR?

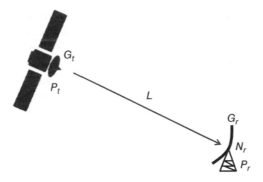

Answer

- EIRP $= P_t + G_t = 61$ dBm
- Path loss $= 32.4 + 20\log(36\,000) + 20\log(12\,000) = 205$ dB
- Received signal power:

$$P_r = \text{EIRP} + G_r - L = 61 + 30 - 205 = -114 \text{ dBm}$$
$$\text{Received SNR} = P_r - N_r = -114 - (-120) = 6 \text{ dB}$$

Earth Orbits

Earth orbits are classified based on their distance (altitude) from the Earth. As articulated by Arthur C. Clarke, at a specific altitude, the orbital period exactly equals the time it takes for the Earth to rotate once around its axis or approximately 24 h. A satellite at such an altitude and orbiting within the equatorial plane appears stationary to an observer on the ground. This is known as the GEO and satellites orbiting here are called GEO satellites.

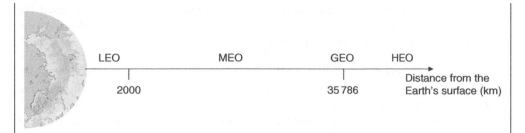

Figure 11.7   Earth orbits

The GEO satellite's altitude for the length of a sidereal[2] day of 86 164 s is calculated to be equal to 42 164 km from the centre of the Earth or 42 164 − 6378 = 35 786 km from the surface of the Earth.

Orbits higher than GEO are collectively referred to as high Earth orbits (HEOs). Orbits below GEO are called medium Earth orbits (MEOs) or low Earth orbits (LEOs) if they are lower than 2000 km, as illustrated in Figure 11.7.

Examples of satellite systems at these orbits include GEO long-distance communications satellites from operators such as Intelsat; a number of LEO mobile satellite systems orbiting in altitudes of 700–800 km and the Global Positioning System (GPS) operating in the MEO of 20 200 km.

Broadband Satellite Telecommunications Systems

A number of satellite telecommunications of interest to broadband services exist, and these may operate in a range of orbital altitudes. Directional antennas with strong gain can be used to communicate with a GEO satellite as these are 'stationary'. However, due to the long distance between the satellites and the Earth, path loss is very high and in the order of 200 dB or more. This means that systems with smaller antennas cannot provide high capacity transmission rates. Moreover, signal delays can be excessive for many applications as end-to-end conversational delay is in the order of 0.5 s. Yet another limitation is the maximum number of satellites that can be placed in the GEO altitude and plane: for example if a 2–3° separation is needed the maximum number of satellites will be 120–180.

GEO Satellite Systems

As stated in the Historical Note, the first GEO satellite system was operated by Intelsat, an intergovernmental organisation under the auspices of the United Nations. It soon launched satellites to cover the entire Earth and operated as a long distance voice and TV

---

[2] A sidereal day is the time it takes for the Earth to once rotate fully around its axis, and is slightly shorter than a 24-h 'solar' day. This is because the earth travels approximately 1/365th, or almost 1°, of its solar orbit a day. For the same point to directly face the sun, the Earth needs to rotate around its axis slightly more than 360°. That is, a 24-h day is slightly longer than the time it takes for the Earth to rotate once around its axis. A sidereal day is approximately 23 h, 56 min and 4 s (86 164 s).

traffic carrier. Soon another international satellite operator, Inmarsat, was formed for maritime (ship–shore) communications, again as a body under the United Nations. As the need grew for communications from remote locations such as mines or oil rigs, both Intelsat and Inmarsat started ad-hoc satellite network services using portable earth stations. These systems were mostly intended for digital voice communications, but gradually low rate data communications services were also offered. The main traffic over the GEO satellites remained voice and broadcast TV: these were known as fixed satellite services (FSS). With the growth in the mobile telephony market, a niche market appeared in global MSS coverage in the late 1980s.

With the roll out of fibre optic cables, most of Intelsat's long-distance telephony and broadcast TV business was lost to these cheaper and higher quality media by the early 1990s. Moreover, cellular mobile technologies substituted most of Intelsat and Inmarsat MSS and reduced their revenues. Both companies filed for bankruptcy protection and went through restructure. Both have now emerged focusing on the provision of fixed broadband satellite telecommunications services to remote and underserved areas using very small aperture terminals (VSATs), as well as MSS.

Very Small Aperture Terminals

GEO satellites can be used for ad-hoc telecommunications network set-up using portable terminals. These systems operate using VSATs to digitally communicate connecting remote locations to the public telephone network and the internet. A VSAT terminal consists of a transmitter/receiver (transceiver) to communicate with the satellite and a portable antenna with a size ranging from a few tens of centimetres to a few metres. The antenna size generally determines the received SNR and therefore the channel capacity and transmission rate. A typical VSAT terminal is illustrated in Figure 11.8.

Figure 11.8   A typical VSAT terminal antenna connection. Source: Axlsite (own work) CC BY-SA 3.0 (https://commons.wikimedia.org/wiki/File%3ATooway_satellite_antenna_drawing.png)

Very Small Aperture Terminal Multiple Access Methods

Multiple users access the resources of a satellite though multiple access protocols. These function typically belong to layers 2 and 3 of the OSI model, or the media access control (MAC) and logical link control (LLC) layers. These specifications determine how the limited resources of a satellite (i.e. the available spectrum) may be simultaneously used by many VSAT users. VSAT architecture as shown in Figure 11.9 requires a central hub which determines which users transmits to the satellite and at what time. The process requires a user to first send a request to the satellite when it needs to transmit data in the uplink. In response the satellite hub allocates resources in the time and frequency domains so this transmission will not collide with other users' traffic. Resource allocation, or channel multiplexing, may be done in a number of ways, but it generally follows four domains: frequency, time, space and code. The first two are discussed below, space multiplexing was shown in Figure 11.4, and code multiplexing will be described in the next chapter as it is more commonly used in terrestrial mobile telecommunications.

Frequency Division Multiple Access

In Frequency Division Multiple Access (FDMA), a portion or the whole of a transponder bandwidth (Figure 11.6) is allocated to a VSAT subscriber to use. This set-up is useful for major hubs which need constant connectivity to a network, and which themselves divide this connectivity among a number of local users (this may be a wholesale buyer who provides services to its retail subscribers.)

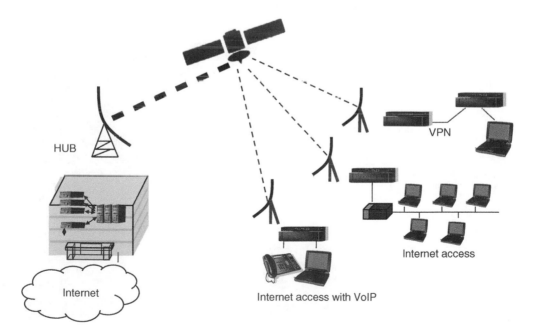

Figure 11.9 VSAT system architecture. VoIP, voice over internet protocol; VPN, virtual private network

Time Division Multiple Access

In Time Division Multiple Access (TDMA) transmission capacity on one or more transponders is allocated to a VSAT user for a certain period of time. Several VSAT users may consecutively gain access to a satellite, each within a well-defined time slot. In general these VSAT subscribers' needs are intermittent and therefore constant access to a VSAT is not necessary.

Frequency Reuse

The same transponder carrier frequency may be reused in a number of different regions if these are sufficiently separated so that their signals do not interfere. This technique can be used to increase the total transmission rates a satellite may support in the downlink. This is illustrated in Figure 11.10, where different shades indicate the carrier frequency used in a region. As can be seen, the same frequency can be used multiple times, thereby increasing the system's capacity several fold.

Frequency reuse requires multiple spot beams and/or multiple polarisation beams. Mutual geographical isolation ensures that little interference is perceived for ground receivers.

VSAT systems can provide data communication rates from a low of 9.6 kilobits per second (kbps) to upwards of 50 megabits per second (Mbps) [10]. They are used in a number of applications: from conflict-zone reporting, to remote-field exploration data connectivity, to internet service provision to rural areas. Rural and underdeveloped regional internet connectivity are sometimes funded through philanthropic institutions such as the Belinda and Bill Gates Foundation.

Inmarsat has announced plans to provide services branded as the Broadband Global Area Network (BGAN) using spectrum in the K_a band. The spectrum in this band is available to

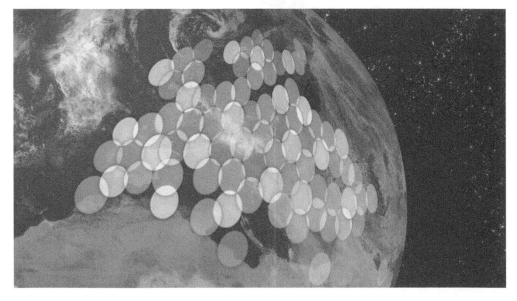

Figure 11.10   Frequency reuse in satellite communications [10]

both civilian and military applications, and as a result Inmarsat toggles services using the bands to offer services to both the public and government [11, 12].

Example 11.3: Australian National Broadband Network

The Australian NBN roll-out is expected to provide connectivity to some 3–4% of Australians using a K_a band satellite system. This system is expected to use transponders with a 300–600 MHz bandwidth to provide services using a VSAT architecture. Transmission rates in the downlink are expected to be in the order of 12 Mbps, and the systems is expected to support a total traffic of 80 gigabits per second (Gbps) [13]. The network architecture is expected to follow that of VSAT systems as illustrated in Figure 11.9.

Mobile Communications Satellite Systems

Usage of GEO satellites for international voice and TV broadcast service requires large capacity links and therefore high-gain antennas which can be directed to the satellite. However, satellites may provide other applications which do not require large capacity, and therefore the link budget can be met with smaller, less directional antennas. Such receivers on the ground may be non-stationary, and therefore incapable of maintaining the direction of an antenna towards a GEO satellite. These satellites may also support non-real-time applications: a spy satellite may collect information from a point on the Earth and download the data to a ground station during its next pass over.

Two classes of satellites relevant to telecommunications exist beside GEO. One class is the LEO satellite systems, with an orbital altitude of less than 2000 km. LEO satellites are primarily used in mobile communications systems. The other class is the MEO satellites with an orbital altitude of between 2000 km and 35 786 km. Some MEO satellites are used for broadband telecommunications and another set for positioning services such as in GPS. Since the altitude of these two classes of satellite systems is lower than for the GEO satellite, the path loss experienced by their signals is smaller and therefore smaller antennas can be used for achieving their link budget. Moreover, the required total capacity/transmission rates are significantly smaller than that of GEO satellites for most services offered.

Low Earth Orbit Satellites

The LEO satellites orbit the Earth at altitudes of less than 2000 km. Because of this their orbiting time is significantly shorter. For example, a LEO satellite orbiting at 800 km experiences an orbital time of around 100 min. Such a satellite would not be very suitable for a real-time telecommunications system since the receiver on the ground can only view the satellite for a short period: the satellite rises from one horizon and sets in the opposite horizon on average in less than 50 min. As a result, continuous connectivity to one satellite is clearly impossible. However, a constellation of such satellites can be designed to maintain constant connectivity. Such a constellation helps so when one satellite sets, a new one has already risen to maintain connectivity. In this way, a call is relayed through a network of several satellites to a ground station.

In LEO systems the communications distance is short and therefore the path loss is much lower compared with GEO satellites. For example a LEO satellite operating at the 4 GHz band orbiting at 800 km altitude experiences the following path loss to the Earth:

$$L = 32.4 + 20 * \log(800) + 20 * \log(4000) = 162.5\,\text{dB}$$

The link budget is improved by 33 dB, or 2000 times, compared with GEO satellites which means smaller antennas can be used. In practice such an antenna can be mounted on a mobile phone (pictured).[3]

Several LEO satellite systems are presently in operation or on the design board. Iridium is one such system. It consists of a constellation of 66 satellites in 6 orbital planes as shown in Figure 11.11. The system is designed so an Iridium satellite is always visible from any point on the Earth, including the poles.

Iridium and another LEO system, Globalstar, were unsuccessful commercially at their launch in the late 1990s (see Case Study 11.1). Their business models

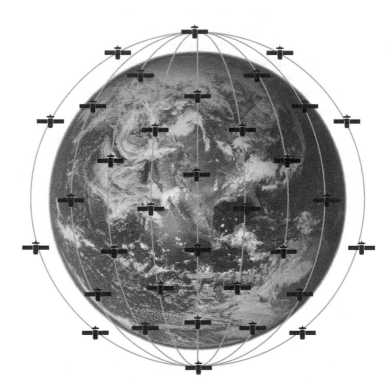

Figure 11.11   Iridium LEO satellite constellation

---

[3] An Iridium satellite phone. Photograph taken by Mark Pellegrini in the Steven F. Udvar-Hazy Center (Smithsonian Air and Space Museum extension in Dulles, VA, USA). Modified by ivob.

Table 11.2 Details of the main mobile and broadband satellite systems

System	Number of satellites	Orbital height (km)	Frequency band	Downlink spectrum	Uplink spectrum	Transmission rate down/up
Iridium	66	780	L/K$_a$ band	1616–1626.5 MHz		28.8 kbps
Globalstar	48	1400	S/L band	1610–1618.75 MHz	2483.5–2500 MHz	9.6 kbps (telephony)
Teledesic	288	1400	K$_a$ band	28.6–29.1 GHz		720/100 (Mbps)
O3b	8	8062	K$_a$ band	17.7–20.2 GHz	27.5–30.0 GHz	24 Mbps

have since been restructured to serve specific niches such as mining operators and military applications: indeed the conflicts in the Middle East have been a boon to the satellite communications companies in both the GEO and LEO fields. Another system, Other 3 billion (O3b) is being launched and is expected to provide global connectivity to the other half of the human population who do not have access to the internet. Note that the O3b orbital height of 8000 km classifies it as a MEO satellite. Details of various mobile and broadband satellite systems are shown in Table 11.2.

Example 11.4: Iridium System Capacity

The Iridium system uses the frequency spectrum between 1616 MHz and 1626.5 MHz. This 10.5 MHz spectrum is divided into 240 channels of 43.75 kHz, and an effective channel of 41.67 kHz with guard band as illustrated in Figure 11.12.

Each satellite uses 48 spot beams within its coverage area, with frequency reuse factor of 12 as shown in Figure 11.13. Because of this frequency reuse each cell contains 240/12 = 20 frequency channels, which are time division multiple accessed by 4 users. Therefore, each satellite can support 20 * 4 = 80 telephone links/spot beam. Overall, each satellite can carry 48 * 20 = 960 links. As the Iridium system is made of 66 satellites, a total of 3168 spot beams are available. However, because of uninhabited regions around the world, only 2150 may be useful, which means a total of 2150 * 80 = 172 000 voice links are available through the Iridium system. Some of these links may be used for data communication which reduces the overall voice capacity.

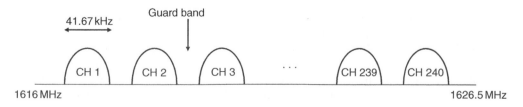

Figure 11.12 Iridium frequency spectrum usage

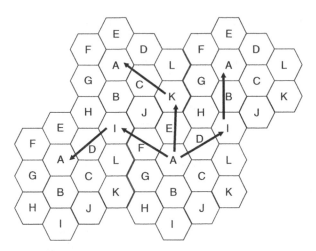

Figure 11.13   Iridium frequency spectrum reuse

Clearly the limited spectrum made available to the Iridium system limits the overall system capacity. Iridium is also planning on launching new services using spectrum in the K_a band in order to provide up to 8 Mbps data communications services [14, 15, 16].

Case Study 11.1: Iridium and Globalstar

Using satellites to provide mobile connectivity gained currency in the late 1980s as the first generation of mobile telephony systems were rolled out. At the time the mobile service coverage was patchy and confined mainly to large metropolitan areas. Even within these cities the coverage was not perfect and calls would regularly drop out. Moreover many countries did not have a mobile phone service, and no coverage existed outside populated areas. Furthermore, mobile phones were expensive with handset prices of more than $2000 and phone call rates of more than $3 per minute. There was a business opportunity to improve the cost/performance of these systems. One response was the development of mobile telecommunications using GEO satellites, however these systems were generally intended for trucks and trains where large antennas could be installed to meet the required link budget.

Mobile communications using a constellation of LEO satellites solved the problem of large antennas as the path loss could be reduced by more than 30 dB. Three LEO mobile satellite projects were launched in the early 1990s, Teledesic was primarily funded by Microsoft, Globalstar was funded by Qualcomm, and Iridium by Motorola. Of the three, Teledesic was the most ambitious with plans to launch some 840 satellites and provide transmission rates of 100 Mbps and downlink of 720 Mbps. The plans of Globalstar and Iridium were more modest, calling for 48 and 66 satellites in their constellations, respectively, and providing transmission rates sufficient to support digital voice. While Teledesic demurred, Globalstar and Iridium started to build and launch a global mobile communications service based on the cost assumptions of the early 1990s.

By the mid-1990s, the second generation of mobile communications systems had been rolled out and were providing better coverage and higher transmission rates. The cost of devices and calling had also significantly fallen from the early days, and many operators were fully subsidising the cost of a mobile. Regardless, both Iridium (picture [17])[4] and Globalstar followed with their satellite constellation build up, apparently based on the earlier business assumptions. As the service launch neared, the target market had to shift from the general public to the businessman on holiday. Despite the shrinking market the system build up continued.

By the time Iridium and Globalstar entered the market, the second generation of mobile phone systems had become very popular. Mobile phone sizes were reduced to smaller than 100 g and 100 cm^3, and the costs of devices and calling rates were a fraction of those in the early 1990s. Both businesses failed to attract enough customers and had to file for bankruptcy protection. The main supporters of the companies, Qualcomm and Motorola, also suffered. Of these, Motorola faced other problems and ultimately had to sell its handset business to survive.

Case Study Questions

- What were the initial motivations for LEO mobile satellite communications?
- How did the costs and revenue items compare at the start of these projects?
- What were the risks?
- What do you see as the extent of the target market at product launch?
- Is there a market for mobile satellite to make this a viable business?

Business Ecosystems

Broadband satellite systems create an infrastructure over which connectivity services and content may be delivered.

The industry ecosystem structure is heavily influenced by the high cost of building and rolling out the infrastructure, and therefore the fixed initial cost. While national satellites may operate by licensing frequency from a national agency, these players usually find most local markets too small for sufficient return on their investments. Most of these services are therefore entirely funded or subsidised as part of a government broadband roll-out strategy. International satellite infrastructure players on the other hand need

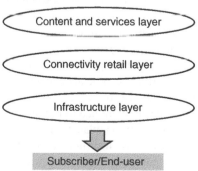

---

[4] Iridium Extreme Satellite Phone, Iridium Satellite Communications.

to secure the necessary frequency spectrum through the ITU within specifically set aside bands. Again, the high cost of network build-up, combined with small-sized, low-income markets, means that the return on investment is low. As a result there is not much interest in these systems from commercial operators. Regardless, there are a number of players each active in a specific segment of the market, from LEO to MEO and GEO. This ensures competition, and price stabilisation for service subscribers.

At the retail layer, most infrastructure players provide capacity on a wholesale basis to national operators. This is mostly due to the international character of infrastructure businesses and the necessity to comply with service provision policies of national governments. At times satellite services are packaged with terrestrial service through a national operator: for example Telstra resells Iridium services in Australia alongside other products.

The contents and services layer consists of applications served to a general broadband market, as well as those for the satellite niche market. Iridium for example maintains a specific weather application for the maritime market. Shipping and mining sectors are major customers of satellite services and use customised applications.

The choice of a nationally owned satellite infrastructure, for example the Australian government-owned NBN, vis-à-vis a privately owned international infrastructure such as O3b or Inmarsat depends on many factors. The technology–business–policy (TBP) framework can be used to make a decision as different satellite configurations and technologies lead to different cost and performance figures. The range of services required by a government, from nationwide broadband telecommunications to security and military applications, are important in the final decision.

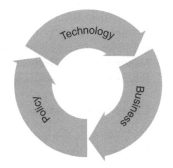

Positioning Satellites

One's coordinates on the ground in combination with accurate maps have long been of great importance to commerce and governance. Accuracy of positioning measurement grew in importance from the 15th century as shipping for trade and conquest gathered pace. Ships at sea did not have any reference point to accurately measure their position and this in many instances led to shipwrecks when coming to shore.

Any point on Earth can be uniquely specified by its latitude (its north–south position) and longitude (its east–west position) with respect to a reference point [18]. Latitude can be accurately measured with reference to the sun, however longitude measurement had to rely on the positions of the moon, planets and stars and its accuracy was insufficient. In fact the British parliament enacted a prize in 1714

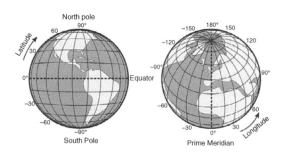

to the value of £20 000 (£2.5 million in 2014 money) to anyone who could determine the position of a ship to within 56 km accuracy. This was the latest in similar prizes that had been offered by Spain and the Netherlands in 1567 and 1598, respectively, for the same purpose.

The Longitude Prize was finally won by John Harrison, a clockmaker who managed to build a very accurate chronometer. The premise of his solution was that if one can measure the time accurately, then the time difference at noon with respect to a reference point will yield the longitude difference. As the Earth rotates 360° in 1 day, each hour is equal to 15°, and therefore the difference in degrees between one's position and the reference point, Greenwich Village near London, can be calculated from the time difference. This accurate time measurement revolutionised navigation and gave Britain significant technological and military advantage.

The emergence of satellites in the 1960s, and the US military's need for accurate positioning measurement started a project in 1973 that led to the development and launch of GPS in 1993. In essence, GPS satellites are very accurate clocks in the sky, which broadcast this time and their exact position to receivers on the ground. The time is maintained according to an atomic clock with an accuracy loss of 1 s every 2.7 million years. The receiver can then calculate its distance relative to four of these satellites and triangulate its position based on these data as well account for inaccuracies of the local clock.

Positioning satellites may be positioned at any orbit around the Earth. The only requirement is that they broadcast their present position and time. GPS operates at a MEO of 20 200 km using a constellation of 24 satellites orbiting at 20 200 km. These satellites orbit in 6 orbital planes each with 4 satellites as shown in Figure 11.14. The system is designed so that at any time a receiver on the ground can measure its distance from four of these satellites.

Position measurement is accomplished by determining the time it takes for a satellite signal to be received at a GPS device by measuring the signal propagation time from four or more visible satellites, as well as the satellites' physical x–y–z location in space as broadcast. This information is continuously updated and transmitted from each GPS satellite, and can be decoded to within a specified accuracy by a GPS receiver. The propagation delay and the satellites' physical positions enable the receiving GSP device to calculate its position to within a designed accuracy.

GPS applications are widespread in location services such as emergency assistance, and in location-based business services using smart phones.

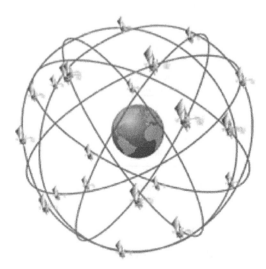

Figure 11.14 GPS satellites. Source: https://commons.wikimedia.org/wiki/File:GPS-24_satellite.png

Case Study 11.2: Google Maps

Google is one company that has taken a leading role in taking advantage of the GPS infrastructure. A number of companies offered on-line maps as the World Wide Web usage grew in the mid- to late 1990s. Companies such as Map-quest provided location search and display on the World Wide Web. Alternatively, some mobile manufacturers integrated maps into their handset platforms and provided directional assistance as applications. These services were prominently offered on the Japanese mobile handsets with GPS receivers. Yet a number of other players manufactured GPS devices for use in car navigation with audio turn-by-turn instructions, a market that grew significantly in the late 1990s and early 2000s. Initially these devices were integrated in luxury cars and later on offered as accessories to all vehicles. Prices for these car navigation devices were in the order of $2000.

Google was a relatively late-comer to this field with the purchase of a little-known Australian start-up company called 'Where 2 Technologies' in 2004. They soon integrated this on-line map into a web application and started providing maps in association with their search results. This service has now become an integral part of Google services in both fixed and mobile domains and a source of advertising revenue.

Google has further enhanced on-line maps with street view and Google Earth, both augmented by satellite generated information. Street view and Google Earth information collections are linked with the GPS-provided coordinates, so these can seamlessly be integrated into the map searches. Google has further enhanced GPS accuracy with terrestrial mobile network and local area network information. These enhancements enable delivery of many location-based services in connection with search result provision. By all measures Google Maps has been a great success: it constantly ranks as the most used smartphone app in the world [19]. Traffic and driving instruction information as well as audio augmented services have largely removed the need for GPS navigation devices.

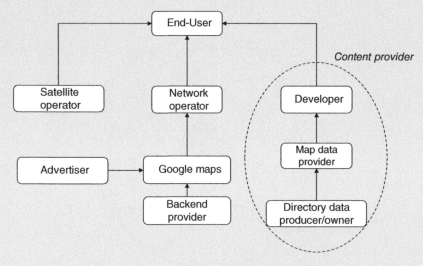

Figure 11.15   Google map business system diagram

The Google map business system diagram can be drawn as shown in Figure 11.15. It greatly depends on the data gathered from GPS and other satellite services and when the information is retailed to the users of Google services [19]. As a player in the content/service provision layer, Google Maps is an important part of the overall company's strategy [20].

Case Study Questions

- Why did Google acquire Where 2 Technologies? What is the value addition of maps for search results?
- How do satellite systems feature in Google Maps?
- Who are the infrastructure, retail and content/service players in an on-line map business?
- How costly are enhancements to on-line maps and how have these impacted the overall navigational business' competitive landscape?
- How does Google address the limitation of GPS systems?

Table 11.3   Orbital altitude of satellite navigation systems

System	Owned by	No. of satellites	Orbital height (km)
GPS	USA	24	20 200
GLONASS	Russia	24	20 000
Galileo	EU	30	23 222
BeiDou-1	China	35	35 786
INRSS	India	7	35 786
Quasi-Zenith	Japan	4	35 786

A number of other countries have launched, or are in the process of launching, positioning satellites to provide a national/global infrastructure alternative to that of the US military run GPS. Among these is the European Union's €5 billion Galileo programme which has launched a number of the constellation satellites and is expected to be fully operational by 2019 [21]. Other systems include the Russian GLONASS, Chinese BeiDou2, Indian regional IRNASS, and Japanese quasi zenith satellite system. Some of these systems use the MEO while some operate in the GEO. The principal behind all these systems is the same: a receiver needs to detect the position of 4 satellites and its distance from them so it can calculate its own coordinates.

While duplication of this infrastructure is inherently inefficient, these countries clearly consider a positioning system as a strategic asset worth the cost of launch and operation. For many of these countries, military applications require accurate positioning to sub-metre levels, presently considered to be available only to the US military through GPS. Some of these systems are already operational such as Russia's GLONASS, and the rest are expected to become operational by the turn of this decade (2020). Details of these systems are listed in Table 11.3. The GEO-based systems (BeiDou, INRSS and Quazi-Zenith) are generally national or regional, whereas the MEO-based systems are global.

Satellite telecommunications systems are finding a new life in the provision of broadband services and contents. On the one hand they are essential to GPS, which in turn is used in the provision of many location-based services. On the other hand they are the most economic, as well as the quickest, roll-out option for provision of broadband internet in remote, under-served areas of the world. While the latter services are mostly government subsidised, new service provision business models are approaching competitive pricing points. Satellite systems have become an important part of the broadband telecommunications landscape.

Review Questions

1. What is a geostationary orbit?
2. What is a sidereal day?
3. Why is link budget calculation necessary?
4. Why did Intelsat's business shrink?
5. How can real-time communication be possible with non-geostationary satellites? Give an example.
6. For a VSAT operator, list cost and revenue items.
7. A GEO satellite transmits a signal in an 11700 MHz downlink channel. What is the path loss the signal experiences? If the EIRP is 40 dBW, and the receiver noise level is -125 dBm, what is the received SNR if the earth station antenna gain is 60 dB? Can this system support high quality analogue voice communications if the required SNR is 30 dB?
8. Explain two reasons behind the commercial failure of LEO satellite mobile communications companies (such as Iridium).
9. How was Intelsat governed considering its international ownership?
10. How do satellite systems complement broadband service provision?
11. What are the relative advantages and disadvantages of using an internationals service such as Intelsat or O3B for a national broadband network?
12. Compare the O3B and Intelsat broadband services using the TBP framework.
13. What are the drivers behind the roll out of a service such as O3B?
14. What were the drivers behind the roll out of GPS services?
15. Why did the US policy makers allow for GPS services to be used in civilian applications?
16. Why are there so many satellite navigation programmes in the pipeline?
17. What are the policy issues regarding spectrum allocation for LEO satellite mobile telecommunications services?
18. What were the business drivers behind Iridium's network development? What were the miscalculations?

References

[1] https://en.wikipedia.org/wiki/Sputnik_1#/media/File:Sputnik_asm.jpg, accessed 17 August 2015.
[2] http://www.intelsat.com/about-us/company-facts/, accessed 26 October 2014.
[3] http://www.acma.gov.au/theACMA/mobilesatellite-services-band-plan-2-ghz-overview, accessed 23 October 2014.
[4] IEEE. (1984) IEEE Standard Letter Designations for Radar-Frequency Bands, IEEE Standard 521-1984.
[5] http://www.ntia.doc.gov/files/ntia/publications/spectrum_wall_chart_aug2011.pdf, accessed 23 October 2014.

[6] http://www.ntia.doc.gov/files/ntia/publications/redbook/2014-05/4b_14_5.pdf, accessed 23 October 2014.

[7] http://www.ceduna.sa.gov.au/page.aspx?u=495#OTC, accessed 24 October 2014.

[8] http://commons.wikimedia.org/wiki/File:Ceduna_OTC,_South_Australia.jpg, accessed 24 October 2014.

[9] http://www.jsati.com/why-satellite-how-Spacesegment4.asp, accessed 19 October 2014.

[10] ITU. (2015) Satellite Broadband Comes of Age. https://itunews.itu.int/En/2727-Satellite-broadband-comes-of-age.note.aspx, accessed 27 October 2014.

[11] Christensen, J. ITU Regulations for Ka-band Satellite Networks. (2012) http://www.itu.int/md/dologin_md.asp?id=R12-ITURKA.BAND-C-0001!!MSW-E, accessed 27 October 2014.

[12] Padhye, A.V., Li, A., Sattler, C.A., Mohan, N., Liu, R. and Patil, R.P. (2013) There and Back, Inmarsat's Satellite Technology and Beyond. Telecommunications Management Team Project Report, Carnegie Mellon University Australia.

[13] Booker, M., Zhang, M., Shen, M., Vembuli, S. and Khan, R. (2011) An Analysis on Ka-Band Satellite for NBN. Telecommunications Management Team Project Report, Carnegie Mellon University Australia.

[14] http://www.argo.ucsd.edu/sat_comm_AST13.pdf, accessed 26 October 2014.

[15] https://www.iridium.com/About/IridiumNEXT.aspx, accessed 26 October 2014.

[16] Shameel, A., Johnson, S., Dasgupta, T. and Li, H. (2013) Iridium," Telecommunications Management Team Project Report, Carnegie Mellon University Australia.

[17] https://www.flickr.com/photos/iridium_communications_inc/6119847235/in/photostream/, accessed 17 August 2015.

[18] http://geographyworldonline.com/tutorial/instructions.html, accessed 26 October 2014.

[19] http://www.statista.com/chart/1345/top-10-smartphone-apps-in-q2-2013/, accessed 17 August 2015.

[20] Alohali, A, Lao, D., Shao, H.-W., Gao, L., Yang, M., Lertanuntasuk, M., and Arora, V. (2013) Google Maps. Telecommunications Management Team Project Report, Carnegie Mellon University Australia.

[21] Taverna, M.A. (2011) Completing Satnav System to Cost $2.5 Billion. http://aviationweek.com/awin/completingsatnav-system-expected-cost-25-billion, accessed 19 October 2014.

12

Personal Wireless Communications Systems

Preview Questions

- What are unlicensed frequency spectrum bands and why are they used?
- How are personal communications systems regulated?
- How can telecommunications systems network without the presence of a central node?
- What are telemetry and sensor networks and how do they communicate?
- What is visible light communications and its import?

Learning Objectives

- Unlicensed band wireless communications systems technologies
- Mesh networking systems
- Wireless local area networking technologies
- Wireless personal area networking technologies
- Visible light networking technologies

Broadband Telecommunications Technologies and Management, First Edition. Riaz Esmailzadeh.
© 2016 Riaz Esmailzadeh. Published 2016 by John Wiley & Sons, Ltd.
Companion Website: www.wiley.com/go/BTTM

Historical Note

As discussed in previous chapters, the radio frequency spectrum is generally allocated by a national body to provide specific telecommunications services to end-users. Governments further design policies to ensure the spectrum is licensed to those organisations which can efficiently provide these services.

The discovery of dielectric heating at microwave frequencies for medical usage (such as in physiotherapy) and industrial use (such as microwave ovens, pictured [1])[1] in the 1940s necessitated frequency allocations for non-telecommunications applications. However, these devices emitted substantial electromagnetic energy which potentially could interfere with nearby telecommunications equipment operating using the same frequency band. To ensure zero mutual interference a number of frequencies were set-aside worldwide for industrial, scientific and medical (ISM) applications by the International Telecommunications Union (ITU) in 1948 [1]. These are known as ISM bands, and devices can operate in these frequencies subject to transmission power constraints to minimise interference. As no government licence is required for devices operating in these frequencies, ISM bands are also known as *unlicensed bands*. ISM bands were reserved for non-telecommunications applications until 1985 when the US Federal Communications Commission (FCC) approved their usage for telecommunications applications such as cordless telephones [2].

With increased usage of computer terminals and personal computers (PCs) in the workplace and the move to paperless offices, the mid- to late 1980s also witnessed a move towards wireless intra-office communications. The goal was to untether the office, and use wireless links to connect computers to each other and to servers and printers. However, this was not a trivial technological undertaking and research projects on short range, indoor telecommunications were faced with many obstacles. One main obstacle was the lack of available frequency spectrum: while public mobile telephony operators had licensed spectrum from relevant government authorities, usage of such licensed bands within a private office was not possible. One reason was technological and business related in that mobile operators were not (yet) capable of providing service coverage within all buildings and homes. Another reason was policy related and associated with the complexity of spectrum allocation to multiple numbers of private building owners for telecommunications applications in confined spaces. A non-licensed personal communications system was therefore more suitable, and therefore the ISM band was deemed most appropriate. The release of ISM band for personal communications systems by FCC in the mid-1980s resolved many business and policy concerns [3, 4]. Accordingly, a number of standards were developed to facilitate wireless local area networking, including the now common IEEE 802.11 series.

Popularity of mobile telephones in the 1990s led to the emergence of a set of new devices. Wired hands-free headsets (speaker and microphone), gave way to wireless headsets using

---

[1] http://transition.fcc.gov/sptf/files/E&UWGFinalReport.pdf. Used with permission of Raytheon Company.

Bluetooth technology in the mid-1990s. Again these personal communications devices were designed to use the ISM band. Both 802.11 and Bluetooth standards were designed to form ad-hoc peer-to-peer networks, without the need to connect to a central node such as a base station.

A need for telemetry, especially remote reading of utility usage meters gave rise to the ZigBee standard. Again, this standard uses an ISM band to facilitate remote electricity, gas or water meter readings, as well as enabling smart meters and devices.

The above technologies, as well as a number of others such as near field communications (NFC) and visible light communications (VLC) devices mostly augment the broadband services discussed in previous chapters.

Wireless Personal Communications

Personal communications, also known as consumer communications, devices and standards generally complement (although at times compete) with public wireless standards such as 3G and 4G. These systems generally operate within the ISM unlicensed spectrum bands and are an important component in the delivery of broadband telecommunications services. Similar to other telecommunications systems, their success in the market place has depended on addressing technological, business and policy issues.

A number of technological challenges exist for systems operating in non-licensed bands. Historically these frequency bands are utilised by a large number of devices using a number of different standards. By definition, these devices operate in an uncoordinated fashion, and therefore mechanisms have been developed to ensure they can work under multi-user, multi-standard interference. Another challenge is network establishment in the absence of a central node. This has resulted in the design of special protocols for peer-to-peer and ad-hoc networking.

A major technological challenge is associated with the fading characteristics of the indoor channels where most personal systems operate. Solid walls, desks and chairs, and movement of workers lead to significant signal power loss as well as the existence of a large number of reflections and transmission paths. These lead to large signal variations and multi-path fading. As discussed in Chapter 10, fading results in highly degraded quality of service, and must be compensated by power control and diversity reception techniques. However, compensation techniques are complex and are not readily practical for inexpensive, simple personal communications systems. Moreover, wireless digital communication techniques were nascent and signal processing requirements for fading compensation well beyond what early devices could deliver.

From a business point of view, personal consumer communications devices needed to be of a small size and low cost to allow for integration with and/or within other consumer devices such as mobile phones and PCs. To ensure maximum market size, international standards were needed. Therefore, while most of these standards were developed based on proprietary technologies, standards and global consortia had to be developed. International standards further required harmonisation of unlicensed bands, which required governmental policy cooperation.

Considering all these challenges it is surprising that working solutions could be developed in a very short space of time. The first IEEE 802.11 standard for a wireless local area network (LAN) was published in 1997 and inexpensive access points and data cards were available in the market as early as 1998. The development of Bluetooth technology by Ericsson in the mid-1990s was another solution for inter-device wireless connectivity and soon commercial devices were offered to the market. A number of technologies have been developed since for short- and mid-range wireless communications using a shared spectrum. Wireless LAN and Bluetooth remain the most successful, although a more recent addition, ZigBee, is finding a niche.

IEEE Networking

Our wireless connectivity needs can be defined over a wide range of distances and applications: from near-field communications over a distance of a few centimetres for keyless entry to office, to wireless headset for a mobile phone over a distance of a metre, to communicating with a fixed satellite for remote internet access. One way to classify these different wireless telecommunications systems is based on the distance between the transmitter and receiver. Alternatively, the required rate of transmission and the required transmission power may be considered.

The Wireless World Research Forum classifies terrestrial public wireless telecommunications systems based on the service needs of an end-user. Table 12.1 illustrates a model based on this classification, with an individual body and personal, immediate and wide area connectivity applications. Personal and body area network (PAN and BAN) address applications include networked glasses, watches, medical monitoring and fitness sensors. A number of such devices exist in the market, for example Google Glass and Samsung Galaxy S5 Gear. These devices wirelessly connect to a central device over a typical distance in the order of a few metres. The next level is the local area, where connection is to networked appliances (e.g. TV, fridge, PC, printer). Wireless LAN devices are the most commonly used devices in this category. Connectivity is also required over wide areas as one needs to connect to other people and sources of information, and smart phones are a prominent example.

A number of technologies have been developed to support these networking applications. Each category may emphasise a parameter of interest, such as transmission rate, power and range depending on the application.

As noted above, personal (and body) area networking is for applications within a range of a few metres. Two IEEE standards have emerged to address this market: one is the 802.15.1 series, better known as Bluetooth. Bluetooth applications are for short range connectivity at speeds of one to a few megabits per second (Mbps) and transmission power of up to 100 mW. The other is 802.15.4, better known as ZigBee. ZigBee devices transmit at power levels between 1 mW and

Table 12.1 A summary of networking levels

Personal/body area networking	Biometric devices (Fitbit), smart glasses (Google Glass), smart watches (Apple Watch), keyless car entry, etc.
Local area networking	Wireless office (wireless LANs), smart devices (fridges, air conditioners, TV), home security, etc.
Wide/metropolitan area networking	Mobile telecommunications networks

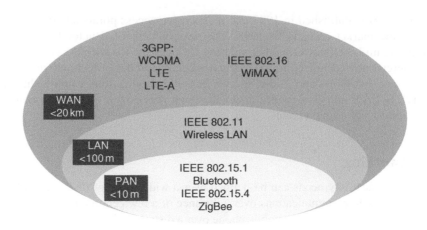

Figure 12.1   IEEE depiction of networking technologies and their range

100 mW, and have transmission rates in the order of 20–250 kilobits per second (kbps). ZigBee applications are mostly in telemetry and sensor networking, with a range of 20–100 m.

A number of technologies were developed for local area networking applications in the late 1980s and 1990s. One such standard was HyperLAN which was developed under the auspices of the European Telecommunications Standards Institute (ETSI). Several other proprietary standards were also developed. However, the surviving standard is the IEEE 802.11 series, popularly known as Wi-Fi. IEEE 802.11 standard specifications cover a range of transmission rates and ranges. They have also been adopted for a range of applications from indoor office to vehicle-to-vehicle and vehicle-to-road communications. Although transmission range can be extended to several hundred metres, the common range is 50–100 m. Both IEEE 802.15 and 802.11 families of standards operate using the ISM frequency band.

Wide area networking has been facilitated by a number of standards such as GSM, CDMA and LTE, many of which were discussed in Chapter 10. Gradually there has been a convergence towards a single technology and the fourth[h] and firth generation of wide area network (WAN) standards are centred on LTE and LTE Advanced. These standards are generally developed under the auspices of ITU, although one competing standard was initially developed by the IEEE. It is possible that the IEEE 802.16 standard, better known as WiMAX, may still serve a niche market of fixed wireless access in a number of markets. However, it is likely that one technology will capture the lion's share of the market. All WAN standards operate within licensed bands and are not considered as personal communications systems (Figure 12.1).

Industrial, Scientific and Medical Frequency Band

As discussed above, ISM frequency bands were initially set aside internationally for industrial, scientific or medical applications. Originally no communication was allowed within these bands to prevent mutual interference between devices. Applications included microwave ovens (2.45 GHz), plastic welding and physiotherapy diathermy machines. ISM bands cover a range of frequencies in different spectrum bands: from a 14 kHz bandwidth at 13.553–13.567 MHz, to a very wide bandwidth of 2 GHz at 244–246 GHz. Since the 1980s

Table 12.2 ISM frequency bands for telecommunications

Frequency range	Bandwidth (MHz)	Region
433.05–434.79 MHz	1.74	EMEA
902.0–928.0 MHz	26	Americas
2.4–2.5 GHz	100	Worldwide
5.725–5.875 GHz	150	Worldwide
61.0–61.5 GHz	500	Subject to acceptance

Source: Electronic Code of Federal Regulations [5].

five ISM bands have been opened internationally to telecommunications applications as shown in Table 12.2. These are used as unlicensed bands where personal communications systems may operate. The first two at 400 and 900 MHz are used by analogue cordless telephony systems. PAN and LAN standards such as Bluetooth, ZigBee and Wi-Fi, operate within the 2.4 and 5 GHz bands. The millimetre wave frequency of 61 GHz is a newly proposed addition.

Although devices can operate in the ISM band without license, they still need to abide by power emission regulations issued by the relevant authorities in different countries. For example in the United States the Federal Communications Commission (FCC) mandates that the transmission power from the devices operating in the 902–928, 2400–2483.5 and 5725–5850 MHz bands to be less than 30 dBm (1 W). Moreover, if the antenna gain (directionality) is more than 6 dB, for every 3 dB, the maximum transmission power is to be reduced by 1 dB. For example if the antenna gain is 12 dB, the maximum transmission power is reduced by $(12 - 6) / 3 = 2$ dB to 28 dBm [3].

IEEE 802.11 Standards

Wireless LANs or Wi-Fi equipment are one of the most common personal communications devices in use: Wi-Fi access points are found in homes, offices, coffee shops, public squares, shopping centres and so on. Similarly Wi-Fi chipsets are included in smartphones, laptop computers, tablets, consumer electronics, cameras, printers, fridges, air conditioners and so on. North River ventures reports that the number of Wi-Fi enabled devices is expected to reach nearly 8 billion by 2015 (Figure 12.2). To connect these devices, a network of public and private Wi-Fi *hotspot* access points has emerged in many cities around the world. As shown in Figure 12.3 the number of such hotspots is expected to reach 8 million by 2016 as reported by Informa [6].

The IEEE 802.11 series of standards was first released in 1997 with the 802.11a and 802.11b specifications. These early releases have been enhanced with new features including faster transmission rates, better security, and more efficient usage of frequency resources. Recent releases of the standards use wider bandwidths, higher order modulations and usage of advanced antenna techniques (such as multiple-input and multiple-output, MIMO) to enhance transmission rates [8]. A summary of some commonly implemented releases, their transmission rates and technology usage is shown in Table 12.3.

DSSS, Direct-Sequence Spread Spectrum; OFDM, Orthogonal Frequency Division Multiplexing.

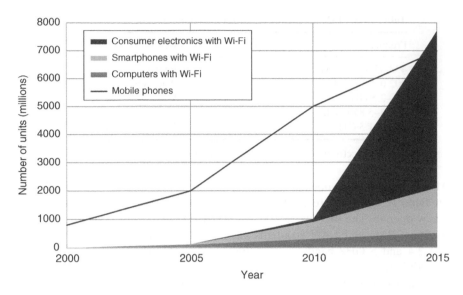

Figure 12.2 Wi-Fi enabled device shipment. Source: Ref. [7]

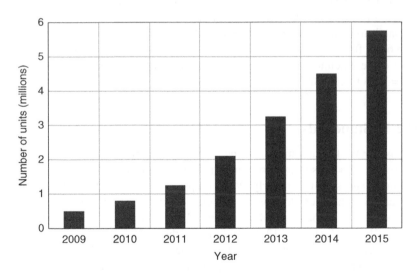

Figure 12.3 Global number of public hotspots. Source: Ref. [8]

With the exception of the 802.11b standard, all other 802.11 standards are based on OFDM for modulation. The carrier frequency bandwidth is typically 20 MHz, which fits in well within the ISM bands of interest. A typical device may use up to 3 subcarriers at any time, as illustrated in Figure 12.4. The subcarriers are used in a time division duplex (TDD) fashion, with two-way communications duplexed in the time domain [as there does not exist a central device such as a base station, a frequency division duplex (FDD) mode cannot work]. OFDM subcarrier bandwidth is 312.5 kHz, and therefore 64 subcarriers can fit in a 20 MHz carrier. Of these 64 subcarriers 52 are used: 48 for data and 4 for pilots. The remaining 12 subcarriers

Table 12.3 Features of IEEE 802.11 standards

802.11 Protocol	Release	Frequency (GHz)	Bandwidth (MHz)	Data rate (Mbps)	MIMO	Modulation
–	June 1997	2.4	20	1, 2	1×1	DSSS/OFDM
a	Sep. 1999	5	20	6~54	1×1	OFDM
b	Sep. 1999	2.4	20	1~11	1×1	DSSS
g	June 2003	2.4	20	6~54	1×1	DSSS/OFDM
n	Oct. 2009	2.4/5	20/40	7.2~150	4×4	OFDM
ac	Jan. 2014	5	20/40/80/160	87.6~866.7	8×8	OFDM
ad	Dec. 2012	60	2160	Up to 6750	1×1	Single-carrier/OFDM

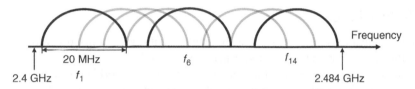

Figure 12.4 802.11 carrier allocation with the ISM band

are left unused and act as guard bands. Data modulation may use any of BPSK, QPSK, 16-QAM, 64-QAM or even 256-QAM. Moreover, MIMO techniques have been specified for the 802.11n standard to further increase the possible transmission rate. Wider carrier bandwidth of 40, 80 and 160 MHz (104, 208 and 416 subcarriers) have been specified for the 802.11n and 802.11ac standards, with a corresponding 2, 4 and 8 times increase of transmission rate.

IEEE 802.11 standard carrier allocation within the 2.4 GHz ISM band is illustrated in Figure 12.4. As noted above, the ISM band is shared with other ISM equipment, as well as other Wi-Fi devices. Multiple carrier allocation allows a Wi-Fi device to avoid interference from other devices by choosing a subset of carrier frequencies which are least occupied. The available transmission rate is calculated from the transmission rate supported by a subcarrier.

Example 12.1

Calculate the maximum transmission rate for the 802.11g standard using the system parameters shown in Table 12.4.

Table 12.4 Some IEEE 802.11n physical layer parameters

Modulation	BPSK, QPSK, 16-QAM, 64-QAM
Channel coding rate	1/2, 2/3, 3/4
OFDM symbol length	3.2 ms
Guard interval	0.8 ms
Carrier bandwidth	20 MHz
Number of subcarriers	52
Pilot subcarriers	4

Answer

The maximum transmission rate is when all subcarriers are transmitted using the highest order modulation and highest coding rate.

- Total number of data subcarriers = 52 − 4 = 48
- Total OFDM symbol length = 3.2 + 0.8 = 4 ms; and therefore total symbol rate = 1 / 4 ms = 250,000 symbols/s
- Higher order modulation = 64-QAM, which supports 6 data bits per symbol
- Highest coding rate = 3/4, which means of every 4 transmitted bits 3 are data bits
- Total maximum transmission rate = 250 000 x 6 x 48 x 3/4 = 54 Mbps

Wi-Fi Networking

Mobile communications systems such as 3G generally use a centralised control system for resource allocation. For example a radio network controller allocates wireless resource to users as the need arises. It ensures multiple users access a network's resources in a fair and efficient manner. This centralised resource allocation can lead to maximal network usage efficiency. In addition, other centralised nodes keep track of where a user is located so calls can be forwarded.

In contrast most personal communications systems operate without centralised nodes. Call set-up is carried out in an ad-hoc manner, and collisions are avoided through special mechanisms. Resource access in Wi-Fi systems is carried out using a channel sense multiple access (CSMA) with collision avoidance (CA) technology. In CSMA-CA, a wireless LAN device wishing to send a packet to another device first listens to see if the channel is currently being used. If the channel is silent, the device sends a short message 'Request to Send – RTS' to the intended destination. If the destination device is free, a reply 'Clear to Send' is transmitted. Upon completion of this 'hand-shaking', packetised data communications between the two devices commence. This process is illustrated in Figure 12.5.

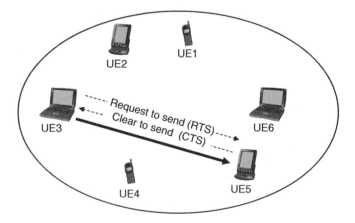

Figure 12.5   CSMA-CA process in wireless LAN systems

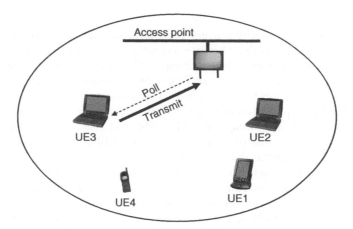

Figure 12.6 Wi-Fi transmission through access point wireless LAN centralised control

In systems with a centralised access point, the access point acts in a similar way to a base station regulating which user devices may transmit at any particular interval through a polling process. When a device is in the proximity of an access point, it responds to the polling signal when it has traffic to send. Based on this signalling channel resources are shared among all active Wi-Fi equipment. This process is illustrated in Figure 12.6. This centralised control mechanism is now the most common 802.11 network configuration: while the peer-to-peer networking mechanism is fully specified, it is rarely utilised.

Case Study 12.1: Wi-Fi Applications and Roll-outs

Although Wi-Fi systems are generally designed to operate for short range communications, they can and have been proposed for long range and compete with mobile telecommunications systems such as GSM or CDMA. The main driver in the early 2000s was the significant cost advantage that Wi-Fi systems enjoyed in comparison with GSM or CDMA base stations. The difference in cost was at times several orders of magnitude. Moreover, the transmission rate capability of IEEE 802.11a or 802.11g was up to 54 Mbps which was several times higher than what a GSM or CDMA standard base station could support. A number of operators considered rolling out Wi-Fi networks to provide a broadband infrastructure. Indeed a number of city-wide networks have been and continue to be rolled out. These systems rightly belong to the infrastructure layer of the telecommunications ecosystem.

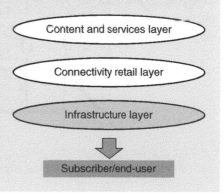

Wi-Fi systems did not become a universal infrastructure layer technology for a number of reasons. One was the inherent short range of communications due to usage of the CSMA-CA resource allocation technique. Moreover, higher transmission rates were possible over short ranges: no performance advantage existed on comparable distances of a few kilometres typical of GSM/CDMA macro cells. Rolling out of a large number of small cells presented other challenges: each Wi-Fi access point needed to connect to a fixed network through an optical fibre link. Such connections were not readily available and the cost of rolling out an optical fibre backhaul was generally prohibitively large, as illustrated in Figure 12.7.

Latest technologies facilitate a network roll-out topology where only a fraction of the access points need a fixed line connection. The remaining access points can connect to these using wireless links. This topology is at times referred to as multi-hop or mesh networking as broadband traffic hops over several access points until it reaches an access point which is connected to fixed backhaul. For example only a subset of all access points in Figure 12.7 are connected to the backhaul network by wire (those

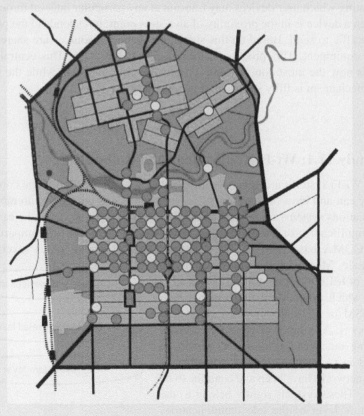

Figure 12.7   A typical city-wide wireless LAN roll-out (Adelaide)

shown in lighter grey). The rest of the access points connect to these, and to the back-haul network wirelessly.

A number of municipalities have rolled out Wi-Fi networks to provide coverage over their central business districts. Many such municipal wireless networks exist: from Johannesburg, to Hong Kong, to Adelaide, to San Francisco. However, most of these systems have been publicly funded, and most have failed to produce sufficient revenue to cover their operating costs. As a result, many cities have stopped their free Wi-Fi services.

A number of cities have turned to a public–private partnership in order to roll out and continue operating city-wide Wi-Fi networks. Adelaide City Council has partnered with Internode, a locally based internet service provider to operate a free Wi-Fi network of more than 300 access points. A mesh topology of 3 to 1 ratio is used to reduce the number of fixed connections: about 100 access points are connected to a fibre or xDSL backhaul. The partnership means that the capital expenditure cost of network roll-out is borne by the City, and operating costs by Internode. While the revenues are zero, the Council benefits from information on people movement in the central business district, as well as providing a free service to attract tourists. Internode benefits from having a free Wi-Fi infrastructure over which it can offer premium services to its customers.

Case Study Questions

- Why have many city-wide Wi-Fi networks failed to operate successfully?
- What are the technology, business and policy aspects of Adelaide free Wi-Fi?
- Is there a case for a local mobile operator to partner with Internode to use the Wi-Fi network?
- Are there any possible revenue streams? What are the associated costs with these business models?
- The contract between Adelaide City Council and Internode is up for renewal in 2020, after National Broadband Network roll-out is scheduled. Should the contract be renewed?

Bluetooth

Bluetooth is a personal area wireless communication technology first developed by Ericsson in the mid-1990s for wireless headsets (pictured).[2] The technology has now been adopted for a number of applications such as wireless mouse and game consoles which require transmission rates in orders of 1 Mbps over distances of several metres. Bluetooth has been further evolved  to combine with a WLAN interface to provide transmission rates in the order of tens of Mbps. It is the second most widespread personal wireless technology after Wi-Fi, and ABI research

---

[2] Bluetooth headset for mobile phones: https://en.wikipedia.org/wiki/Bluetooth#/media/File:Bluetooth_headset.jpg.

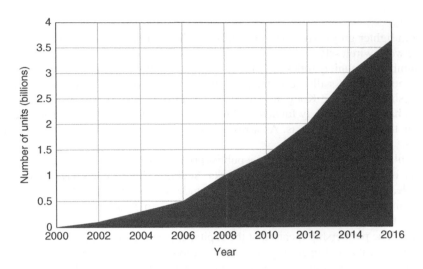

Figure 12.8 Bluetooth enabled device shipment. Source: Ref. [10]

reports that by 2016 there will be some 4 billion Bluetooth enabled devices worldwide (Figure 12.8). Applications include wireless headsets, mouse, keyboard, as well as game consoles of Sony PlayStation and Nintendo Wii, wireless speakers, and so on. Bluetooth technology is also used to connect wearable devices, such as heart monitors and pedometers, to smart phones.

Bluetooth technology is specified through IEEE under the 802.15.1 family of standards. These devices operate in the same 2.4 GHz ISM band as Wi-Fi devices, and similarly need to overcome interference emitted by other devices operating in the same frequency band. Similar to Wi-Fi systems, interference can be caused by other Bluetooth devices, Wi-Fi devices and other ISM equipment active in the same frequency band. In order to operate under this interference, the Bluetooth physical layer uses a technology designed in the 1940s for military communications to avoid enemy jamming. The tech-

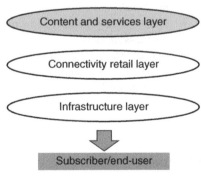

nology is spread spectrum frequency hopping, in which the carrier frequency is regularly switched over a range of available carrier frequencies. In this way, even if one or more of the carrier frequencies experience jamming (interference), time-averaged interference can remain small.

Bluetooth IEEE 802.15.1 standard specifies 79 carriers of 1 MHz bandwidth over which frequency hopping may be made. The transmitter/receiver pair uses any one of these 79 carriers once every 1/1500 s. A hop is then made to another carrier frequency based on a pseudo-random sequence arranged between the pair, as illustrated in Figure 12.9. Interference from other Bluetooth transmitter/receiver pairs can be small as the probability of a collision (i.e. two pairs using the same frequency carrier at any time) is 1/79, or approximately −18 dB. As the number of

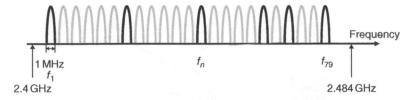

Figure 12.9   Bluetooth frequency hopping in the 2.4 GHz ISM band

Table 12.5   Summary of different Bluetooth standard releases

Version	Data rate (Mbps)
1.0	1
2.0	3
3.0	24
4.0	24

interfering pairs in a small area increases, the overall probability of interference also increases, ultimately resulting in unacceptable performance. However, the probability of many devices transmitting at the same time in close proximity (within a room or an office) is not very high. Interference from other equipment is also similarly mitigated by the frequency hopping process.

A Bluetooth device's maximum transmission rate is limited by the carrier bandwidth of 1 MHz. Several versions of Bluetooth standard have been released and the transmission rate varies between releases. Bluetooth 1.0 specifies a 1 mega symbol/s transmission rate in combination with frequency shift keying technique to deliver a maximum data rate (including channel coding and other overheads) of 1 Mbps. The actual application level data rate however is in the order of 700 kbps. Bluetooth 2.0 uses enhanced data rates through higher order modulation techniques differential QPSK and differential 8-PSK. The latter increase the transmission rate to 3 Mbps. Again application level data rate is smaller and in the order of 2.1 Mbps. Bluetooth 3.0 uses media access and call set-up technologies of Bluetooth in combination with transmission technologies associated with wireless LAN to increase transmission rates to 24 Mbps [10, 11].

Bluetooth 4.0, released in 2010, combines a number of features from previous versions including high speed and low energy protocols. It is mainly intended for long range, low transmission power, low transmission rate applications, such as sensor networking, telemetry and automation. It still maintains a maximum transmission rate of 24 Mbps by incorporating features from Bluetooth 3.0. Transmission rates for different Bluetooth releases are summarised in Table 12.5.

Bluetooth Networking

Bluetooth devices pair and communicate in an ad-hoc manner. Two peer user devices form what is known as a piconet with one device acting as a 'master' and one device a 'slave' during the connection. Up to eight active user devices may form an ad-hoc piconet with one device acting as a master and others as slaves (Figure 12.10). The master controls hopping

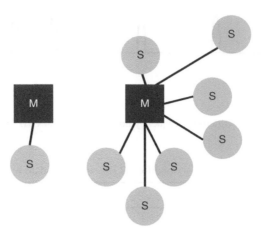

Figure 12.10   Bluetooth piconets

sequences and clock timing for all slaves. All communications flow through the master and slaves do not communicate directly. Generally any device may act as a master. However, in most present applications the smart phone is the master.

In contrast to Wi-Fi devices which mostly use a centralised access point configuration, Bluetooth devices are generally one-to-one master–slave networks. This is useful in applications such as wireless headsets, printing from mobile devices, sharing files between two computers and so on. These work well regardless of whether any other wireless infrastructure such as 3G/4G or Wi-Fi is available.

Example 12.2: Tazzle iT

A number of Bluetooth devices were developed in response to the need for ad-hoc peer-to-peer communications for file transfer or printing applications in the early 2000s. Most computers and printers were shipped without Bluetooth or Wi-Fi connectivity, and therefore stand-alone PC cards or USB dongles were offered to the market. One such device was Tazzle iT, a Bluetooth dongle which was designed to connect Blackberry smart phones to any PC. Bluetooth equipped Blackberry phones could pair with a Tazzle dongle and send files to computers for printing or sharing. A competitive advantage of Tazzle was that all files transferred from a Blackberry device resided on the dongle and therefore full security could be ensured.

Shoot a video

The arrival of iPhones and later on Android smart phones on the market largely eroded the share of Blackberry devices. Tazzle dongle software has evolved to work with new smart phones again to facilitate secure file transfer where other networking options are unavailable or undesired. A Tazzle device sets up the peer-to-peer networking connection using Bluetooth technology, and transfers files using the Wi-Fi air interface. Applications range from printing to sharing files, pictures and videos from a smart phone with a laptop device. Tazzle claims one-click file transfer as its software is programmed to work with the Bluetooth interface.[3]

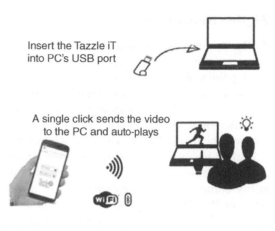

Insert the Tazzle iT into PC's USB port

A single click sends the video to the PC and auto-plays

ZigBee

The ZigBee system is a competitor technology to Bluetooth for short range communications and is similarly specified under the IEEE 802.15 family of standards [12]. It is managed through ZigBee Alliance, a consortium established in 2002 and consisting of more than 400 companies, including major global telecommunications and control companies. ZigBee applications focus on short to long range machine-to-machine (M2M) sensor networking, using the 900 MHz and 2.4 GHz ISM bands. In Europe ZigBee may operate in the 860 MHz band, which is set aside for unlicensed sensor networking applications.

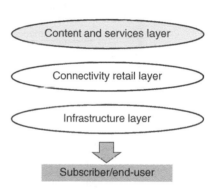

Content and services layer

Connectivity retail layer

Infrastructure layer

Subscriber/end-user

The market for ZigBee devices has been growing at a very fast rate and it is expected that by 2019 there will be more than 220 million devices shipped worldwide [13] (Figure 12.11). Initial applications have been in the utility sector, with sensors used for telemetry, automated and connected home and energy management. The total market size is expected to be $4.3 billion according to the ZigBee Alliance.

The ZigBee physical layer uses the same DSSS technology used in the 802.11b and Wideband CDMA (W-CDMA) systems. Frequency channel bandwidth is 5 MHz, and 16 channels may be used in the 2.4 GHz ISM band. BPSK modulation is used in the 868 and 915 MHz bands, and a form of QPSK in the 2.4 GHz band. Transmission rates are deliberately kept low as sensor networking and automation applications need longer battery life.

---

[3] Photograph and illustration reproduced by permission of Tazzle iT (tazzleit.com).

ZigBee Networking

ZigBee systems use the same CSMA-CA networking mechanism as that of Wireless LANs. The networking topology may be mesh, star or cluster tree as illustrated in Figure 12.12. In a star topology all ZigBee devices connect to a central device, called a coordinator, which may be connected to a network using a wireless or fixed link. Mesh topology allows all devices to communicate with each other and to a coordinator which again passes/receives data to/from a destination network. A cluster tree topology allows some nodes to act as a coordinator to sensors which have low activity factors or are remote.

ZigBee has been successful in finding a niche in smart meters: electricity, water or gas utilities. These meters are connected to a coordinator which passes the usage data to a utility company using telephone lines or a mobile telecommunications standard such as GSM or W-CDMA. ZigBee Alliance is also active to promote usage of its technology in the sensor-based world of the Internet of Things (IoT). ZigBee technology competes with

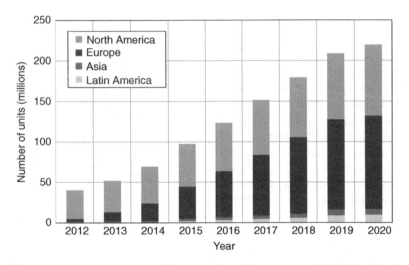

Figure 12.11 ZigBee-enabled installed devices in several regions in the world 2012–2020, based on analysis by Navigant Research. Source: Ref. [14]

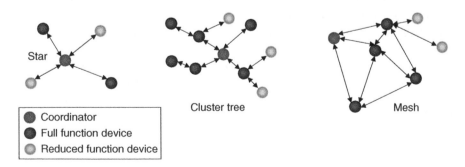

Figure 12.12 ZigBee networking topologies. Source: Ref. [15]

Table 12.6 A comparison of the different personal wireless communications standards

	Bluetooth	Bluetooth 4.0	ZigBee	Wi-Fi
IEEE standard	802.15.1	802.15.1	802.15.4	802.11 (b, g, n)
Frequency (GHz)	2.4	2.4	0.868, 0.915, 2.4	2.5 and 5
Maximum bit rate (Mbps)	1–2	1	0.250	11, 54, 600
Typical throughput	0.7–2.1	0.27	0.2	7, 25, 150
Maximum range outdoors (m)	10-100	50	10–100	100–250
Relative power consumption	Medium	Very low	Very low	High
Battery life	Days	Months to years	Months to years	Hours
Network size	7	Undefined	64 000+	255

Bluetooth 4.0 in serving low power, low transmission rate, medium range applications, such as those of the IoT. Table 12.6 compares the personal wireless communications technologies discussed above. There appears little that separates the Bluetooth 4.0 low energy and ZigBee devices.

Example 12.3: Itron

A major market for ZigBee devices is in utility smart meters. Itron is a leading manufacturer of intelligent metering and data collection. It provides management solution to 8000 utility companies in 130 countries.

Itron OpenWay smart meters (pictured)[4] allow for remote reading and control of service. Electricity, water or gas flow can be monitored and managed through the OpenWay platform. The platform can also record detailed information on what time utilities are used and by what equipment. Such customer information can be used for security and service enhancement. Depending on network load, electricity service flow may be stopped or rationed during periods of peak usage to selected or all individual devices within selected homes.

OpenWay smart meters use ZigBee to communicate with information collection and control hubs. These communication devices require relatively small transmission speeds and can therefore operate using batteries for a specified period of 2 years. OpenWay meters also communicate with a display monitor which allows a customer to choose when to use a service: for example to run a dishwasher at off-peak for lower charges. The smart meter may also control the air conditioning if the overall load on the network is unsustainable. Such smart metering and control can help reduce the overall cost of utility delivery as well as save resources through a centralised communications platform [16].

---

[4] Dwight Burdette (own work) CC BY 3.0 (https://commons.wikimedia.org/wiki/File%3AItron_OpenWay_Electricity_Meter_with_Two-Way_Communications.JPG).

Near Field Communications

The NFC standard is a short-range communications standard, based on the ISO 14443 proximity-card standard associated with radio frequency identification (RFID) systems. An RFID device is an electronic tag (pictured)F[5] associated with a good, which can be read by an electromagnetic transceiver. The tag may be battery powered and be capable of transmitting its unique ID number. Alternatively it is a passive device which is activated when it comes in the proximity of a reader. In passive operation, the reader transmits an electromagnetic wave which induces an electrical current in the tag, which is then used to transmit the ID number. The range of transmission may be several metres in the case of an active device or 3–100 mm in the case of a passive device. An RFID reader usually acknowledges a successful reading of an ID by playing a sound: this operation is illustrated in Figure 12.13.

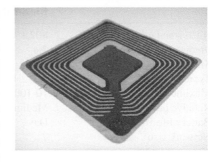

RFID tags are used to identify individual goods and are a direct competitor to technologies such as one-dimensional and two-dimensional barcodes. Other applications include contact-less credit cards (such as Visa paywave and Mastercard paypass), vehicular electronic toll payment systems and contact-less ticketing for public transport systems.

The NFC standard was developed through a consortium formed by Sony, Philips and Nokia in 2004. It uses the operating frequency 13.56 MHz in the ISM band with a bandwidth of 14 kHz, and supports a transmission rate of 106–424 kbps. NFC systems are generally associated with mobile phones and find applications in mobile commerce. These include payment systems, electronic tags, contact sharing, electronic keys and so on, and may be classified within the content/service layer of the broadband telecommunications ecosystem [18, 19, 20, 21, 22].

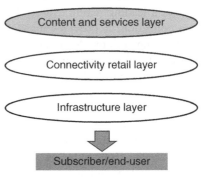

The growth of NFC technology has greatly depended on the growth of an NFC platform and associated networks. One network is manufacturers who embed NFC technology in their products, another is merchants who install NFC readers, and another is the financial institutions which facilitate the transactions. These associated networks provide value to a network of consumers who pay for these services. As the platform has demonstrated its value, the size of the associated networks has grown, reinforcing the original value proposition. This is illustrated in Figure 12.14.

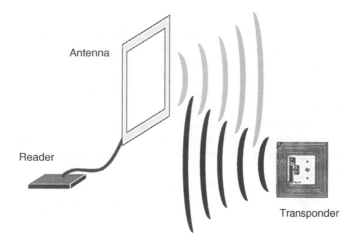

Figure 12.13 RFID operation

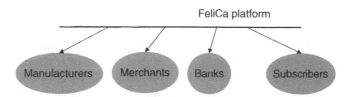

Figure 12.14 A typical NFC network-mediated platform

Case Study 12.2: FeliCa

The history of integrated circuit cards (IC cards, later called smart cards), a forerunner to RFID systems, dates back to the late 1980s when the cards were produced by a number of Japanese manufacturers led by Sony under the product name *FeliCa*. These cards were originally intended for automatic ticketing systems in public transport and employee ID systems.
A number of systems were rolled out in Hong Kong and Singapore before a large roll-out by the East Japan Railway (JR East) company under the brand name of Suica (pictured)[6] in November 2001. These were read–write cards and could be charged up to 10 000 yen, which could then be used to pay for train usage and JR East soon equipped platform kiosks and vending machines with Suica readers to allow micropayment and further enhance the card's value. Soon other merchants such as convenience stores installed readers, and other railway companies introduced similar cards. The convenience

---

[6] Source: East Japan Railway.

of the transport card as well as the popularity of its micropayment function attracted many customers and a large market share in a short time.

The success of the smart card platform coincided with the growth of mobile telephony and the Japanese operator Docomo's i-mode platform (as well as other operators' own data communications platforms). Japanese operators' mobile telecommunications systems of the early 2000s offered many features such as browsing capability and access to an application market place, and the mobile phones are considered as forerunners of smart phones such as iPhone. It was natural for Docomo to consider incorporating Sony's FeliCa platform within i-Mode. The phones could then be used in public transportation as well as micropayment.

The advantages for the users were the security of the device (a passcode could be set for payment acceptance) and integration as a single device could be used for both telephony and as an electronic wallet. There were also a number of benefits for the operators in developing the mobile FeliCa market. The inclusion of a payment system in a mobile would improve customer acquisition and retention. A bigger advantage was entrance into the credit and payment market, where an operator could collect a commission on all such sales. The operators considered their business to gradually move from being a telecommunications company to a financial institution. One such move was Docomo's credit card brand iD.

Mobile FeliCa was a great success and by December 2007 there were 47 million subscribers with FeliCa enabled handsets (more than 50% market penetration). As for the iD credit card, Docomo had 17 million credit card customers as of October 2012, who could use their cards in 560 000 merchants in Japan and, in association with Mastercard paypass, in more than 41 other countries [19,20,21,23].

Case Study Questions

- How did the success of Suica help the roll out of the FeliCa platform?
- Why was Docomo interested in FeliCa? What are the values it brings?
- FeliCa spent $200 million to subsidise the installation of FeliCa readers by small merchants. Why?
- What is the revenue model for FeliCa networks?
- Despite a number of attempts, the FeliCa platform was not successful outside Japan. Why?

Visible Light Communications

Infrared light emitting diodes have been used for applications such as remote control devices since the 1980s. At around the same time, visible light emitting diodes (LEDs) found applications in calculators, computer screens, and later on colour TVs, and automotive, commercial and residential lighting. With the emergence of fast switching LEDs, their application for short range telecommunications was proposed by Masao Nakagawa and members of his laboratory at Keio University in Japan in 1999–2000 [24, 25, 26, 27]. Nakagawa's vision was that white LED devices can be used simultaneously for transmission of information in addition to their usual lighting use, thereby reducing the energy necessary for communications to zero. He coined the phrase visible light communications (VLC) to describe these systems (Figure 12.15).

Other sources of LED could also have a dual telecommunications utility: for example, traffic lights can be used to convey traffic information to cars, gallery lights for communicating background information on paintings, home lighting used for local area networking and so on. Moreover, due to the fact that telecommunications are carried out using visible light, one can naturally see if a link is active.

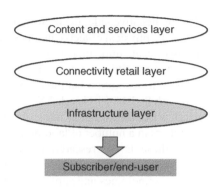

There are many advantages to VLC systems. As noted above, transmissions are embedded in visible light and therefore telecommunication power usage is nearly zero. Light has a very wide bandwidth and therefore can support very high transmission rates. Since visible light frequency is very high, it does not interfere with devices operating in radio frequency bands. Moreover, as light is blocked by walls inter-room interference does not exist. VLC systems are also inherently secure as communications remains invisible outside a room. Furthermore, because the sources of lighting are generally fixed VLC technology can be used for locating and controlling objects such as robots with high accuracy in indoor environments.

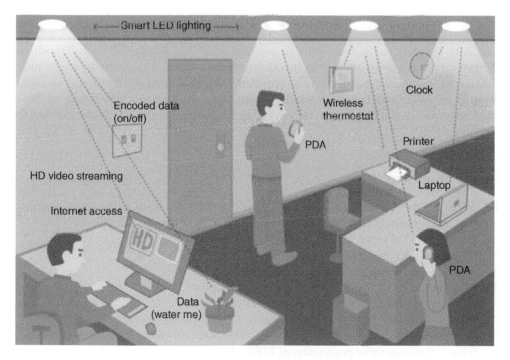

Figure 12.15 A visible light communications system [27]

On the negative side, the high frequency of operation makes it very susceptible to channel conditions: physical objects can block the light and interrupt any communications. Networking can also be an impediment: while lighting in new homes can be wired using power-on-Ethernet, retro-fitting homes can be costly and impractical. Internet over power lines has been suggested as a solution but the transmission rate over these power lines is much slower compared with 802.11 wireless LANs.

High VLC transmission rates of several gigabits per second have been demonstrated by researchers in a number of laboratories, and many applications have been proposed. Foremost among these is the research work carried out in Harald Haas' laboratory at Edinburgh University (e.g. precise location, gigabit/s local area networking, robotics, etc.).A good overview of technologies and applications may be found in [28].

Example 12.4

An interesting application of VLC technology is in diving and snorkelling communications [29]. Electromagnetic waves do not propagate well in salty water, and sonar systems have very low capacity. In contrast VLC systems can support high transmission rates. They can also facilitate a natural way of communications between divers since they can be seen by divers. Marine Com Ryukyu, a company based in Okinawa, Japan, manufactures and markets i-MAJUN, an underwater VLC device. According to the company's website, i-MAJUN can support underwater voice communications of up to a distance of 30 m [30].

Ubiquitous wireless connectivity may be accomplished in a number of ways. Broadband wireless standards may be adopted to provide machine-to-machine connectivity over a range of distances. In practice connectivity is provided through a number of different standards operating in unlicensed frequency bands. These *personal communications* systems are a complementary prominent feature of broadband telecommunications landscape service provision.

Review Questions

1. What are the IEEE specified technologies for personal area networking/local area networking/metropolitan area networking? Which of these have you utilised?
2. What is the ISM frequency band and how is it regulated?
3. Where are IEEE 802.11 standards used?
4. What are the component technologies for wireless LAN?
5. What have been the main drivers for Wi-Fi adaptation?
6. Why is Wi-Fi not suitable for large area roll-out?
7. Wi-Fi systems can be both competitive and complementary to 4G networks. How?
8. Use the technology–business–policy (TBP) framework to compare Wi-Fi and 4G systems.
9. What is the main value proposition of Bluetooth technology?
10. What are the component technologies for Bluetooth?

11. Draw a network-mediated platform for Bluetooth technology and explain why it has grown.
12. Why is peer-to-peer networking easier in Bluetooth networks compared with Wi-Fi?
13. What is the main application of ZigBee technology? In your opinion why is ZigBee used instead of Bluetooth in this application?
14. Are there other applications for ZigBee? How can this platform grow?
15. Where does ZigBee feature in the broadband communications ecosystem?
16. ZigBee and Bluetooth can be considered as competitive. Use the TBP framework to compare the two systems.
17. How are NFC devices powered?
18. What is the value to mobile operators from the growth of mobile payment systems? How do they benefit?
19. What are the main values of VLC? What are the applications?
20. What are the challenges for wider adaptation of VLC systems?

References

[1] Reytheon Microwave Oven. http://www.wired.com/images_blogs/thisdayintech/2010/10/Radarange_first.jpg, accessed 25 August 2015.
[2] http://www.itu.int/dms_pub/itu-s/oth/02/01/S020100002B4813PDFE.pdf, accessed 25 August 2015.
[3] http://transition.fcc.gov/sptf/files/E&UWGFinalReport.pdf, accessed 25 August 2015.
[4] http://transition.fcc.gov/sptf/files/E&UWGFinalReport.pdf, accessed 25 August 2015.
[5] http://www.ccfr.gov/cgi-bin/retrieveECFR?gp=&SID=b52a0c0dbe58a975c37ec6ef1d6e8787&r=PART&n=47y1.0.1.1.16#47:1.0.1.1.16.3.236.31, accessed 25 August 2015.
[6] http://www.itu.int/ITU-R/terrestrial/faq/index.html#g013, accessed 25 August 2015.
[7] http://vr-zone.com/articles/smartphone-and-tablets-expected-to-drive-global-increase-in-wi-fi-hotspots/13915.html, accessed 25 August 2015.
[8] http://www.mobilitytechzone.com/topics/4g-wirelessevolution/articles/2012/12/19/320276-wi-fi-just-part-carrier-infrastructure.htm, accessed 25 August 2015.
[9] http://standards.ieee.org/about/get/802/802.11.html, accessed 25 August 2015.
[10] http://www.bluetooth.com/Pages/SIG-Membership.aspx, accessed 25 August 2015.
[11] http://scholar.lib.vt.edu/theses/available/etd-11182002-115502/unrestricted/Chapter2.pdf, accessed 25 August 2015.
[12] http://www.av-iq.eu/avcat/images/documents/dataSheet/AMX.ZigBee.White.Paper.pdf, accessed 25 August 2015.
[13] https://www.abiresearch.com/press/installed-base-of-802154-enabled-devices-to-exceed/, accessed 25 August 2015.
[14] http://newmore.blogspot.com.au/2013/04/setting-smart-meter-standard.html, accessed 25 August 2015.
[15] http://www.eetimes.com/document.asp?doc_id=1273396, accessed 25 August 2015.
[16] Burdette, D. (2012) Itron OpenWay Electricity Meter with Two-way Communications. https://commons.wikimedia.org/wiki/File:Itron_OpenWay_Electricity_Meter_with_Two-Way_Communications.JPG#/media/File:Itron_OpenWay_Electricity_Meter_with_Two-Way_Communications.JPG, accessed 25 August 2015.
[17] Maschinenjunge. (2008) RFID Chip. https://commons.wikimedia.org/wiki/File:RFID_Chip_004.JPG, accessed 25 August 2015.
[18] http://www.nfcworld.com/2012/10/11/318353/ntt-docomo-to-take-japanese-mobile-wallet-global/, accessed 25 August 2015.
[19] http://www.sony.net/Products/felica/NFC/index.html, accessed 25 August 2015.
[20] http://www.sony.net/Products/felica/business/data/RC-S888_E.pdf, accessed 25 August 2015.
[21] http://www.radio-electronics.com/info/wireless/nfc/near-field-communications-modulation-rf-signal-interface.php, accessed 25 August 2015.
[22] Patauner, C. et al. (2007) High speed RFID/NFC at the frequency of 13.56 MHz. The First International EURASIP Workshop on RFID Technology.

[23] Bradley, S. P. *et. al.* (2006), "NTT Docomo Inc., Mobile Felica", Harvard Business School case study.

[24] Tanaka, Y., Haruyama, S. and Nakagawa, M. (2000) Wireless optical transmissions with white colored LED for wireless home links. Proceedings of PIMRC, 18–21 September, London, UK.

[25] Horikawa, S., Komine, T., Haruyama, S. and Nakagawa, M. (2003) Pervasive visible light positioning system using white LED lighting. *IEICE Transactions*.

[26] https://www.academia.edu/6996573/CSE_Study_Paper_on_._Li-Fi_Technology_The_latest_technology_in_ wireless, accessed 25 August 2015.

[27] Alejosemejiagp. (2014) Li-Fi. https://commons.wikimedia.org/wiki/File:Lifi.jpg, accessed 25 August 2015.

[28] Haas, H. (2015) Latest Li-Fi Research News. http://www.see.ed.ac.uk/drupal/hxh, accessed 25 August 2015.

[29] Ito, Y., Haruyama, S. and Nakagawa, M. (2006) Short-range Underwater Wireless Communication Using Visible Light LEDs. http://www.slideshare.net/guestcd295/short-range-underwater-communication-using-visible-led-presentation, accessed 3 November 2015.

[30] http://www.mcrvlc.jp/, accessed 25 August 2015.

13

Network Topologies, Design and Convergence

Preview Questions

- What are macrocells, microcells, picocells and femtocells?
- How is a wireless network designed?
- How do heterogeneous technologies coexist in wireless service delivery?
- What is fixed-mobile convergence?
- What are the factors for network self-organisation?

Learning Objectives

- Steps in design of a wireless network
- Different wireless networking topologies
- Characteristics of next generation networks
- Convergence of public and private wireless networks
- Self-organising networks

Broadband Telecommunications Technologies and Management, First Edition. Riaz Esmailzadeh.
© 2016 Riaz Esmailzadeh. Published 2016 by John Wiley & Sons, Ltd.
Companion Website: www.wiley.com/go/BTTM

Historical Note

A fundamental mobile telecommunications system configuration consists of a central device, usually known as a base station, which simultaneously transmits and receives signals from several end-user devices. The area covered by the base station is called a 'cell', giving rise to the term cellular communications. This configuration still remains the way all public wireless systems are designed and rolled out. Nevertheless much has changed since the first generation systems.

The first generation system networks were designed around fixed line operator exchanges. Most exchanges had a telecommunications tower for point-to-point relay connectivity and on these the first cellular system antennas were installed. These exchanges were already a part of the public-switched telephone network (PSTN) and readily switched mobile calls to their destination. Moreover, exchanges served a certain number of fixed subscribers, usually 10000–20000, and were therefore ideally located to be the centre of a cellular network. Although the capacity of base stations was limited by the available frequency spectrum and the reuse factor, the number of initial subscribers was low and this early configuration worked quite well. Nevertheless, network coverage was not comprehensive and calls frequently dropped due to weak signal reception.

The introduction of the second generation systems in the mid-1990s and the growth in the number of subscribers necessitated the expansion of cellular networks and roll out of new base stations. Moreover, new operators sprung up in many markets each rolling out a new parallel network. Competition led to price and quality comparison between different operators, which in turn required high quality coverage, at a minimal cost. This required determining the total expected traffic offered to the wireless network, where base stations must be located, and how the network should grow as more subscribers join. Initially these tasks were done by radio frequency (RF) planning groups within the operator but soon specialist companies grew to design and plan the network and roll them out. To streamline these processes, specialised software was developed by planning companies as well as network equipment manufacturers such as Ericsson. Network planning and RF engineering companies (and engineers) were in great demand and charged high premiums for their services in the late 1990s.

Several new network configurations were used to improve coverage inside buildings, underground train stations and crowded areas. These base stations had smaller coverage and served fewer users. To distinguish these from larger coverage base stations, the terms macrocell and microcell were coined. Many new terms have since been coined to represent ever smaller cell sizes such as picocells and femtocells. The combination of all these cell configurations work together to provide comprehensive coverage of populated areas.

Another factor of complexity has come about due to advanced antennas. From the mid-1990s sectored antennas were used to increase the system capacity. Later on adaptive array antennas, and multiple-input and multiple-output (MIMO) antennas further increased the amount of data that could be communicated over the air.

With the introduction of third generation (3G) and fourth generation (4G) standards, many systems have had to coexist in one base station site to reduce capital and operational expenditure. Co-location of base stations from the same operator or even competing operators has become commonplace. These have added a new layer of complexity in network design and operation. Newer broadband telecommunications systems are designed to automatically assign a user to the most optimally placed network, whether second generation (2G), 3G, 4G or Wi-Fi. System design and operation has greatly evolved over a period of less than 25 years.

Cellular Topologies

Mobile telecommunications technologies can provide broadband connectivity to mobile user equipment over broad geographic areas. They are one of the fastest and least expensive means of broadband service provision. However, since capacity is shared among many users and since different places have different topographies, network design is complex. A wireless network operator needs to decide on many technology, business and policy issues before a network is rolled out. This chapter deals with how a wireless network is designed, planned, installed, optimised and maintained.

Wireless networks can cover broad areas. However, coverage is generally made over populated areas, or where people travel through as it is unnecessary to provide 100% geographical coverage of a land mass. It is also difficult to provide coverage of a country's *100%* population as some people live in very remote areas where economical service provision is impossible. Furthermore, different services and technologies have different coverage range business factors that determine which service may be commercially viable at a particular region. This is illustrated in Figure 13.1: Telstra's 3G network covers 99.3% of Australia's population while its network covers 2.3 million km of Australia's land mass (28% of total). Figure 13.1 is for illustrative purposes only and is current as at August 2015 [1].

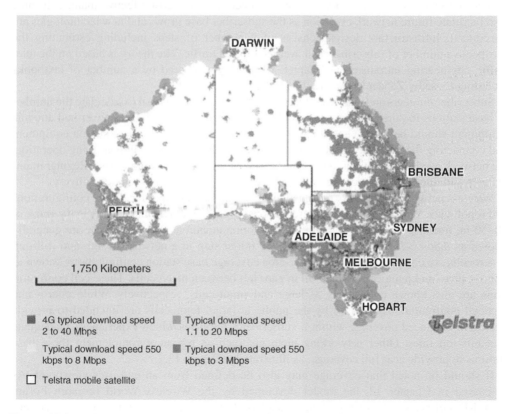

Figure 13.1 Telstra's network coverage in Australia. Reproduced by permission of Telstra Ltd & MapData Services Ptd Ltd [2]

Universal coverage is at times a contractual obligation placed on an operator by the government. However, the cost of such universal coverage is high as many areas may be sparsely populated, and the revenue per these particular base stations can be less than the cost of the service provided. Regardless of its contractual obligations, an operator may still decide to provide wide coverage. This ensures the provision of high grade of service and wide-area user connectivity. Indeed, the extent of coverage is primarily a business decision; lack of widespread coverage has been a primary cause of failure for a number of technologies. A primary example of this is the PHS service in Japan. While the service was initially successful when introduced in 1994, PHS operators could not compete when a wide-coverage 2G service was rolled-out. The main reason was the inferior coverage, and as a result the PHS service has now been nearly phased out.

For highly populated areas such as cities, it is important to provide full geographic coverage to ensure good service reputation. However, this is not easy as a good measure of signal to noise ratio (SNR) for a minimum service capacity is complex to maintain. Physical objects – both natural such as hills and valleys and manmade such as buildings – block the propagation of radio signals, and result in insufficient received signal power. Complex coverage strategies and configurations are needed to provide high quality universal signal reception. However this high quality coverage must be balanced with an operator's need to minimise infrastructure costs.

Wireless telecommunications network infrastructure design, planning and installation is a specialised task and is aimed to service a populated area in a cost effective manner. It must also facilitate future network evolution as the subscriber base grows and new technologies are introduced. Infrastructure design consists of a number of steps including estimating the prospective number of subscribers and average offered traffic. The theory is based on the teletraffic engineering introduced in Chapter 3 [3], and is covered by a number of textbooks including those by Zander [4] and Laiho *et al.* [5].

Subscriber numbers and average offered traffic parameters are used to calculate the number of base stations required, where they should be installed, and which transceiver and antenna equipment should be used. Then comes the actual installation of the base station equipment and connecting them to core network control nodes. As part of initial and on-going operation, the network needs to be optimised to ensure best coverage to the customer base. Regular maintenance and upgrades are necessary after a base station has started its service delivery.

Wireless networks are usually designed using a number of different cell size configurations and topologies. Wide-area coverage base stations, usually with antenna heights in the order of 30–50 m, form the backbone of a wireless telecommunications network. These are generally known as *macrocell* base stations and are the initial step in a network design and roll-out. Macrocells are complemented by smaller-area coverage base station configurations known as micro-, pico- and femtocells, which fill in gaps left between macrocells. These cell configurations are also known collectively as large and small cells, respectively. While micro- and picocell configurations are used by the public at large, femtocells are intended to provide exclusive enhanced coverage within a subscriber's premises which experiences inadequate transmission rates. Other networking elements such as repeaters complement these base stations to provide near full coverage, as illustrated in Figure 13.2.

It should be noted that coverage may also be defined from an end-user viewpoint. As discussed in Chapter 12, the model developed by the Wireless World research Forum (WWRF) defines wireless networking needs based upon an individual user and the services which he/she may use on a personal, immediate, and wide-area basis as discussed in Chapter 12 [6, 7]. This approach can be analysed using the content and services layer of the

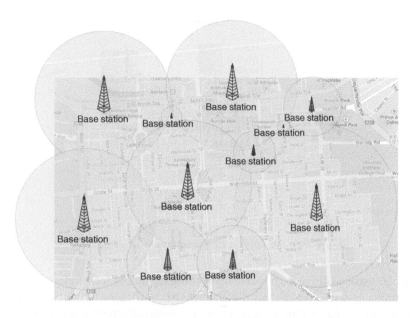

Figure 13.2 An example of full coverage through multi-level cells for central Adelaide

broadband telecommunications ecosystem. The network planning and design approach on the other hand concerns the infrastructure layer. These two approaches to network design bring together device and network technology fields in an attempt to provide seamless connectivity for an end-user over the broadband telecommunication ecosystem. The focus of this chapter is on how the wireless infrastructure layer is designed and rolled out, and content and services layer issues are addressed in Chapter 14.

A complete network needs to be fully connected to fixed telephony and internet. Furthermore, an operator's base stations should be connected to each other and to other operators' networks. In practice all base station nodes and wireless local area networks (LANs) connect to the core broadband telecommunications network through gateways and control equipment. Where possible, connections are made using optical fibre networks. The remaining links are usually wireless: point-to-point relay links are used in rural area base stations, and wireless mesh networking for wireless LAN access points.

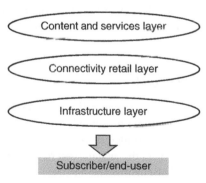

Figure 13.3 illustrates a few nodes in a 3G network and how they are connected to other network nodes. The mobile station connects to a base station through an air interface. Several base stations then connect to a central control node, known as the Radio Network Controller (RNC) in 3G standards. An RNC is served by several databases such as a Home Location Register (HLR) which checks if a mobile device is a valid subscriber. Several gateways connect voice and data traffic to the fixed network.

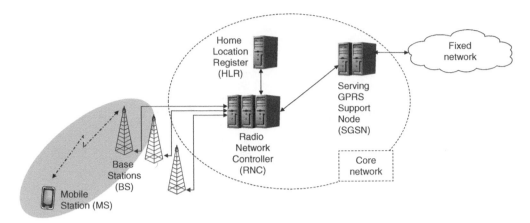

Figure 13.3   Wireless node connection to a core network

This chapter lists and discusses possible networking topologies for wireless broadband tele-communications systems. We discuss the characteristics of these topologies and how they perform considering the specific challenges of broadband telecommunications systems. We also discuss the convergence of wireless and fixed networking technologies and how network coverage may be automatically optimised as the number of base stations and traffic levels change in different geographical areas.

Cell Structure

The classic wireless network design is based on a star topology, where a central base station serves mobiles in a circular area with the base station at its centre. As shown in Figure 13.2 several cell structures can be defined based on the capacity of the base station and the size of the coverage area. Different cell size configurations are further described below.

Macrocell

A macrocell is generally a wireless network area with a radius of several hundred metres to tens of kilometres served by a single base station. The macrocell base station utilises powerful antennas installed atop a tower or a tall building at a height of 30 m or more. Transmission power from the antenna can be up to tens of watts, although it typically is between 5 W and 10 W (37–40 dBm). The base station transceiver equipment is generally housed in a room inside the building or within a box at the base of the tower where an appropriate temperature may be maintained. With recent reductions in base station equipment size, they are at times placed near the antenna on the rooftop. An example is shown in Figure 13.4. Figure 13.5 shows antennas for two typical macrocell base station installations.

A macrocell is usually the first element used in broadband wireless communications systems because it allows for blanket, albeit imperfect, coverage of a populated area. This serves two goals: an operator can claim blanket coverage in its marketing material. It can also minimise

Figure 13.4 Ericsson's Radio Base Station 6301 (size: 1115×415×536 mm). Reproduced by permission of Ericsson AB [8]

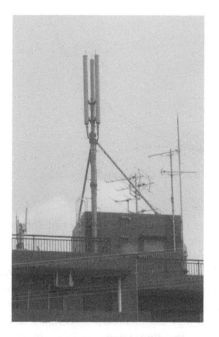

Figure 13.5 Two macrocell base station antennas

initial capital expenditure, and grow the network with the number of subscribers and traffic. For example the initial network roll-out in the central business district (CBD) of Adelaide may look as shown in Figure 13.6. This roll-out requires minimal investment and as the initial number of subscribers to a new system is low, service quality is acceptable. Subsequent introduction of smaller cells may be made in targeted areas as user subscriber traffic grows.

The radius of a macrocell depends on two factors: the amount of expected offered traffic at peak time, and the maximum coverage distance of the technology. The former is a function of

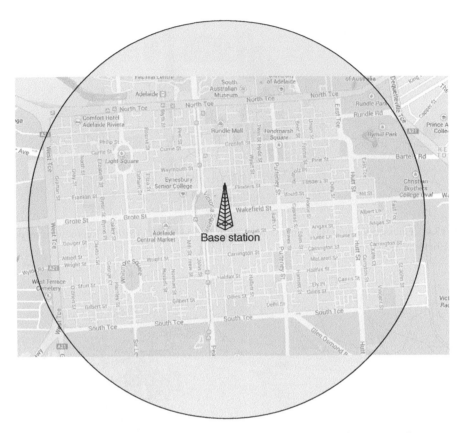

Figure 13.6   A possible initial coverage of Adelaide CBD using macrocells

expected subscriber density. More traffic capacity is required as the number of subscribers increases, and as each base station can service a limited number of users at any one time, the coverage area of a cell in a densely populated urban area is usually small. In suburban and rural areas, population density is relatively smaller and therefore a base station can cover a larger area. The cell radius therefore ranges from several hundred metres in dense urban areas, to several kilometres in suburban areas and to tens of kilometres in rural areas. The limit of coverage in rural areas is determined by the technology and standard constraints. Usually sector antennas are used to divide a cell into several (often three) parts, and thereby increase the capacity of a base station site by a corresponding factor of several (three) as illustrated in Figure 13.7.

The height of base station antennas are a function of cell topography and building characteristics. A summary of macrocell designs is as follows:

- Dense urban
  Downtown areas in major cities, characterised by high-rise buildings and high population density. The cell radius is in the order of 300–500 m. Typical antenna heights are 35 m or higher. Cell sites and antennas are usually placed on rooftops.

Figure 13.7   Sector antennas

- Urban
  Town centres, characterised by medium height buildings. Cell radius is in the order of 1 km and base station antenna heights are 25–40 m. Again cell sites and antennas are usually placed on rooftops.
- Suburban
  Urbanised areas, with single to multi-storey buildings. Cell radius is in the order of a few kilometres and antenna height is 10–25 m. Cell sites are usually rooftops or in specially built towers.
- Rural
  Sparsely populated areas with little traffic. Here the cell radius is determined by a system's link budget (i.e. how far the signal can travel before received SNR falls below a required threshold). The cell radius also depends on the technology's coverage limitations, which is typically tens of kilometres.

Smaller Cell Configurations

Smaller cell configurations are used to complement macrocell coverage. One such configuration is a microcell. A microcell is quite similar to a macrocell with the main difference being a less powerful smaller antenna and a smaller transmit power setting. Microcells are generally employed to fill gaps left after macrocell roll-out and/or to enhance capacity. They are also installed in areas where coverage cannot easily be provided using macrocells. For example, in highly built-up urban areas, buildings block the signals and create areas with low received signal levels. Another example is isolated areas such as underground train stations or shopping centres. Microcell designs may also be utilised in high population density areas such as concert arenas or stadiums to increase system capacity. In these configurations smaller cells usually underlay a macrocell as shown in Figure 13.8. A microcell base station is usually connected directly to the core network using an optical fibre link, but at times the traffic may be routed through a macrocell base station with a wireless point-to-point link.

Microcells and other smaller cell configurations are nowadays integrated into building and street fixtures for aesthetic reasons. One example is a joint design by Ericsson and Philips marketed as 'zero site', shown in Figure 13.9 where the base station equipment and antennas are integrated into street lighting poles.

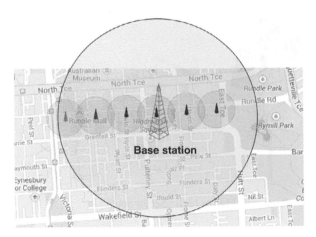

Figure 13.8   Smaller cells underlay a macrocell to cover a high subscriber density area (Rundle Mall in Adelaide)

Figure 13.9   An Ericsson and Philips small cell design 'zero site' integrated into an electricity pole. Reproduced by permission of Ericsson AB [9]

When a small cell underlays an existing macrocell service area, the signals from the two base stations may interfere with each other and network planning needs to find ways to minimise intra-system interference. In systems where the carrier frequency is reused such as GSM (Global System for Mobile communications), the overlapping configuration uses different carrier frequencies. In 3G and 4G standards, dynamic resource allocation techniques are used to minimise mutual interference.

Picocell

Another small cell configuration with a coverage area smaller than that of a microcell is a pico-cell. The difference from the micro-cell however is not only in size: picocell base stations are usually installed in isolated, confined areas. This can be in an auditorium, inside an aeroplane, or on an underground train platform. A picocell base station capacity is also usually smaller than that of a microcell as the expected traffic is quite low. Accordingly, the transmission powers and antenna sizes are also smaller. Picocells are designed to fill in gaps left by macro- and microcell configurations. As they are generally well isolated, frequency planning is not a significant issue: low transmit powers will further ensure minimal interference to its surroundings.

Again, connections to the core network can be via a macro- or microcell base station, or through dedicated fixed links. The connection may itself be wireless, fibre or cable. At times the picocell base station may act as a subscriber to the overlaying cell structure's larger cell, which relays traffic to the core network. Other wireless connections may be fixed wireless, or even satellite links as in the case of aeroplane picocells.

In practice large and small cell configurations usually overlay to cover an area. However, still there may exist areas where coverage remains weak. Femtocell configuration may be used to provide connectivity in these areas as described in Case Study 13.1.

Case Study 13.1: Femtocells

Femtocells are another small area cell configuration, and are similar to picocells. The main difference is that femtocell equipment is installed at a subscriber's premises and connected to an operator core network over ordinary fixed (such as cable) lines. They are utilised where a subscriber's premises has poor reception and therefore cannot transmit or receive at a desired rate. As such, femtocell equipment are personal consumer devices which fill in a gap in an operator network reception weakness. The range of coverage is tens of metres to fill in the coverage gap at a home or in an office. Femtocells were first introduced in the mid-2000s for GSM, and later on for 3G and 4G standards. A range of manufacturers and original equipment manufacturers produce these devices and they are usually sold under operator brands. Examples are Optus' Home Zone [10] or Docomo's My Area (discontinued in February 2012) [11]. Manufacturers may also market small cells which connect into a building's Ethernet network. An example is Ericsson's Dot shown in Figure 13.10.

Figure 13.10   Radio Dot by Ericsson. Reproduced by permission of Ericsson AB [12]

The femtocell main value proposition for a customer is better mobile network coverage. This translates into higher 3G/4G data transmission rates and better voice quality calls. Furthermore, subscribers can continue to be connected to the network using their mobile phone when they are inside a home/office and therefore removes the need for a home/office phone. They are targeted particularly at businesses housed in large buildings where reception quality degrades the further one moves inside. Connection charges over a femtocell can be at a discount or free as traffic travels mostly over fixed lines rather than operators' wireless links. This fits well with a recent trend called *bring your own device* (BYOD) has become very popular with many companies. BYOD generally means an employee uses his/her personal device (smartphone, tablet, PC) also at work. Under BYOD regimes high quality reception is a significant requirement everywhere in the office, and in many instances can only be delivered through femtocell configurations.

The value proposition for the operator is high quality network coverage at lower cost, as a similar level of quality necessitates more macro/micro base stations. Operators have generally followed two main different femtocell roll-out models: one where the equipment remains a property of the operator; and another where it is sold to the subscriber. In the former, an operator incurs the cost of the equipment and needs to also service and maintain the device if it breaks down. In the latter, a subscriber purchases the femtocell equipment, and is responsible for its installation and repairs. This may be considered unfair as the customer is burdened with the cost of repairing an operator's coverage gap. However, owners of commercial buildings may be incentivised to install femtocell equipment and provide high quality coverage to prospective tenants. In either case, the femtocell traffic is carried over a fixed [digital subscriber line (DSL), cable] link to the operators' core networks. This traffic may count towards a subscriber data allowance and be accordingly charged.

Femtocell roll-outs have been attempted by a large number of operators around the world. Some operators have in contrast decided against using these devices. At times an operator's marketing of femtocells has been considered as evidence of defects in its network coverage. One operator referred to a competitor's usage of femtocells as a 'means of compensating for poor coverage' [13]. Subscribers have also rightly asked why they should bear the cost of equipment which fixes an operator's lack of high quality coverage.

During the course of writing this book, many operators were still using femtocells. There is little marketing for the service, and the equipment cost is usually borne by the operator. As network coverage matures it is expected that the need for femtocells will diminish. However, universal high quality coverage is very difficult to achieve and there remains a business case for femtocells in the foreseeable future.

Case Study Questions

- Why are femtocells necessary?
- How are coverage imperfections discovered? Can it be automatic?
- Why are femtocells usually marketed and sold under operator brands?
- How can a mobile operator compensate for the traffic carried over a subscriber's fixed network?
- Why are femtocells an economical solution for coverage defects?

Repeaters

Two other cell figurations are used in practice. One is a repeater as illustrated in Figure 13.11. Repeaters are simple devices consisting of an outdoor receiving/transmitting antenna, an amplifier, and an indoor transmitting/receiving antenna. The function of a repeater is to increase the signal power within a confined area in the same way that a microcell or a pico-cell base station does. However, a repeater is simpler as it does not carry any signal processing itself. It only bridges a well-covered outdoor environment with an isolated, often under-ground, environment through a receiver and transmitter antenna pair. They are also used to increase signal power inside buildings. One example is the Mobile Smart Antenna offered by Telstra [14].

Femtocells and repeater configurations show that no coverage can be perfect, and 'band-aids' are necessary to fix remaining areas after a network has been rolled out.

Distributed Antenna Systems

Another solution for indoor building coverage is illustrated in Figure 13.12. Here the radio signal is sent inside a building using fixed links such as fibre or cable. The radio signal is connected to distributed antenna systems throughout the building. This ensures that the coverage within the building remains consistent and of high quality to enable desired signal power levels and connectivity. Femtocells and distributed antenna systems are the two possible solutions for businesses wishing to move to a BYOD regime.

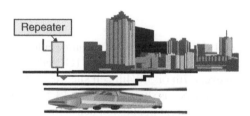

Figure 13.11 A repeater connecting an underground area to street level coverage

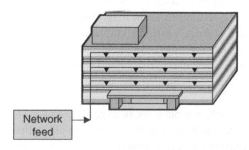

Figure 13.12 Indoor building coverage using distributed antenna systems

Network Planning

The design, roll out, optimisation and maintenance of a wireless network is a complex task, and must consider a large number of factors at its different stages of planning and implementation. The most basic factors are the range of coverage as determined by the base station maximum transmit power, and the subscriber density and expected offered traffic. Usually the former determines cell size in sparsely populated areas such as rural or suburban. Offered traffic is on the other hand the determining factor for cell size in urban and densely urban areas. The initial plan is based on these factors.

In the early days of mobile communications, network planning was carried out by a specialist group within the operator. With the explosive growth of mobile communications in the mid- to late 1990s, many specialist network design companies entered the market. Network planning followed by implementation, roll out, optimisation and maintenance functions are time-consuming and human-resource intensive. Furthermore, after the network is designed and rolled out, and service is introduced, the operator needs to regularly test the network to ensure a high quality of service is maintained. Moreover as the customer base grows, the operator needs to improve the service, revise its network plan, add new cell sites, and retune the network.

In addition, there are business strategic aspects to the network design stage of the business. The operator must decide which markets to target; how broadly to cover a country/region; where to the place base stations or access points; which network topology to use; how to connect the base stations or access points to a core network; how to facilitate further evolution of the network's design and so on. It must also decide whether or not to partner with other operators to share network resources such as base station sites and antennas.

Frequency Reuse

Another factor impacting cell design is spectrum resource allocation and frequency reuse factors as discussed in Chapter 10. The first generation (1G) and 2G standards reused the allocated frequency spectrum in every 7th or 3rd/4th cell site, respectively. While this process is easy to design and implement on paper as shown in Figure 13.13, designing for a real city with buildings, hills and valleys, and different population density requires different size and shape of cells. Frequency allocation was a main task of cell design companies.

The Code Division Multiple Access (CDMA)-based 3G standards simplified frequency allocation as the same frequency could be used in all cells. Cell design for 3G systems however required a good management inter-cell multi-user interference. The signal from any base station represented itself as interference to all neighbouring cells and therefore antennas had to be tuned to ensure signal leakage outside a cell was minimised. Figure 13.14 shows how two neighbouring cell antennas are tuned. As the signal from a base station appears as interference to the neighbouring cell's users, the antennas of base stations are tilted in order to minimise the mutual interference. Adjusting the antenna tilts is one factor in controlling and tuning the network.

Another challenge arises when a network is expanded and new base stations need to be introduced into the network. Such introduction is not trivial as there is no suitable place to install a new base station and create a cell around it, as can be seen from Figure 13.15. Antennas and networks need a retune to ensure optimised performance.

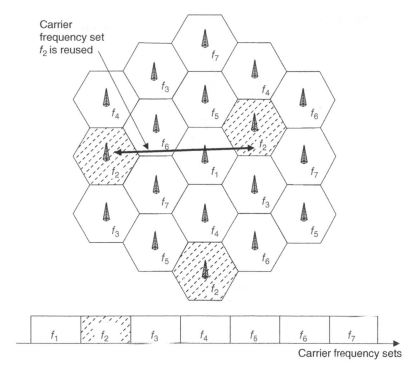

Carrier frequency sets

Figure 13.13   Frequency reuse factor of 7 for 1G systems

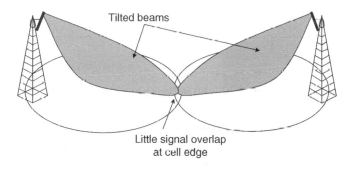

Figure 13.14   Antenna tilting to optimise coverage

The 4G standards present yet a different set of challenges as the Orthogonal Frequency Division Multiple Access nature of resource allocation cannot work with a frequency reuse factor of 1. However, dynamic resource allocation ensures that each time/frequency slot is used as often as possible. In practice, a fractional frequency reuse factor of 1 and 3 as demonstrated in Figure 13.16 is used [15]. In these systems, a cell's central area uses all available frequencies, but at cell-edge frequencies are dynamically allocated. Under full system utilisation, the cell edge is likely to have a frequency reuse factor of 3.

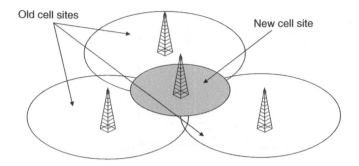

Figure 13.15   Introduction of a new base station

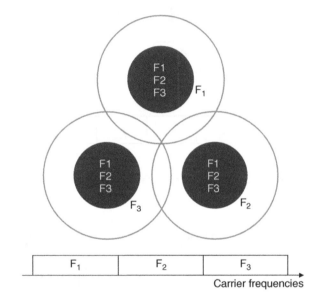

Figure 13.16   Fractional frequency reuse in 4G systems

In practice, frequency resource allocation for a 4G system is dynamic and time/subcarrier slot allocation is coordinated between advance base stations. This simplifies the design of cell configuration and allows for heterogeneous networks to coexist and operate.

All these tasks are resource intensive, and many are outside the core competency of many operators. Therefore most operators outsource network planning to the specialist companies. A number of different organisations work together to plan and manage mobile communications infrastructure. These include software companies which develop signal network design programmes, building and installation companies, real-estate management companies and so on. A business system diagram of a network planning process is shown in Figure 13.17. Network planning companies which carry out or manage the above functions have become an integral part of infrastructure roll-out and therefore of great value to mobile operators.

As Case Study 13.2 demonstrates, infrastructure and network planning companies' fortunes have been mixed: high demand led to large profits and high valuations in the late 1990s. As demand subsided, and as more competitors entered the market, margins and stock valuations

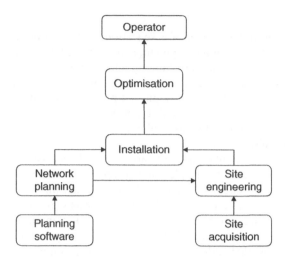

Figure 13.17 Cell design value chain

sharply declined, a trend accentuated by the general downturn in the information technology industry after the year 2000. The supply and demand have balanced and the rates charged are comparable with other professional services companies. The core competency of these companies has also changed over the past two decades as will be described further in this chapter.

Case Study 13.2: Cell Design Companies

The mid- to late 1990s witnessed a fast growth in the mobile telephony market. The success of the 2G standards such as GSM and CDMA enabled operators to offer services at competitive rates. Subsidised handsets and economic call plans led to the growth of the number of subscribers and mobile telephony traffic. As a result, a large number of new cell sites had to be rolled out in response to this increase in demand. While many incumbent operators had a cell design division in-house, the size of this work force was not adequate for a fast roll out of new sites.

At the same time deregulation of the telecommunications industry had allowed for multiple players to apply and be granted licenses to operate mobile networks. The majority of these new entrants did not have in-house expertise in designing and rolling out mobile networks. A great demand emerged for specialist companies who could design, plan, install and optimise wireless networks. A number of these companies came into existence in the mid- to late 1990s and grew significantly. Network planning jobs naturally required expert human resources as well, and RF engineers were in great demand. Many communications engineers found employment in these companies with inflated salaries: one company even offered a brand new BMW as a sign-up bonus. These engineers were in turn tasked to design and build operators mobile networks at high premiums. The valuation of network planning companies grew significantly, with many of them priced at several billion dollars at the turn of the 21st century.

The value proposition of network planning companies was their turn-key solution provision. Many of these companies designed the cellular network based on the expected voice traffic, determined the number of base stations and locations, built the infrastructure and installed the base station equipment, and optimised the coverage by tuning the antennas. They also offered continuous tuning and optimisation as the subscriber base grew. Many of these tasks were carried out in-house but some were also outsourced. A business model diagram of the cell design companies is shown in Figure 13.17. Of particular value was the network planning software which simulated a base station signal coverage based on three-dimensional maps. Expertise in RF engineering such as site engineering was also of value.

As more accurate network planning software entered the market, the main component of the value proposition of network planning companies became an experienced work force. However, as more engineers entered the job market this value proposition fell. Moreover, much of the design and simulation work was carried out remotely and in lower cost countries. The entrance of lower cost network planning companies into the market and competition for operator's business drove the prices down. On the other hand, the mobile network build-up phase was gradually completed and therefore the demand side cooled. By the mid-2000s most of the demand was in network tuning and optimisation, and significantly smaller numbers of engineers were needed at site. Consequently much of the high valuation of network planning companies disappeared, and most were taken over by rivals or fixed-line networking companies.

Case Study Questions

- What were the main drivers behind the high valuation of wireless network planning companies?
- What value did wireless planning companies provide to mobile operators? Were these values inimitable and enduring?
- What was the value of the work force in this business and how has it changed?
- How did technology, business and policy factors impact this business?
- What were the reasons for the relative decline of network planning companies?

Wireless Local Area Network Cellular Structure

As discussed in Chapter 12 IEEE 802.11-based wireless LANs are a personal communications element in overall delivery of broadband telecommunications services. A possible cell structure for these systems is similar to the macrocell definition above. Although several technologies have been developed to increase the coverage area of wireless LAN systems, due to the channel sense multiple access nature of their operation, coverage areas are likely to be small and more of a small cell nature. The difference here is that there exists no overlaying wireless LAN macrocell structure to provide large area coverage.

Systems based on IEEE 802.11 technologies therefore require a large number of cell sites to provide continuous coverage. Present business models of public wireless LAN operators are mostly aimed at the so-called hot-spot coverage, such as within coffee shops, train stations, airports and so on. City-wide wireless LAN services also have been provided for a number of years and use the fibre and cable infrastructure present in a CBD. In addition

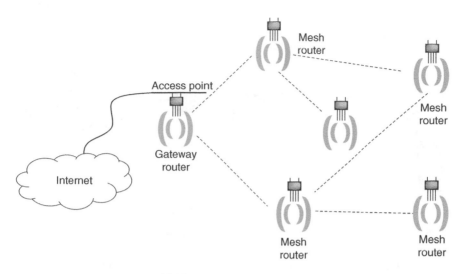

Figure 13.18   Wireless LAN system roll-out

wireless LAN multi-hop technologies enable a wider roll-out of these services as only a fraction of access points need to be connected to a fibre/cable fixed network. An example of a roll-out is shown in Figure 13.18.

Wireless Local Area Network Off-loading

With the explosive growth in mobile broadband data communications due to widespread usage of smart phones, operator networks are under significant loading constraints. The amount of capacity required is growing at a very fast rate and networks are struggling to cope with this demand. One strategy in reducing this traffic is to off-load traffic to the wireless LAN systems. In this configuration, wireless LAN systems provide connectivity to mobile devices in dense subscriber demand areas such as shopping centres and central business areas. As a user moves away from an area covered by a wireless LAN, connectivity is provided by the mobile network using a 3G/4G technology as illustrated in Figure 13.19.

Many operators have partnered with wireless LAN operators or rolled out one of their own to cope with the increasing demand on their mobile network. One such example is the Japanese operator Docomo who has a network of 150 000 hotspots throughout metropolitan areas [16]. The network subscribers can connect to any of these networks at a small monthly fee.

Hierarchical and Overlay Cell Design

As different distributions of population density, and traffic needs lead to cell designs with various degrees of cell radius, a mixture of macro-, micro- and picocells are required to design a network with adequate coverage and user capacity. These structures may use different standards (2G/3G/4G) and frequency spectrum. When the same frequency and technology are used, dynamic resource technologies can be used to minimise interference. This approach is

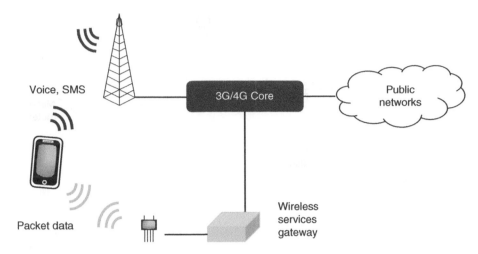

Figure 13.19　Hybrid networking using macrocells and wireless LAN access points

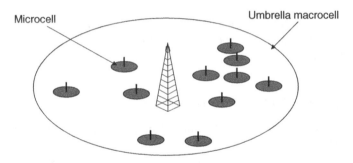

Figure 13.20　Umbrella cell structure

useful in a dynamic roll-out of a wireless network: as the number of subscribers and traffic increases, more cell sites may be deployed in areas of large traffic concentration. One solution is to use an *overlay* structure, as illustrated in Figure 13.20, where different cell configurations overlap to provide service to end-users. Such an approach may also use wireless LANs in its overall service provision strategy.

Handover

An overlay network needs to facilitate a very fast handover process. In general, a mobile device travelling within a cellular system connects to the most suitable base station. Cellular telecommunications systems are designed to facilitate handover of a user from one base station to another as the channel conditions vary. Handover mechanisms have been specified for almost all cellular standards since analogue 1G systems. As illustrated in Figure 13.21, a mobile stops communicating with a base station and starts with another which has a higher

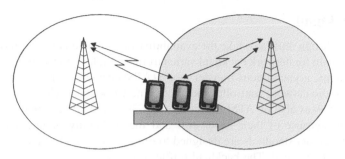

Figure 13.21   Handover process

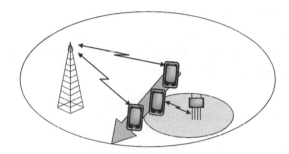

Figure 13.22   Hard handover in multi-structured cell topologies

signal strength. In Figure 13.21, the mobile continues to communicate with both base stations for some time, until the received signal from one is lower than the other by more than a designed margin. This process is known as a soft handover. Some standards allow a mobile device to be connected to only one base station at any one time. The process is known as a hard handover.

In 3G and 4G packet-switched broadband telecommunications systems a hard handover process is used. This is mainly due to the difficulties of managing packet transmissions from multiple base stations. As illustrated in Figure 13.21 a mobile station in an overlaid, multi-structured base station system with large- and small-cell topologies, can connect to any of the base stations or access points near it. The decision on which base station to use is based on the cost and rate of transmission as well as network traffic conditions. It is desirable that the most efficient connection is used at any time, and handovers are made as soon as required. However, the delays associated with handover processes make frequent handovers difficult. Furthermore as packets are directed to and from a central network, as illustrated in Figure 13.22, only a limited number of base stations can be associated with a mobile station at any time. This is further restricted by the fact that a central network controller is connected to a limited number of base stations, and at times inter-network handover is not readily possible. The smaller the cell size, the more frequently a mobile station crosses boundaries from one cell to another. This can result in handover failure, and dropped packets.

Network Backhaul

Finally, network design must also take the availability of backhaul network into account. The transport mechanism for network backhaul varies from technology to technology. In older 1G, 2G and 3G systems, asynchronous transfer mode (ATM) devices were used to connect base stations to their radio network controller and to other network nodes. Connection to the public telephone network was also on ATM devices. Latest 3G standards use internet for backhaul packet transport, because of their lower cost and the scalability of internet protocol (IP) switches. Moreover, 4G standards are designed to operate on an end-to-end IP system where all traffic is packet switched. The backhaul traffic travels mostly over fibre networks. Very-high-bit-rate DSL (VDSL) and cable connections are other options, and where all these are not available, fixed wireless links are used.

Next Generation Networks

The broadband packet-switched telecommunications system of 3G and 4G standards can be categorised as Next Generation Networks (NGNs), and philosophically follow the vision of the International Telecommunications Union as:

> a packet-based network able to provide telecommunication services and able to make use of multiple broadband, QoS-enabled transport technologies and in which service-related functions are independent from underlying transport-related technologies. It enables unfettered access for users to networks and to competing service providers and/or services of their choice. It supports generalized mobility that will allow consistent and ubiquitous provision of services to users. [17, 18]

In other words, NGNs will provide an all-IP network which transports multiple services (voice over IP, video, etc.) in both fixed and wireless systems. Moreover the services will be broadband and use 4G, VDSL and fibre to the x to transport content to end-users. The systems are to have stringent quality of service (QoS) constraints to ensure delay and transmission rate sensitive services are supported, regardless of whether transmission networks use fixed or wireless links. Furthermore, NGNs propose to provide QoS control through recognising the application and routing accordingly. This can be accomplished by protocols which enable the router to recognise what kind of data are included in a packet, and switch according to a set priority.

An important part of the vision of NGNs is independence between application provision and data transport mechanisms. This may be interpreted in two ways: one is that a network operator should not discriminate against services and applications that use its infrastructure. This is also referred to as net-neutrality, a concept which will be discussed in Chapter 14. The other interpretation is that the infrastructure and retail strata of broadband telecommunications service providers are separated. We have used this model in building the broadband telecommunications ecosystem model.

Infrastructure Convergence

The objective of the Australian government in construction of its National Broadband Network (NBN) has been to create a monopoly infrastructure entity for fixed (as opposed to mobile) services. The NBN manages the infrastructure and leases capacity to retailers who manage service provision. However can the NBN fixed network model apply to wireless networks?

Most operators have spent billions of dollars in rolling out 3G networks and are spending similar amounts on 4G networks and beyond. Infrastructure expenditure is the main part of a wireless network operator's fixed cost and any reduction can boost overall business performance. Moreover, there is a shortage of infrastructure roll-out engineers and fewer installations can speed up service roll-out for all operators. Clearly there are economic efficiencies by rolling out a single set of wireless infrastructure for all service providers.

One may envision a future where only one set of network infrastructure exists, which is owned or managed by a third party. All operators in this model will be in effect virtual network operators – although they will still own spectrum. This is the complete opposite of the model where every operator owns and manages their wireless infrastructure. While the joint ownership of the infrastructure provides significant cost saving, it also removes any competitive advantages an operator may enjoy through having a superior coverage. The likely scenario will therefore be somewhere between the two: operators may jointly own and manage part of their infrastructure and separately own other parts.

Unfettered access further envisions a future characterised by net-neutrality where service providers treat all content streams equally. It should be noted that the NGN vision of ITU is aspirational and not necessarily mandatory. Different governments may legislate to enforce any of these desired outcomes.

Fixed Mobile Convergence

Another aspect of NGN is supporting organised mobility and seamless access to broadband infrastructure in both fixed and mobile form. This is known as fixed mobile convergence (FMC) and envisions delivery of packets to/from an end-user device using the most efficient route (e.g. through a wide area network such as LTE, or via LAN or wireless LAN, or through even a personal area network). A single device IP address will remain unique and is recognised and located by the delivery network. This is illustrated in Figure 13.23.

As packet transport is over multiple networks, each possibly owned by a different service provider, tariff calculation will be complex. Two methods of tariff calculation may be envisioned. One is when the end-user pays for the consumed content from a content provider, who in turn compensates transport providers. The other method is a monthly or per-packet contract

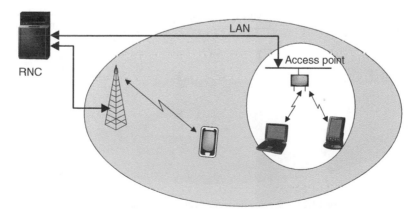

Figure 13.23   Packet delivery over heterogeneous networks

between the end-user and a broadband service retailer, who distributes revenue among fixed and mobile network owners.

FMC requires traffic to flow over an all-IP backhaul. This means that all service will be packet switched including telephony. Moreover, QoS constraints will be implemented using protocols such as Multiprotocol Label Switching (MPLS) to facilitate traffic type differentiation.

Many broadband wireless operators have integrated wireless LANs into their 3G and 4G networks. The motivation is mostly to offload wide area network traffic to local networks. Also, BYOD regimes require seamless transition between 3G/4G, femtocell and wireless LAN systems. These are early manifestations of fixed and mobile convergence.

ITU's NGN vision faces challenges from a number of fronts. Many operators are vertical integrated and own network infrastructure, retail service and deliver content from a subsidiary. For example, Telstra has ownership in content delivery companies such as BigPond Movies and Foxtel, in addition to infrastructure ownership and retail arms. This may be a competitive advantage to Telstra and opening up the network to treat all content similarly may not be in their interest. Such companies have little incentive to provide an open platform.

Other complexities arise from charging for services and percentage of revenues which goes to each player. Furthermore, such operators will not be incentivised to provide services to sparsely populated areas as revenues will not justify the costs. Without government intervention a 'digital divide' will come to exist within a country between urban and rural areas. Transmission speeds and capacity are another challenge as traffic growth is soon expected to exceed wireless network abilities.

Self-organising Networks

As discussed above, broadband service provision in the NGNs will be through several different coexisting technologies. This is illustrated in Figure 13.24. The technology landscape includes 3G and 4G mobile as well as wireless LAN standards. Network configuration will

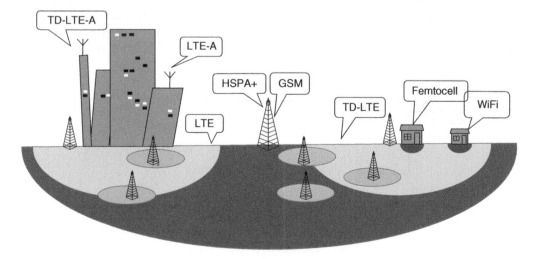

Figure 13.24   Service delivery over heterogeneous networks

also include from large cell to small cell. Traffic is routed through the best positioned base station or access point. A cost function is constructed for each user based on their profile and service level agreement. This function is used to find the optimal method of packet delivery to/from user equipment [19].

Clearly handover issues are complex as a mobile device travels through the coverage area. Fast handover is required especially for delay-sensitive services such as voice or streaming audio and video. More complex is network planning as different entities may own and operate parts of the network and coordination may not be practical.

New networks designs are expected to be self-configuring, self-optimising and self-healing. That is, initially umbrella cover macro base stations are installed within an overall plan. Further addition of smaller cells requires little planning as these cells are designed to be self-configuring in that they communicate with neighbouring base stations and dynamically allocate resources to minimise mutual interference. The new cell configurations then self-optimise, in that their signal coverage is coordinated together to minimise areas of overlap and insufficient received power. This reduces the requirement for drive testing and saves transmission power. Furthermore, as offered traffic to the area varies, self-optimisation allows base station load sharing and load balancing. This also allows for self-healing when one cell site goes off-line. Under such a situation neighbouring cells detect the outage and balance their load to heal the network.

Another aspect of self-organising network (SON) design is analysis of network signalling for optimisation. As noted above, driving tests were needed to examine the quality of coverage within a cell. These driving tests were effective to some extent, but could not give an accurate picture of reception quality inside buildings. However signalling mechanisms within 3G and 4G standards allow for reception signal strength to be fed back to the base station and to the mobile core network. A SON allows for this information to be analysed and used to identify and heal areas where reception is weak.

Example 13.1

Celcite is a vendor of SON software for operating a heterogeneous network (HetNet). The software identifies operator's 2G/3G/4G network nodes and maps a multidimensional network coverage picture. Real-time network load is then analysed to balance the node of different cells. SON software is also used to allocate frequency and time-slot resources of the 4G network to ensure best resource utilisation. The software needs to analyse a large collection of information from base stations and subscriber signal strength feedback to map an optimal resource allocation solution for the users and the operator [20, 21, 22].

Network planning and roll-out, optimisation and maintenance has therefore become a problem for big data field, where an 'optimum' solution for a network at any particular time may be found. The role of network planning companies is minimised as cell planning and tuning can be done using vendor (e.g. Ericsson, Huawei) software installed within network equipment. The need for drive tests to analyse the network is also replaced by data analysis software based on information collected from subscribers.

Wireless service provision to a populated area is a function of subscriber density as well as standard specifics, as operators build the necessary infrastructure. It comprises of acquiring sites for installing network equipment and antennas and ensuring that the resulting signal coverage provides efficient, high quality service. It is a dynamic process involving many players: from networking planning to determine suitable positions, to real estate suppliers for physical space (land or rooftops) to measurement and network optimisation suppliers to ensure high quality coverage. It is a costly and complex dynamic process which is further complicated by the interplay of technology and business issues.

Review Questions

1. What are the technology issues related to the roll out of a mobile telecommunications infrastructure?
2. What are the business issues related to the roll out of a mobile telecommunications infrastructure?
3. What are the policy issues related to the roll out of a mobile telecommunications infrastructure?
4. What are the reasons behind different cell configurations (macro, micro, etc.)?
5. How is connectivity addressed from the infrastructure structure layer and the content and services layer points of view?
6. What are four different macrocell configurations and how do they relate to population density?
7. How does population density impact on the size of a cell?
8. Why have femtocell, repeater and distributed antenna systems been developed?
9. What are the complexities involved in the design of a mobile network?
10. What values do network planning companies offer to a mobile operator?
11. What roles do wireless LANs play in a mobile infrastructure?
12. What is wireless LAN off-loading and what is its importance?
13. What are the characteristics of the NGNs?
14. Why is an IP end-to-end network considered for NGN?
15. What is fixed-mobile convergence and why is it of importance to the roll out of a mobile network?
16. What do SONs accomplish?
17. How is SON of value to 4G network roll-out?
18. How is subscriber connection control information used to optimise mobile network operation?
19. What are the advantages and disadvantages of a monopoly infrastructure provider (an entity that owns all base stations and core network and which sells wholesale capacity to retailers) vs individual infrastructure ownership?
20. What are the policy considerations to multiple operator license provision? What are business considerations?

References

[1] http://www.itnews.com.au/News/372659,telstra-wins-legal-battle-over-optus-ads.aspx, accessed 25 August 2015.

[2] Telstra. Our Coverage. https://www.telstra.com.au/coverage-networks/our-coverage, accessed 25 August 2015.*

[3] Kleinrock, L. (1975) *Queuing Systems*, John Wiley & Sons, Ltd.

[4] Zander, J. (2001) *Radio Resource Management for Wireless Networks*, Artech House Publishers.

[5] Laiho, J., Wacker, A. and Novosad, T. (eds) (2002) *Radio Network Planning and Optimisation for UMTS*, John Wiley & Sons, Ltd.

[6] Tafazolli, R. (ed.) (2004) *Technologies for the Wireless Future: Wireless World Research Forum (WWRF)'* John Wiley & Sons, Ltd.

[7] Wireless World Research Forum. http://www.wireless-world-research.org/, accessed 25 August 2015.

[8] http://www.ericsson.com/ourportfolio/products/base-stations?nav=productcatagory006, accessed 25 August 2015.

[9] http://www.ericsson.com/ourportfolio/products/zero-site?nav=productcatagory006%7Cfgb_101_0516% 7Cfgb_101_0548, accessed 25 August 2015.

[10] http://www.optus.com.au/shop/support/answer/home-zone-what-is-it?requestType=NormalRequest&id=1509 &typeId=5, accessed 25 August 2015.

[11] https://www.nttdocomo.co.jp/info/news_release/2012/02/03_00.html, accessed 25 August 2015.

[12] http://www.ericsson.com/news/1731153, accessed 25 August 2015.

[13] http://www.smh.com.au/digital-life/mobiles/optus-makes-customers-pay-to-fix-its-blackspots-20110411-1da6b.html, accessed 25 August 2015.

[14] https://www.telstra.com.au/my-offer-summaries/download/document/my-offer-summary-smart-antenna.pdf, accessed 25 August 2015.

[15] A. Stolyar, A. and Viswanathan, H., (2008) "Self-organizing dynamic fractional frequency reuse in OFDMA systems," IEEE Conference on Computer Communications INFOCOM.

[16] http://visitor.docomowifi.com/en/area/, accessed 25 August 2015.

[17] Bertin, E., Crespi, N. and Magedanz, T. (eds) (2103) *Evolution of Telecommunication Services, The Convergence of Telecom and Internet: Technologies and Ecosystems*, Springer Verlag.

[18] http://www.itu.int/ITU-T/studygroups/com13/ngn2004/working_definition.html, accessed 25 August 2015.

[19] http://www.emf.ethz.ch/archive/var/pres_matthias_eder.pdf, accessed 5 November 2015.

[20] http://pipelinepub.com/cloud_and_network_virtualization/SON_D-Son_and_C-Son, accessed 25 August 2015.

[21] http://www.lightreading.com/spit-(service-provider-it)/customer-experience-management-(cem)/amdocs-shines-son-acquisitions-on-ran/d/d-id/707637?f_src=lightreading_sitedefault, accessed 25 August 2015.

[22] http://www.amdocs.com/site/ran/pages/default.aspx, accessed 25 August 2015.

14

Content Delivery and Net Neutrality

Preview Questions

- What are the characteristics of the traffic that flows over broadband telecommunications networks?
- What are the roles of network operators and content providers?
- How do different players in the content provision business collaborate/compete?
- Is there a need for regulation of anti-competitive behaviour when the network is not neutrally served?
- Can a network operator within its right discriminate against content from a competitor?

Learning Objectives

- Content over a broadband telecommunications network
- Entities that facilitate content delivery
- Network operators' revenue trends
- Content provision revenue sharing
- Vertical integration and network neutrality

Broadband Telecommunications Technologies and Management, First Edition. Riaz Esmailzadeh.
© 2016 Riaz Esmailzadeh. Published 2016 by John Wiley & Sons, Ltd.
Companion Website: www.wiley.com/go/BTTM

Historical Note

The value proposed by telecommunications networks has until recently been largely *connecting people*. Telephone operators have provided an infrastructure over which two subscribers can communicate. Both telephone and telegraph systems were in essence 'pipes' over which information flow took place. Very few third parties were involved with any extra 'information' in this value chain. The only information provided over these networks besides that coming from the calling and called parties was directory assistance provided by the operator. This service was usually complimentary and did not contribute significantly to operators' revenues.

With the introduction of automatic switches in the late 20th century, a number of third-party service providers emerged. Initial services included weather or stock price information, news, and some entertainment. These were offered to subscribers on a fee-per-service or subscriptions basis. Some of these services originated from within an operator's network, while others were provided by external parties. Thus content provision emerged as a new source of revenue for operators – charging for the service or taking a commission – albeit the amount so raised was very small compared with the revenue from the 'pipe'.

With further advancement of telephone switches towards the end of the 20th century, more *sophisticated* content and services were offered. Voice boxes, call transfer, intelligent networks, conference calls and so on are a few of such services provided by operators. Horoscope, counselling, adult calls are examples of services/content provided by third parties over telephone lines. Gradually combinations of these 'non-pipe' services became a significant revenue source for operators, still small compared with the revenue from the 'pipe'. Furthermore, most if not all of these services were provided on the circuit-switched, analogue telephone network.

The advent of the World Wide Web and abundance of on-line content and services transformed the telecommunications ecosystem landscape at the turn of the 21st century. News, audio, video, games and so on started to flow over fixed-line internet, raising significant revenues for content providers such as newspapers, iTunes, YouTube and Netflix among many companies. Furthermore, services such as search engines and social networks, as well as peer networking, further provided many service companies with unforeseen income streams. Meanwhile, the telecommunications infrastructure, built and maintained by operators, formed the 'pipe' over which this traffic was flowing. Unfortunately for these operators, very little leverage existed to capture a portion of the content revenue. The reason was that operators did not have any exclusive 'gateway' capability to ensure content providers could be charged for a commission. However, there was one exception to this. The Japanese mobile operators, led by NTT Docomo, routed most traffic through their mobile content portal platforms (such as i-Mode) and charged a commission on the content providers' revenues [1].

Mobile portal platforms were of great interest to operators and many tried to replicate the i-mode model. However, none managed to successfully replicate the entire ecosystem. The arrival of iPhone and emergence of the smart phone industry shifted the platform from operators to operating system developers: iOS, Android, Windows Mobile. Fixed-line and mobile operators have been mostly locked out of the content/service layer, except for those who are vertically integrated and have a stake in content producers/aggregators [2].

Content and Service Provision

Broadband telecommunications infrastructure is used to provide a variety of services and content to a range of end-users. These services range from human–human, human–machine or machine–machine information transfer. Infrastructure operators' and retailers' value proposition is facilitating this information transfer to the end-users.

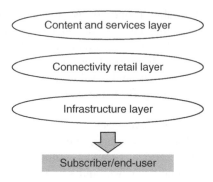

The operators' business can be exemplified as constructing a network of 'pipes' over which information may flow. The 'pipes' are then *rented* on a temporary or semi-permanent basis to customers who will pay according to the value such services provide. The service business system diagram is shown in Figure 14.1, where the wholesaler and retailer roles are separated. The two need not be a separate entity: many retailers own and operate their wholesale network which they use exclusively or share with other retailers. In practice a number of wholesalers need to work together to provide the overall service. For example, an international telephone call traverses several networks owned and operated by several infrastructure owners and associated wholesalers. The business system diagram is completed by the equipment manufacturers and retailers chain.

The infrastructure value includes provision of a telecommunications link with a designed level of service quality to a subscriber. The link may be fixed and exclusive to one user, as in an FTTP (fibre to the premises) link. Or it may be shared, as in mobile telecommunications systems. The telecommunications ecosystem may be defined as a 'pipe' business, where the information transferred over the network is user-generated. Examples may be a phone call between two parties, or an email exchanged through two computers, as illustrated in Figure 14.2. The role of infrastructure/retail operators is to provide a link to carry this end-user generated content, and charge subscribers for the connectivity service. Historically, the vast part of operators' revenue has come from this 'pipe' business. Different revenue models exist: the fee may be calculated per minutes of a voice call or the size/number of transferred data packets. Other sources of revenues may include a

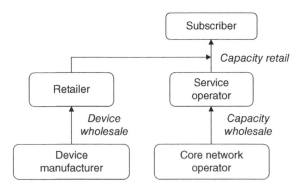

Figure 14.1   Business system diagram for a telecommunications infrastructure business

fixed monthly subscription fee. These added together and averaged for all subscribers result in a widely used metric for operators' performance, known as average revenue per user (ARPU). This metric is commonly used to evaluate the performance of a fixed or a mobile operator both in an absolute fashion as well as relative to their competitors.

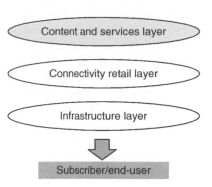

Another kind of traffic that flows over a telecommunications infrastructure is associated with third party content and services. Such content may originate from within the operator network or be provided by a third party as illustrated in Figure 14.3. As discussed in the Historical Note, content provision has evolved significantly over the past decades. Earlier content included typical information a subscriber may need,

Figure 14.2   Pipe service provision for voice and data service

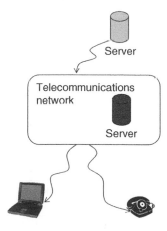

Figure 14.3   Value added service and content provision over a telecommunications network

for example weather, stock, lottery, directory assistance, horoscope and so on. Initially such information originated within the network and was provided on a complimentary basis. As telephone switching became computerised and more sophisticated the extent of this content grew, and premium third-party service provision became more common. These services can be classified into two groups – information and entertainment, based on which the term *infotainment* was coined. In addition, many value added services (VASs) were also offered by the operator, including voice mailbox, call forwarding and call holding. Furthermore, special services were offered to business customers such as ring back tones, music/radio play back, contests and voting and interactive voice response (IVR). All these additional services, most of which exist today, were value additions, and intended to increase operators' ARPU through extra revenue or better customer retention. These types of broadband telecommunications revenues differ to those associated with the 'pipe' provision, and are associated with the content and services layer.

Service Provision

Value added service provision business has evolved from its telephone/data-centric nature in a number of significant directions. One early direction was associated with operators and retailers moving into telecommunications service management provision for corporate clients. In the 1990s many companies started outsourcing their information technology divisions to specialist companies. These included telecom service retailers, such as Telstra and Optus in Australia. Such service provision created a new revenue stream for players in the retail layer of the broadband ecosystem.

Another direction which emerged in the mid-2000s is commonly known as over-the-top (OTT) services. OTT companies use the telecommunications infrastructure network to provide services such as telephony, video telephony, messaging and so on. Prominent examples include Skype, FaceTime and WhatsApp. These companies substitute the services of a retail layer telecommunications service provider, and are therefore *competitors* to these companies. The business system diagram associated with OTT services is relatively simple and is shown in Figure 14.4.

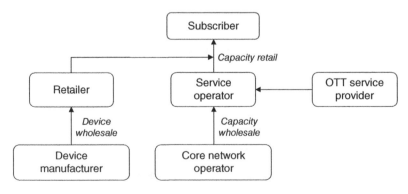

Figure 14.4   Business system diagram for an OTT service business

Case Study 14.1: Mobile Messaging and Voice Applications

The popularity of smart phones and the introduction of broadband data plans in third generation (3G) and fourth generation (4G) mobile subscriptions, made it possible for a number of third party applications to provide telephony and short-messaging services. These *apps* competed with 'monopoly' mobile telecommunications services which accounted for most of operators' revenues. Early apps were evolved from personal computer (PC) voice over IP (VoIP) applications such as Skype and Viber which allowed for primarily low-rate international, and mobile-to-fixed line calls, as well as instant messaging. Furthermore, communication between apps was free. Additional services such as video telephony, multi-party conference, file sharing and gaming were provided over the platforms associated with these apps. The apps were however less successful in substituting mobile-to-mobile telephony and messaging as the call rates were not significantly lower than those charged by mobile operators. Furthermore, Skype and Viber worked mainly with usernames instead of a telephone number which made it difficult to communicate with many subscribers.

A second generation of mobile apps entered the market in early 2010 which allowed for short messaging across a mobile telecommunications network and used telephone numbers instead of usernames. A number of these apps have greatly grown in popularity. In particular Skype, WhatsApp and WeChat count their subscriber numbers in the order of several hundred millions. This has led to multi-billion dollar valuations for these companies as shown in Table 14.1 [3, 4, 5, 6, 7].

A common feature of these apps is short messaging, but they differ in the features and functions they provide. Some offer telephony and gaming services while others focus on short messaging. Furthermore, some services are offered for free while some are charged. In parallel some smart phones have started to include these OTT services within the operating system, such as iOS Facetime and iMessage. Moreover, a social networking app such as Facebook may also be considered an OTT as it provides many connectivity values traditionally provided by the (mobile) telephony networks.

With the exception of a few global networks (Facebook messaging, iMessage) most OTT apps are concentrated in a few markets; WhatsApp dominates in the US, WeChat in China, Line in Japan and Korea and so on. This may be explained by the *social networking* nature of these apps. We connect with people we know, who are likely to live where we do. While all these apps may be installed freely, we tend to communicate on platforms which are used by our social network. Moreover, the platform created by

Table 14.1 OTT services and valuations

OTT Service	Valuation (year)
Skype	$8.5 B (2011)
WhatsApp	$16 B (2014)
WeChat	$60 B (2014)
Line	$10 B (2014)
Kakao Talk	$8.5 B (2014)

these apps works to ensure we stay within the network for most of our social needs such as gaming, news and photo sharing, trading and so on.

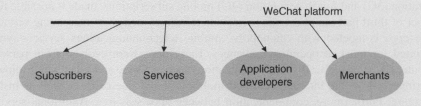

OTT services are global, flow over the local operators' networks and substitute their services. For example, Skype traffic flows over an operator's infrastructure and replaces its long-distance telephone service. The revenue loss has been significant: Ovum reported that SMS revenue declined by some $13.9 billion in 2011 [8] because of applications such as WhatsApp, and has estimated that OTT service providers will erode network operators revenues by up to $479 billion by 2020 [9].

In some countries the impact has been so great that mobile operators have started their own OTT apps (e.g. mVoIP by Korea Telecom) [10]. Some other mobile operators block their subscribers from accessing these apps unilaterally [e.g. Skype is blocked by two internet service providers (ISPs) in the United Arab Emirates] [11]. However, this appears to be a losing battle if the growth of OTT companies and their market share is a guide.

Case Study Questions

- What do you see as the main reason for the growth of Skype in the early 1990s?
- Why did messaging apps such as WhatsApp prosper?
- What do you see as the reason for Facebook acquisition of WhatsApp in 2014?
- How does iMessage complement Apple products?
- Is there a chance for retail/infrastructure operators in their battle against OTT players? Why?

Content Provision

The business system associated with the delivery of content is relatively complex and involves a large number of third-party entities. As illustrated in Figure 14.5, these entities work together to provide infotainment content that may be carried over a broadband telecommunications network infrastructure. The content is produced, retailed and aggregated by a large number of independent or at times vertically integrated players. Such content is then brought near a user through content distribution networks, and accessed by a broadband service provider. Furthermore, a number of manufacturers produce equipment over which infotainment content may be presented to the end-user. In this way a broadband telecommunications network delivers and facilitates infotainment content delivery so that one may read a newspaper, listen

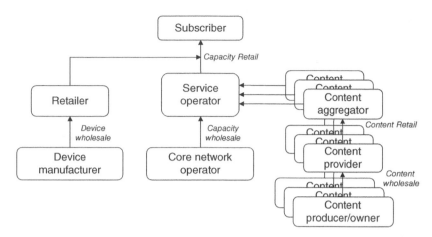

Figure 14.5 Business system diagram for a telecommunications content provision business

to streamed music or watch a movie on-line. While there may exist close collaboration or even cross-ownership between the entities in the parallel links, in general the business chains are independent.

The dynamics of 'pipe' and service/content businesses have undergone a dramatic change since telecommunications industry deregulation and the rise of the internet. Historically 'pipe' services such as local and long-distance telephony, and mobile telephony and messaging accounted for nearly all of ARPU. With deregulation and increased capacity supply due to optical fibres and the new generation of wireless standards the portion of ARPU associated with these services has declined. Fixed telephony ARPU has now been augmented with fixed broadband services such as xDSL and cable. On the wireless side 'data' services have become a substantial addition to the overall ARPU. While the total ARPU has remained stable the share of data has risen from a small fraction to more than half [12]. Figure 14.6 shows the decline in voice ARPU vs increase in data ARPU in the United States. Most of the data ARPU is associated with content services.

In contrast to the general decline in the ARPU and profitability of the 'pipe' business, the size of content business has grown significantly. The ratio of total revenue of the overall telecommunications ecosystem raised by 'pipe' and content businesses may be illustrated as shown in Figure 14.7. In fact it is reported that the revenue associated with data is now surpassing that of voice [13]. As noted above, historically the 'pipe' business accounted for nearly all of the overall revenue and content formed only a very small part. However, after the deregulation the relative revenue share has undergone a dramatic transformation. In particular the internet platform has provided a level playing field so service and content retailers can supply a global customer base with relative ease. Significant revenue may be raised from such content provision: however the infrastructure and retail layer players may not be well positioned to receive a share of this cash flow.

The decline of 'pipe' revenue may be attributed to a reduction in cost-per-bit for transferring information. The large capacity afforded by optical fibres combined with continual reduction in the cost of their roll-out has meant that the cost of transmission for one bit has dropped significantly. Telecommunications industry deregulation on the other hand has meant

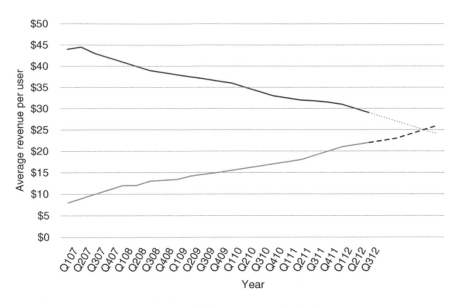

Figure 14.6   ARPU trends in the United States

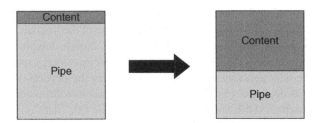

Figure 14.7   Overall trends of 'pipe' and content revenue

more competition, and therefore a shift from value-based pricing to cost-based pricing. Indeed the capacity glut caused by excessive optical fibres drove many long-distance operators to bankruptcy. For instance, the bankruptcy of two long-distance US operators, Global Crossing and Worldcom, occurred in 2002 [14, 15].

The reduced cost of bit delivery as well as open access to a global market place via the internet has allowed single-cast content delivery to become economical. An example is TV content. Before the internet the only economical methods of delivery were TV broadcasting or video rental single-casting. It has now become economical to single-cast video content on-demand over the internet. Content aggregation platforms such as YouTube allow for a global audience to connect to a globally produced body of infotainment.

The infrastructure and retail layers of the broadband telecommunications ecosystem generally act as the 'pipe' that carries the infotainment content. These infrastructure providers are thus faced with a challenge: while they can charge their subscribers for delivering the data streams, they have little leverage to earn revenue from the content provision itself. On the other hand, as these operators control the 'pipe' that carries the traffic they do have some leverage over content providers to ask them to share revenue. For example they may throttle the traffic of a content provider who fails to share revenue compared with a provider which agrees. This however may be illegal as it treats these data streams differently and is not content neutral. We will discuss the issue of net neutrality in more detail later in this chapter.

Traffic Classification

Broadband traffic may be classified as shown in Figure 14.8. On one side operator value proposition is provision of a 'pipe' over which end-user generated information may be transferred. On the other side is third party generated content in the form of infotainment and value added services. Further value operator value proposition is possible: for example quality of service levels, such as end-to-end delay or transmission speed may be enhanced. On the content side similar arrangements may be made with both content provider and subscriber.

A content provider may earn revenue in a number of different ways. The most common are pay per usage (view, song, article, etc.), subscription and advertising sponsored. For example, articles from technical journals such as by the IEEE may be bought for price per article; the video streaming site, Netflix, has a number of monthly subscription plans; iTunes Radio streams music is advertising-sponsored or advertising-free on subscription and so on.

As the profile of service provision has changed from 'pipe' to 'service/content', the relative shares of voice and data traffic has undergone a major transformation. For the fixed network, data traffic only started to become significant in the early 1990s. However data traffic growth was exponential and by the late 1990s its volume had overtaken that of voice. As for mobile networks, global data traffic volume was a small fraction of that of voice as late as 2007 as shown in Figure 14.9. But that also grew significantly and now in 2015 accounts for more than 90% of the total traffic. Meanwhile data traffic is expected to continue its exponential growth for some years as shown in Figure 14.10.

As noted above, data traffic volumes can be classified as those associated with 'pipe' and 'service/content'. A major part of data traffic is associated with audio and video *content* as shown in Figure 14.11. Peer-to-peer 'pipe' traffic accounts for a significant share, and the rest is 'data' which is associated with email service ('pipe') and infotainment and OTT (content/service).

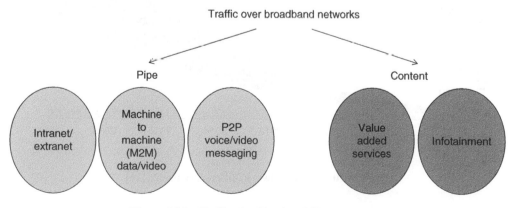

Figure 14.8   Traffic classification. P2P, peer to peer

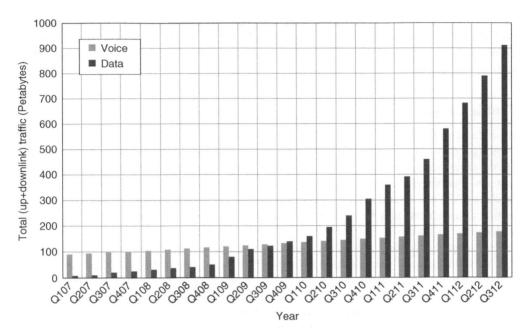

Figure 14.9   Voice and data traffic [16]

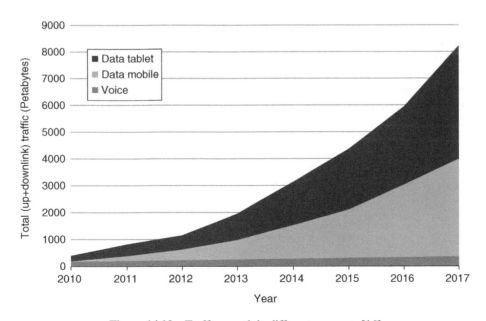

Figure 14.10   Traffic growth in different segments [16]

Revenue Flow

In the 'pipe' business regime, revenue flow was simple and from a subscriber to the operator as illustrated in Figure 14.12. Charges included per usage and subscription fees as well as any content fees. The revenue flow has evolved to include direct transactions between end-user and third-party content providers, indirect transactions through the operator, and other transactions such as special deals between operator and third parties. This is illustrated in Figure 14.13.

In practice a large number of third parties earn a share of what an end-user or an advertiser pays for the delivery of infotainment content. Broadly these include application providers who facilitate the infotainment content delivery and licensors of content. Revenue flow originates from end-users and advertisers as well as the operator as illustrated in Figure 14.14.

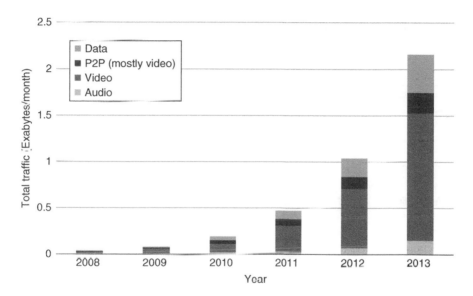

Figure 14.11 Traffic ratios [17]

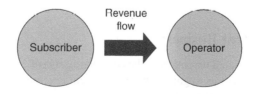

Figure 14.12 Revenue flow in a 'pipe' business regime

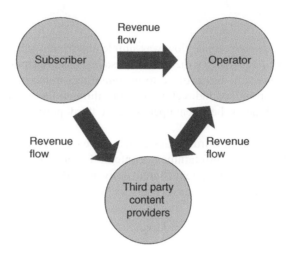

Figure 14.13   Revenue flow in a third party content provider regime

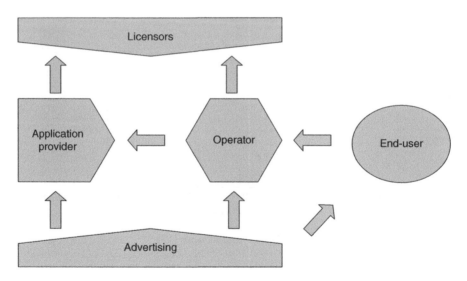

Figure 14.14   Revenue flow in a third-party content provider regime

Operators and Content Business

The decline in 'pipe'-associated revenue and the contrasting rise of content revenue has caused many operators to take measures to gain a share of content revenue. These include starting or buying into content provision companies, creation of a portal over which all content flows, creating a platform to control content discovery and flow, and 'negotiating' with content owners to share revenue through provision of preferential treatment. These measures are described in more detail below.

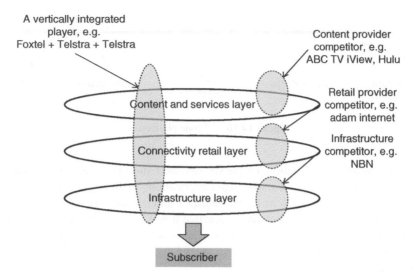

Figure 14.15 A vertically integrated content delivery service provider

Vertical Integration

A number of vertically integrated players have emerged who not only own telecommunications network infrastructure, but also retail these services and run business for delivery of video and other content over the networks. An example is Telstra in Australia which has an integrated content delivery network as shown in Figure 14.15. Telstra owns network infrastructure and is an ISP providing retail service. It also holds a 50% ownership of Foxtel, a pay TV and cable TV operator. Furthermore, it owns and operates several video-on-demand sites, including BigPond Movies which delivers content to its subscribers. In contrast there are players in Australia who are active only in one of the broadband telecommunications ecosystem layers.

Vertical integration may provide an operator with significant competitive advantage vis-à-vis non-integrated players. One example is when streaming video from a competitor is slowed in compared with the operator's own network. Another example is when the content from a competitor counts towards a monthly download limit whereas the operator's associated content does not. Yet another common example is when a vertically integrated service provider bundles services together at a lower cost compared with the sum of the individual services. All these may force an end-user to opt to subscribe to vertically integrated operator services.

Portal Ownership

Processes of discovery and purchase are part of accessing content by an end-user. These processes may be conducted outside of an operator's area of control: one may search for the content and purchase it directly from a retailer. However at times processes of search and purchase are simpler, or mandatory, through a portal owned operator. An example of such a portal is the Japanese mobile operator Docomo's i-mode. This portal was created in 1999

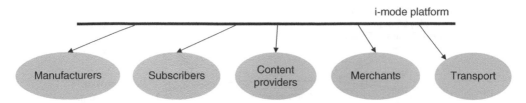

Figure 14.16   i-Mode platform

along with Docomo's introduction to packet-switched mobile data communications. Its purpose was to simplify usage of mobile data applications including sending and receiving email and accessing content such as news, ringtones and screensavers. While it was technically possible to access content directly, in practice it was much simpler to discover and go through the i-mode portal. Since content purchases were made through the portal were billed by Docomo, it could collect a commission, and therefore earn revenue from third-party content provision.

Another example is Telstra's T-Hub. T-Hub provides subscribers with applications where they can access content over fixed telephone lines. Again T-Hub simplifies access and any purchase of content over this platform can potentially be paid through Telstra for which a commission may be charged.

These portal platforms grow as the associated content and services grow. For example the i-Mode network-mediated platform is shown in Figure 14.16. At the height of its success, the i-Mode platform was highly valuable in attracting and holding on to mobile customers. Infotainment content such as games, books, videos, music, stock transaction and so on created a significant revenue stream for the operator platforms. In addition, payment systems using mobile devices with near field communication capability (see Case Study 12.2) created extra revenue for the operators. These in total accounted for some 20–30% of overall ARPU in the mid-2000s.

Retail operator data platforms have lost much ground with the emergence of smart phones. Much of content transactions now go through applications purchased through iOS, Android and Windows App stores, mostly bypassing retail operators.

Content Delivery Networks

Infotainment content needs to be stored at a physical storage device from where it may be accessed. A number of content distribution networks exist which facilitate content access through mirroring them at a storage site near an end subscriber. The content may be streaming video or audio, or software downloads. This is illustrated in Figure 14.17.

At times this storage place is centralised, for example an aggregator server (e.g. YouTube). At other times the content is distributed across many devices and access is controlled through a platform (e.g. BitTorrent). The former paradigm is known as host-centric, where one accesses a 'host' at a particular internet address where the desired content is stored. The latter is known as information-centric where one looks for a particular content (e.g. a movie) on a peer-to-peer platform which then delivers it from multiple storage devices. The information-centric networking paradigm differs from host-centric networking in that the 'internet' address applies

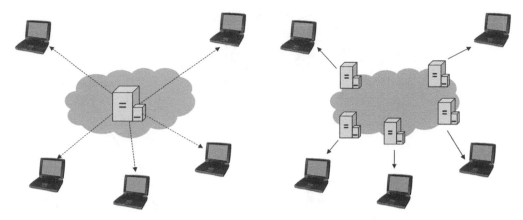

Figure 14.17   Content distribution networking

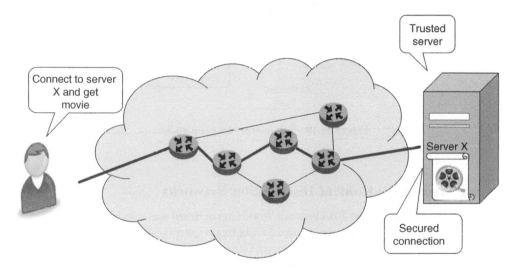

Figure 14.18   Host-centric paradigm

to the content itself and not the address where it is stored. Such content then may be shared by and accessed from the nearest storage place with minimal transportation costs (Figure 14.18 and Figure 14.19). In essence the host-centric paradigm is about *who to communicate with* whereas the information-centric paradigm is about *what to communicate* [18, 19].

Connectivity retail players have a significant role to play in the new information-centric paradigm. Decentralised content delivery may be carried out by these operators where they maintain server and storage facilities which may mirror content from content providers, and compete with content distribution network (CDN) companies such as Akamai Technologies Inc. These retail layer companies are well placed as their mirror services can be co-located with their edge routers and thereby reduce latency as well as cost of transportation. The addressing can be to the content itself and not to the individual who needs to connect.

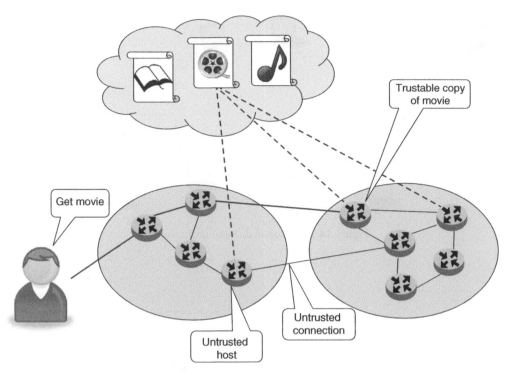

Figure 14.19   Information-centric paradigm

Case Study 14.2: Content Distribution Networks

The semi-final match of the 2014 Football World Cup in Brazil was watched by 8.4 million viewers on-line in Germany (an estimated 25% of total viewers), with 2.35 million of these using a mobile device [20]. The final between Germany and Argentina was watched by an estimated global audience of more than 1 billion, and although information on the global on-line audience is not readily obtainable, a rough estimate is in excess of 100 million. Delivery of a 2.5 h video content over the internet to such a large size audience was a huge telecommunications exercise. This is especially so considering that each viewer received an individually tailored (set up and transmitted) packetised video stream to their PC or mobile device. In contrast, in TV broadcasting the same transmitted video signal can be viewed by anyone who happens to receive the broadcast signal without any limit on the total number.

Clearly streaming the video content individually from the stadium in Brazil is impractical. The challenges are many. Device performance limits mean that switching packets to 100 million subscribers is not possible in terms of energy consumption and speed. This is further complicated as diverse transmission rates are needed because the size of the receiving displays differ: a mobile screen is smaller than a PC and therefore the video quality can be coarser. Given different video rates, the required average transmission rates can be estimated as 1.8–2.5 megabits per second (Mbps), which would require a 180–250×10^{12} bps (Tbps) aggregate transmission rate for 100 million viewers. This would be impossible with present technology. Moreover, carrying all these individual video streams in parallel to approximately the same destination (e.g. the 8.5 million viewers in Germany) is highly inefficient. It is clearly better to copy the content to a local server in, for example, Berlin from where it may be streamed to the viewers in Germany. This clearly saves significant bandwidth in long-distance telecommunications between Brazil and Germany, and results in lower costs. Delivery from a local server also facilitates regulatory and business compliance as some content may not be legally authorised in a particular country in its original format. Furthermore, local delivery allows for value additions such as adding subtitles or dubbing voice-overs.

Bringing the content close to consumers is nowadays a common practice and applies to both content as well as software upgrades. The companies which facilitate this are known as content delivery networks or CDNs. A prominent example is Akamai Technologies Inc. with more than 137 000 servers in 87 countries which mirror data streams from global content providers. The value to content providers is a better quality of service and faster content delivery. They also save significantly from lower long distance packet transmission charges. This value proposition earned Akamai a revenue in excess of $1.5 billion in 2013–2014.

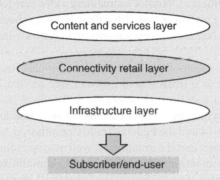

Broadband retailers and infrastructure providers are well positioned to implement and provide content delivery. Their edge switches and routers can be fitted with buffers to mirror content to deliver to their subscribers. This is one way they can earn revenue from content provision.

Case Study Questions

- What are the drivers for video-on-demand content delivery?
- What is the challenge with streaming (video) content to a very large number of end-users?
- How do content distribution companies address these challenges?
- What is a mirror site and what is the value it provides?
- How can retail layer players provide CDN services?

Network Neutrality

Network neutrality or net neutrality is a principle that means an infrastructure/retail operator must remain neutral in respect to all traffic that it conveys, that is all traffic must be processed in a non-discriminatory fashion. It means a consumer can have unrestricted access to all internet content, with respect to any site, content, device, platform and so on.

Net neutrality questions include issues such as whether a company can be forced to carry content against its will (e.g. pornographic content); whether a company can be forced to stop file sharing across its network (e.g. bit torrent); whether a company can be forced to provide equal quality of service to contents from competitors (e.g. Netflix content on a Telstra network); whether a company can be forced to provide unmetered access to video contents to its competitors (e.g. unmetered access by Telstra to ABC iView); whether an infrastructure company can be forced to charge the same wholesale charges to a competitor of its retailer arm (e.g. same wholesale prices by Telstra for Adam Internet and Telstra retail); and so on [21, 22, 23].

A 'pipe' network operator has little reason to discriminate one content stream in favour of another. However, a vertically integrated operator may create advantage for its content provision by throttling a competitor's infotainment stream. An example is how Comcast, a vertically integrated infrastructure, retailer and video cable operator in the US, allegedly slowed Netflix video streams. In late 2013 transmission rates were so throttled that according to Netflix, 'For many subscribers … Netflix's streaming video service became unusable'. The company claimed that some customers 'cancelled their Netflix subscription on the spot, citing the unacceptable quality of Netflix's video streams and Netflix's inability to do anything to change the situation' [24].

Net neutrality discussion means that, for example, Comcast must treat video streaming traffic from Netflix the same as it treats that of its own cable service. The net neutrality policy legal debate has focused on a collection of present US federal and state laws intended to regulate business practices which may restrict competition. These laws include the Sherman Act 1890, the Clayton Act 1914 and the Federal Trade Commission Act 1914. The main objectives of these laws are to restrict cartel formation, as well as stop collusion between companies which may undermine fair competition. The US Federal Communications Commission (FCC) in February 2015 voted to enforce neutrality rules as perceived by the Communications Act of 1934 to 'prevent Internet providers – including cellular carriers – from blocking or throttling traffic or giving priority to Web services in exchange for payment'. Further legislative action in the US Congress to allow infrastructure owners greater control on how they stream traffic is underway, as well as legal action by the companies concerned [23].

Net neutrality discussion also concerns issues such as what content may be legally blocked. For example should an internet service provider block access to sites with child pornography content or those which promote racial hatred and so on. Conversely, can an ISP be sued for *not blocking* such content? Furthermore, peer-to-peer traffic may pose a great burden on an ISP's quality of service: can these streams be delivered at a lower priority? The net neutrality debate is complex and is yet far from resolved.

In particular, the legal fight in the US has been highly politicised. For example, in April 2011 the Republican US House of Representatives passed a resolution mandating FCC to back off from a plan to impose 'net neutrality' rules that would prevent service providers from deliberately slowing or blocking internet traffic. However, the Democratic Senate and President

Obama blocked it [25]. Deregulation of the telecommunications industry and the high degree of competition has made net neutrality less of an issue in many countries including Australia.

While the net neutrality discussion has been generally found to be in favour of content and service providers, nevertheless, many content and service providers have opted to sign agreements with the infrastructure and connectivity retail layer players. The example of Netflix–Comcast was given above. Another example is the agreement between the French telecommunications operator Orange and Google where Google agreed to pay an undisclosed amount for search traffic delivered over Orange's network. It was reported that Orange's dominance in Africa had forced Google to agree to its demands, as they feared that their search results could be slowed compared with those of competitors if such payments were not made [26].

The value of broadband telecommunications systems is in connecting people to each other and to third-party services and content. This value chain has changed significantly over recent decades from one focused on subscriber-to-subscriber connectivity to one focused on subscriber access to third parties. Content and services provision is the fastest growing traffic on the broadband network and clearly of value to end-users. The challenge for traditional operators is how to share in revenues associated with this value provision.

Review Questions

1. What is 'content' traffic and how does it differ from 'pipe' traffic?
2. How has the ratio of 'content' to 'pipe' evolved over recent years?
3. What is ARPU and what is its significance?
4. What are OTT operators?
5. Explain three example of 'services' that may be offered for a broadband telecommunications ecosystem.
6. What services may be classified as 'pipe' and how have these evolved?
7. What are the drivers for OTT businesses and how do they grow?
8. Why are OTT companies so highly valued?
9. What are content distribution networks, and what value do they provide?
10. Why are retail layer operators well placed to become content distribution networks?
11. What are the strategies operators are adopting in order to position themselves to earn revenue from content distribution?
12. What are operator content/services portals and how do they work?
13. Why have such portals lost their power in ensuring broadband retailers earn revenue from third-party content/service provision?
14. What were the business drivers behind mobile operator portals such as i-Mode?
15. How has Apple iPhone replaced mobile operator portals?
16. What is vertical integration within the context of the broadband telecommunications ecosystem?
17. What are the value propositions of the content delivery networks?
18. How can vertical integrated broadband operators discriminate against players without such alliances?

19. What is the net neutrality concept and how does it apply to the provision of broadband content?
20. What are the net neutrality concerns of a broadband ISP?
21. Why do you think the net neutrality discussion is of great interest in the US and comparatively less in most other countries?
22. What do you see as the main policy objective in net neutrality discussion?
23. What are the main business objectives in net neutrality discussion?

References

[1] Natsuno, T. (2003) *i-Mode Strategy*, John Wiley & Sons, Ltd.
[2] Ciferri, L., Koeder, M. and Sugai, P. (2010) The Six Immutable Laws of Mobile Business, John Wiley & Sons, Ltd.
[3] http://www.businessinsider.com.au/numbers-microsoft-skype-deal-2011-5, accessed 9 March 2015.
[4] http://www.cnbc.com/id/101432344, accessed 9 March 2015.
[5] http://techcrunch.com/2014/03/11/if-whatsapp-is-worth-19b-then-wechats-worth-at-least-60b-says-clsa/, accessed 9 March 2015.
[6] http://in.reuters.com/article/2014/07/15/uk-line-ipo-idINKBN0FK0T420140715, accessed 9 March 2015.
[7] http://www.forbes.com/sites/ryanmac/2014/09/24/mobile-master-kakaotalk-creator-becomes-one-of-south-koreas-richest-billionaires/, accessed 9 March 2015.
[8] http://www.fiercewireless.com/europe/story/report-operators-lost-139b-due-ott-messaging/2012-02-22, accessed 14 March 2014.
[9] http://www.ovum.com/ott-voip-to-cost-telcos-479-billion-to-2020/, accessed 9 March 2015.
[10] http://www.mobileworldlive.com/sk-telecom-fights-ott-rivals/, accessed 5 November 2015.
[11] https://support.skype.com/en/faq/FA391/is-skype-blocked-in-the-united-arab-emirates-uae, accessed 9 March 2015.
[12] http://dazeinfo.com/2013/09/04/smartphone-industry-us-q2-2013/, accessed 9 March 2015.
[13] https://gigaom.com/2013/03/13/2013-the-year-mobile-data-revenue-will-eclipse-voice-in-the-us/, accessed 7 March 2015.
[14] http://money.cnn.com/2002/01/28/companies/globalcrossing/, accessed 9 March 2015.
[15] http://money.cnn.com/2002/07/19/news/worldcom_bankruptcy/, accessed 9 March 2015.
[16] http://www.ericsson.com/res/docs/2012/traffic_and_market_report_june_2012.pdf, accessed 25 August 2015.
[17] Pepper, R. (2009) Towards an Exabyte Broadband World. http://www.slideshare.net/stephenmcclelland/session-2-robert-pepper-towards-an-exabyte-broadband-world, accessed 5 November 2015.
[18] Ahlgren, B. *et al.* (2012) A Survey of Information-Centric Networking. IEEE Communications Magazine.
[19] Agyapong, P.K. and Sirbu, M. (2012) Economic Incentives in Information-Centric Networking: Implications for Protocol Design and Public Policy. IEEE Communications Magazine.
[20] http://www.iptv-news.com/2014/07/world-cup-final-breaks-records-worldwide-for-tv-broadcasters/, accessed 3 March 2015.
[21] http://arstechnica.com/business/2015/02/fcc-votes-for-net-neutrality-a-ban-on-paid-fast-lanes-and-title-ii/, accessed 8 March 2015.
[22] http://arstechnica.com/business/2015/03/republicans-internet-freedom-act-would-wipe-out-net-neutrality/, accessed 8 March 2015.
[23] Sirbu, M. (2008) What is the network neutrality debate about? Lecture slides, Carnegie Mellon University.
[24] http://money.cnn.com/2014/08/29/technology/netflix-comcast/, accessed 8 March 2015.
[25] http://www.nytimes.com/2011/04/09/business/media/09broadband.html?_r=0, accessed 9 March 2015.
[26] http://www.theregister.co.uk/2013/01/17/google_orange/, accessed 13 March 2014.

Acknowledgements

I have been guided and assisted by many people in my journey of learning about broadband telecommunications technologies and management. Foremost I am indebted to many educators and professors, especially Professor Masao Nakagawa of Keio University, my PhD supervisor, and Professor Derek Abell, formerly of IMD, my MBA mentor. Both have been great sources of inspiration, guiding me on the topics of technology and management. I am also indebted to Professor Ramayya Krishnan, Dean of Heinz College, Carnegie Mellon University for his pioneering of the Telecommunications Management course. Throughout my career I have also learned much from many colleagues at Telstra, Ericsson, Genista, IPMobile, and Sydney, Keio and Carnegie Mellon Universities. Among these I am particularly grateful to Dr Terrence Percival, Dr Jun Du, Alex Fassel, Kambiz Homayounfar and Hiroshi Takeuchi.

I have been greatly helped by many graduate students and friends in the preparation of this book. Many have assisted with researching and preparing case studies, including Bing Huan Chua, Corey Sattler, Daniel Miller, Isaac White, Manoj Ravi and Murali Ravi. Edgar Anzaldua Moreno designed the cover page and the layout for the e-book version. Chris Rowlinson, Georgina Hafteh, Phil Allan, Sonali Tharwani and Steve Schmid read drafts and made helpful suggestions. I am thankful for all your help.

I am also thankful to the staff at John Wiley & Sons, Ltd including Mark Hammond, Sandra Grayson, Teresa Netzler, Shiji Sreejish and Wendy Harvey, for their help throughout this project.

Most of all I am grateful to my wife and three sons who have patiently endured my preoccupation with this work while at home. To them I dedicate this book.

Case Study Index

Broadband Telecommunications Technologies and Management, First Edition. Riaz Esmailzadeh.
© 2016 Riaz Esmailzadeh. Published 2016 by John Wiley & Sons, Ltd.
Companion Website: www.wiley.com/go/BTTM

Index

Broadband Telecommunications Technologies and Management, First Edition. Riaz Esmailzadeh.
© 2016 Riaz Esmailzadeh. Published 2016 by John Wiley & Sons, Ltd.
Companion Website: www.wiley.com/go/BTTM